The Best
EIT REVIEW
for the
Fundamentals of Engineering (FE) Exam

Ted Huddleston, Ph.D., P.E.
Chairperson & Professor of Chemical Engineering
University of South Alabama, Mobile, AL

Ralph Pike, Ph.D., P.E.
Professor of Chemical Engineering
Louisiana State University, Baton Rouge, LA

Jerry W. Samples, Ph.D., P.E.
Associate Professor of Civil & Mechanical Engineering
United States Military Academy, West Point, NY

Marcia Sullivan, P.E.
Electrical Engineer
Willow Software, Inc., Selden, NY

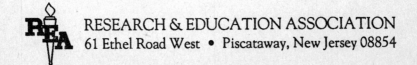

RESEARCH & EDUCATION ASSOCIATION
61 Ethel Road West • Piscataway, New Jersey 08854

THE BEST EIT REVIEW FOR THE
FUNDAMENTALS OF ENGINEERING
(FE) EXAM

Printed in the United States of America

Library of Congress Catalog Card Number 94-65505

International Standard Book Number 0-87891-915-5

Research & Education Association
61 Ethel Road West
Piscataway, New Jersey 08854

REA supports the effort to conserve and
protect environmental resources by
printing on recycled papers.

CONTENTS

About Research and Education Association vii
FE Study Schedule ix

Chapter One
YOU CAN SUCCEED ON THE FE 1
    About Our Authors...3
    About The Test 4
    Test Format 4
    Test Sections 5
    Test Taking Strategies 9

Chapter Two
HOW TO STUDY FOR THE FE EXAM 11

Chapter Three
MATHEMATICS 17
    Algebra..19
    Trigonometry 21
    Linear Algebra 23
    Vector Analysis 29
    Analytical Geometry 33
    Differential Calculus 34
    Integration 38
    Differential Equations 51
    Laplace Transforms 57
    Statistics 60

Chapter Four
ELECTRICAL CIRCUITS 79
    DC Circuits 81
    Current Flow 81

Electric and Magnetic Fields...90
Capacitance and Inductance 99
Transients 106
AC Circuits 114
Power 119
Three-Phase Circuits 123

Chapter Five
STATICS 145
Vector Forces 147
Moments and Couples...150
Two-Dimensional Equilibrium 151
Three-Dimensional Equilibrium 153
Analysis of Internal Forces 156
Distributed Forces, Centroids and Centers of Gravity 159
Friction 161
Moment of Inertia 165

Chapter Six
DYNAMICS.. 181
Kinematics 183
Kinetics 201

Chapter Seven
MECHANICS OF MATERIALS 237
Stress and Strain..239
Shear...241
Tension and Compression..242
Beams...243
Bending Moment Diagram 249
Torsion 252
Columns 253
Combined Stresses 256
Thin-Walled Pressure Vessels 259

Chapter Eight
CHEMISTRY 283
Nomenclature 285
Equations..287
Stoichiometry 288
States of Matter 290
Solutions...295
Periodicity 299

Equilibrium 302
Acids and Bases 306
Oxidation-Reduction 309
Electrochemistry 312
Kinetics 317
Metals and Nonmetals 321
Organic Chemistry 322

Chapter Nine
THERMODYNAMICS 341
Properties 343
Phase Change 345
Thermodynamic Processes 350
Ideal Gases 351
Energy, Heat, and Work 356
Availability-Irreversibility 366
Cycles 369
Air Conditioning and Refrigeration 380
Combustion and Chemical Reactions 386
Heat Transfer 390
Gas Dynamics: Flow Through Nozzles and Blade Passages 393

Chapter Ten
FLUIDS 429
Fluid Properties 431
Fluid Statics 434
Hydraulics and Fluid Machines 441
Momentum 445
Dimensional Analysis and Similitude 448
Pipe Flow and Channel Flow 453
Flow Measurement 459
External Flow 462
Compressible Flow 466

Chapter Eleven
MATERIAL SCIENCE/STRUCTURE OF MATTER 481
Materials and Their Atomic Structure 483
Structures and Structural Defects of Materials 492
Properties of Materials 501
Multicomponent Materials and Phase Diagrams 509
Diffusion and Reactions in Materials 514
Corrosion and Radiation Damage 516
Radiation Alteration of Materials 518

Chapter Twelve
ENGINEERING ECONOMICS 525
  Investment 527
  Time Value of Money 527
  Inflation 529
  Taxes 530
  Risk 530
  Cash Flow 530
  Interest Factors 532
  Minimum Attractive Rate of Return (MARR) 535
  Methods for Project Analysis 536
  Profitability Analysis 540
  Equivalence 545
  Taxes and Depreciation 545
  Sensitivity Analysis 551
  Risk Analysis 552

Chapter Thirteen
ELECTRONICS 563
  Diodes 565
  Bipolar Junction Transistors 580
  Transistor Modeling 587
  Operational Amplifiers 590

APPENDIX 613

ABOUT RESEARCH AND EDUCATION ASSOCIATION

Research and Education Association (REA) is an organization of editors, scientists, and engineers who specialize in various academic fields. Founded in 1959 with the purpose of disseminating the most recently developed scientific information to groups in industry, government, universities, and high schools, REA has since become a highly respected publisher of study aids, test preps, handbooks, and reference works.

Created to extensively prepare students and professionals with the information they need, REA's Test Preparation series includes study guides for the Graduate Records Exam (GRE), the Graduate Records Exam (GRE) in Engineering, the Test of English as a Foreign Language (TOEFL), the Graduate Management Admissions Test (GMAT), and the Medical College Admissions Test (MCAT), and the Law School Admissions Test (LSAT).

Whereas most test preparation books present few practice tests which bear little resemblance to the actual exams, REA's test preparation books present tests which accurately depict the official exams in degree of difficulty and in the types of questions. REA's practice tests are always based on the most recently administered exams and include every type of question that can be expected on the actual exams.

REA's publications and educational materials are highly regarded and continually receive an unprecedented amount of praise from professionals, instructors, librarians, parents, and students. Our authors are as diverse as the subjects and fields represented in the books we publish. They are well known in their respective fields and serve on the faculties of prestigious universities and high schools throughout the United States.

Acknowledgments

In addition to our authors, REA would like to thank the following:

Dr. Max Fogiel, President, for his overall guidance which has brought this publication to its completion.

Stacey A. Sporer, Managing Editor, for directing the editorial staff throughout each phase of the project.

Craig D. Thomason, Editorial Assistant, for preparing the material for print.

Dr. Alkis Constantinides, Arthur Fine, Mark Happe, Jennifer Kovacs, Anne Akemi Ooka, Elizabeth Powell, and Stephen Spitzer for their editorial contributions.

FE
Study
Schedule

STUDY SCHEDULE

The following is a suggested six-week study schedule for the Fundamentals of Engineering Exam. You may want to condense or expand the schedule depending on the amount of time remaining before the test. Set aside some time each week, and work straight through the activity without rushing. By following a structured schedule, you will be able to complete an adequate amount of studying, and be more confident and prepared on the first day of the exam.

Week 1	Acquaint yourself with the FE Review Book by reading the first two chapters: "You Can Succeed on the FE" and "How to Study for the FE." Begin reviewing chapters 3 and 4. As you read the "Mathematics" and "Electric Circuits" chapters, try to solve the examples without the aid of the solutions. Use the solutions to guide you through any questions you missed. Use the practice problems at the end of the chapter to reinforce the concepts you just studied. You may want to leave the last two or three problems for an overall review at a later date.
Week 2	Study review chapters 5 and 6. Take notes as you read the chapters; you may even want to write concepts on index cards and thumb through them during the day. As you read the "Statics" and "Dynamics" chapters, try to solve the examples without the aid of the solutions. Use the solutions to guide you through any questions you missed. Leave the last two or three practice problems for your final overall review.
Week 3	Review any notes you have taken over the last few weeks. Study chapters 7 and 8. As you read the "Mechanics of Materials" and "Chemistry" chapters, try to solve the examples without the aid of the solutions. Use the solutions to guide you through any questions you missed. Save some practice problems for a final review.
Week 4	Study chapters 9 and 10 while continuing to review your notes. As you read the "Thermodynamics" and "Fluids" chapters, try to solve the examples without the aid of the solutions. Use the solutions to guide you through any questions you missed. Continue to leave a few practice problems for your final review.

Week 5	Study chapters 11 and 12. As you read the "Material Science" and "Engineering Economics" chapters, try to solve the examples without the aid of the solutions. Use the solutions to guide you through any questions you missed. Remember to leave a few practice problems for your final review.
Week 6	Study the last review section: "Electronics." Take notes on any material you do not remember from course work, and continue to review your notes from previous chapters. When you have finished reviewing the last chapter, do all of the practice problems that you did not do while studying. Compare your answers to the detailed solutions even if you got them right; it is important to confirm the accuracy of your approach.

Throughout these weeks, you should take as many practice tests as possible. REA's *The Best Test Preparation for the FE(EIT) Exam* provides three practice tests to familiarize you with the format and scope of the FE Exam. Be patient and deliberate as you review; with careful study, you can only improve.

CHAPTER ONE

You Can Succeed On The FE Exam

Chapter 1

YOU CAN SUCCEED ON THE FE EXAM

By reviewing and studying this book, you can succeed on the Fundamentals of Engineering Examination. The FE is an eight-hour exam designed to test knowledge of a wide variety of engineering disciplines. The FE was formerly known as the EIT (Engineer-in-Training) exam. The FE Exam format and title have now replaced the EIT completely.

The FE is now a *supplied reference exam*, and students will no longer be permitted to bring reference material into the test center. Instead, you will be mailed a reference guide when you register for the exam. This guide will provide all the charts, graphs, tables, and formulae you will need. The same book will be given to you in the test center during the test administration.

The purpose of REA's *Best Review for the FE(EIT) Exam* is to prepare you sufficiently for the exam by providing 11 review chapters, including sample problems in each review. These chapters reflect the scope and difficulty level of the actual FE Exam, and provide examples with thorough solutions throughout the text. Along with this book, you may want to use REA's *The Best Test Preparation for the FE Exam*, which provides three simulated FE exams with detailed explanations of answers. While either book is helpful, an effective study plan can incorporate both a review of concepts and repeated practice with simulated tests under exam conditions.

ABOUT OUR AUTHORS

In order to meet our objective of providing reviews that reflect all subjects and subtopics covered on the actual FE, every chapter was carefully prepared by text experts in various fields of engineering. Our authors and editorial review board have thoroughly examined and researched the mechanics of the actual FE exams so that all subject matter is accurate, on the correct difficulty level, and appropriate for an FE review. Our experts are highly regarded in the educational community, holding doctorates and positions in competitive colleges around the United States.

ABOUT THE TEST

The Fundamentals of Engineering Exam (FE) is one part in the four-step process toward becoming a professional engineer (P.E.). Graduating from an approved four-year engineering program and passing the FE qualifies you for your certification as an "Engineer-in-Training" or an "Engineer Intern." The final two steps towards licensing as a P.E. involve completion of four years of additional engineering experience and passing the Principles and Practices of Engineering Examination administered by the National Council of Examiners and Surveying (NCEES). Registration as a professional engineer is deemed both highly rewarding and beneficial in the engineering community.

In order to register for the FE, contact your state's Board of Examiners for Professional Engineers and Land Surveyors. To determine the location for the Board in your state, contact the main NCEES office at the following address:

National Council of Examiners for Engineering and Surveying

PO Box 1686

Clemson, SC 29633-1686

(803) 654-6824

TEST FORMAT

The FE consists of two distinct sections. One section is given in the morning while the other is administered in the afternoon. You will have four hours to complete each section. The morning section consists of 140 questions covering 11 different engineering subjects.

The afternoon section consists of 70 questions arranged in problem set form. Problem set form involves a group of questions based on one diagram or general problem statement. These questions will cover five different engineering subjects. The subjects and their corresponding numbers of questions are shown below.

MORNING SECTION

Subject	No. of Problems
Mathematics	20
Electrical Circuits	14
Fluid Mechanics	14
Thermodynamics	14
Dynamics	14
Statics	14
Chemistry	14

Subject	No. of Problems
Mechanics of Materials	11
Engineering Economics	11
Material Science and Structure of Matter	14

AFTERNOON SECTION

Subject	No. of Problems
Engineering Mechanics	20 (2 problem sets)
Applied Mathematics	20 (2 problem sets)
Electrical Circuits	10 (1 problem set)
Engineering Economics	10 (1 problem set)
Thermodynamics/Fluid Mechanics	10 (1 problem set)

The list of topics covered in the afternoon may seem shorter than the list for the morning section, but keep in mind that the afternoon problem sets will draw on your knowledge of many engineering disciplines, and will not be limited to the subjects specified on the list. For example, the Engineering Mechanics problem set may cover problems involving: Statics, Dynamics, Strength of Materials, and Mechanics of Materials. Remember, all topics on the exam are required; there are no optional sections.

Our review book covers all of these topics thoroughly. Each chapter begins with a short introduction and then progresses to the specifics of the topics. Each topic is explained in detail, with example problems, diagrams, charts, and formulae. Note: The type in the figures may vary slightly from the type in the text. This book is a complete guide to review the topics covered on the exam.

You may want to take some practice exams at various studying stages to measure your strengths and weaknesses. You can obtain sample practice tests and problems from the NCEES, or in a publication that simulates actual FE exams, such as REA's *Best Test Preparation for the FE Exam*. This will help you to determine which topics need more study. Take one test when you finish studying so that you may see how much you have improved. For studying suggestions that will help you to make the best use of your time, see the section of this book entitled "Study Tips."

TEST SECTIONS

As discussed earlier, the FE Exam is divided into two sections: morning and afternoon. These two parts are broken down further into engineering topics. Your registration booklet will explain the breakdown and list the subtopics included under each major heading. The test sections are described in brief in each section that follows.

The topics on the morning section appear as distinct questions, while the afternoon section uses the problem set format. The topics on the AM section will not be specified, nor will this portion be broken down into categories or subtests. The problem sets on the PM section will be labeled according to topic.

Overall, there will be 11 different topics on the FE in both the morning and afternoon sections. The following is a comprehensive summary of the topics included on the exam, along with the types of related questions.

Chemistry (AM: 14 questions, PM: none)

The questions on the morning section involve the following concepts:

Organic Chemistry	Acids and Bases
Nomenclature	Oxidation and Reduction
Kinetics	Periodicity
Equilibrium	Stoichiometry
Equations	States of Matter
Electrochemistry	Solutions

Dynamics (AM: 14 questions, PM: possibly 1 problem set)

These questions will test your knowledge of:

Kinematics	Kinetics

Electrical Circuits (AM: 14 questions, PM: 1 problem set)

The morning section of the FE will definitely contain 14 questions that test your knowledge of Electrical Circuits and related topics. The questions may address the following areas:

AC and DC Circuits	Capacitance and Inductance
Transients	Electrical Fields
Magnetic Fields	Electronics

The afternoon section of the exam may include a problem set dealing with Electrical Circuits, Electronics, or Electrical Machinery. The given diagram or problem statement along with the corresponding questions may test:

Transient and Steady-State Circuits	Electronics
Power and Machines	Electrical and Magnetic Fields

Engineering Economics (AM: 11 questions, PM: 1 problem set)

The morning section of the FE will include 11 questions covering the topic of Engineering Economics. The questions may test your knowledge of any of the following topics:

Time Value of Money	Annual Cost
Present and Future Worth	Capitalized Cost
Break-Even Analysis	Valuation and Depreciation
Depreciation	

The afternoon section of the exam will include one problem set concerning Engineering Economics. In addition to the topics covered in the morning section, the afternoon section may touch upon Present Return.

Fluid Mechanics (AM: 14 questions, PM: 1 problem set)

Fourteen questions in the morning section will test your knowledge of Fluid Mechanics. The questions may deal with any of the following concepts:

Fluid Properties	Fluid Statics
Impulse and Momentum	Dimensional Analysis and Similitude
Flow Measurement	Pipe Flow
Compressible Flow	

The afternoon section of the exam may include a problem set involving Fluid Mechanics. If so, questions will cover the following:

Fluid Statics	Pipe Flows
Momentum and Energy	Dimensional Analysis and Similitude
Hydraulics and Fluid Machines	

Materials Science/Structure of Matter (AM: 14 questions, PM: 1 problem set)

The topics of Material Science and Structure of Matter will be combined on the morning section of the FE exam. Concepts which may be included are:

Physical Properties	Properties
Atomic Structure	Crystallography
Phase Diagrams	Processing and Testing
Diffusion	Corrosion
Materials	

Mathematics (AM: 20 questions, PM: 2 problem sets)

The morning section of the FE will include 20 mathematics problems that test your knowledge in the following areas:

Integral and Differential Calculus   Laplace Transforms

Probability and Statistics   Differential Equalities

Analytical Geometry

Mechanics of Materials (AM: 11 questions, PM: possibly 1 problem set)

Eleven questions in the morning section of the exam will concentrate on the topic of Mechanics of Materials. The questions may test your knowledge of any of the following concepts:

Stress and Strain   Tension and Compression

Shear   Combined Stress

Beams   Columns

Torsion

One problem set pertaining to Mechanics of Materials may appear on the afternoon section of the exam. If so, questions will touch on the following:

Beams   Torsion

Bending   Combined Stresses

Statics (AM: 14 questions, PM: 1 problem set)

Fourteen questions in the morning section of the FE exam will focus on Statics. The questions may deal with any of the following:

Vector Forces   2-dimensional Equilibrium

Concurrent Force-Systems   Centroid Area

Moment of Inertia   Friction

As with Dynamics, Mechanics of Materials and Material Science, a problem set involving Statics may appear in the afternoon section of the exam. Keep in mind that any combination of the two of these four topics will be put on the afternoon portion of the exam. If Statics is one of the two topics, the questions in the problem set may cover the following:

Resultant Force Systems   Equilibrium of Rigid Bodies

Friction Effects   Frames and Trusses

Thermodynamics (AM: 14 questions, PM: possibly 1 problem set)

The morning section of the FE exam will contain 14 questions on Thermodynamics. The questions may test your knowledge of the following topics:

Properties	Phase Change
First Law of Thermodynamics	Second Law of Thermodynamics
Energy, Heat, and Work	Simple Heat Transfer
Availability/Gas Flow	Thermodynamic Processes
Ideal Gasses	Cycles

As mentioned before, either a Fluid Mechanics problem set or a Thermodynamics set will appear on the afternoon section of the exam. If a Thermodynamics problem set is used, it may include:

Cycles	Heat Transfer
Gas Mixtures	Availability-Irreversibility
Chemical Reactions	Nozzles, Turbines, and Compressors

TEST-TAKING STRATEGIES

How to Beat the Clock

Every second counts, and you will want to use the available test time for each section in the most efficient manner. Here's how:

1. Bring a watch! This will allow you to monitor your time.

2. Become familiar with the test directions. You will save valuable time if you already understand the directions on the day of the test.

3. Pace yourself. Work steadily and quickly. Do not spend too much time on any one question. Remember, you can always return to the problems that gave you the most difficulty. Try to answer the easiest questions first, then return to the ones you missed.

Guessing Strategy

1. When all else fails, guess! The score you achieve depends on the number of correct answers. There is no penalty for wrong answers, so it is a good idea to choose an answer for all of the questions.

2. If you guess, try to eliminate choices you know to be wrong. This will allow you to make an educated guess. Here are some examples of what to look for when eliminating answer choices:
    Thermodynamics—check for signs of heat transfer and work
    Fluid Mechanics—check for signs of pressure reading
    Statics—check for direction of forces and compression/tension units.

3. Begin with the subject areas you know best. This will give you more time and will also build your confidence. If you use this strategy, pay careful attention to your answer sheet; you do not want to mismatch the ovals and answers. It may be a good idea to check the problem number and oval number *each time* you mark down an answer.

4. Break each problem down into its simplest components. Approach each part one step at a time. Use diagrams and drawings whenever possible, and do not wait until you get a final answer to assign units. If you decide to move onto another problem, this method will allow you to resume your work without too much difficulty.

CHAPTER TWO

How to Study for the FE Exam

Chapter 2

HOW TO STUDY FOR THE FE EXAM

Two groups of people take the FE Exam: college seniors in undergraduate programs and graduate engineers who decide that professional registration is necessary for future growth. Both groups begin their Professional Engineer career with a comprehensive exam covering the entirety of their engineering curriculum. How does one prepare for an exam of such magnitude and importance?

Time is the most important factor when preparing for the FE. Time management is necessary to ensure that each section is reviewed prior to the exam. Once the decision to test has been made, determine how much time you have to study. Divide this time amongst your topics, and make up a schedule which outlines the beginning and ending dates for study of each exam topic, and includes time for a final test followed by a brief review. Set aside extra time for the more difficult subjects, and include a buffer for unexpected events such as college exams or business trips. There is never enough time to prepare, so make the most of the time you have.

You can determine which subject areas require the most time in several ways. Look at your college grades: those courses with the lowest grades probably need the most study. Those subjects outside your major are generally the least used and most easily forgotten. These will require a good deal of review to bring you up to speed. Some of the subjects may not be familiar at all because you were not required to study them in college. These subjects may be impossible to learn before the exam, although some can be self-taught. One such subject is Engineering Economics; the mathematics are not exceptionally difficult and the concepts are common sense.

Another way to determine your weakest areas is to take a practice FE test. REA's *Best Test Preparation for the FE(EIT) Exam* includes simulated exams which will help you to assess your strengths and weaknesses. These methods will help you find the areas that need the most work, but do not neglect other subjects; do not rule out any subject area until you have reviewed it to some degree.

You may find that a negative attitude is your biggest stumbling block. Many students do not realize the volume of material they have covered in four years of college. Some begin to study and are immediately overwhelmed because they do not have a plan. It is important that you get a good start and that you are positive as you review and study the material.

You will need some way to measure your preparedness, either with problems from books or with a review book that has sample test questions like the ones on the FE. This book contains sample problems in each section which can be used before, during, or after you review the material to measure your understanding of the subject matter. If you are a wizard in Thermodynamics, for example, and are confident in your ability to solve problems, select a few and see what happens. You may want to perform at least a cursory review of the material before jumping into problem solving, since there is always something to learn. If you do well on these initial problems, then momentum has been established. If you do poorly, you might develop a negative attitude as mentioned above. Being positive is essential as you move through the subject areas.

The question that comes to mind at this point is: "How do I review the material?" Before we get into the material itself, let us establish rules which lead to **good study habits**. Time was previously mentioned as the most important asset. When you decide to study you will need blocks of uninterrupted time so that you can get something accomplished. Two hours should be the minimum time block allotted while four hours should be the maximum. Schedule five-minute breaks into your study period and stay with your schedule. Cramming for the FE can give you poor results, including short-term memory and confusion when synthesis is required.

Next, you need to work in a quiet place, on a flat surface that is not cluttered with other papers or work that needs to be completed before the next day. **Eliminate distractions** — they will rob you of time while you pay attention to them. **Do not eat while you study**; few of us can do two things at once and do them both well. Eating does require a lot of attention and disrupts study. Eating a sensible meal before you study resolves the "eating while you work problem." We encourage you to have a large glass of water available since water quenches your thirst and fills the void which makes you want to get up and find something to eat. In addition, **you should be well rested when you study**. Late nights and early mornings are good for some, especially if you have a family, but the best results are associated with adequate rest.

Lastly, **study on weekend mornings while most people are still asleep**. This allows for a quiet environment and gives you the remainder of the day to do other things. If you must study at night, we suggest two-hour blocks ending before 11 p.m.

Do not spend time memorizing charts, graphs, and formulae; the FE is a supplied reference exam, and you will be provided a booklet of equations and other essential information during the test. You can use the supplied reference book as a guide while studying, since it will give you an indication of the depth of study you will need to pursue. Furthermore, familiarity with the book will alleviate some test anxiety since you will be given the same book to use during the actual exam.

While you review for the test, use the review book supplied by the NCEES, paper, pencil, and a calculator. Texts can be used, but reliance on them should be avoided. The object of the review is to identify what you know, the positive, and that which requires work. As you review, move past those equations and concepts that you understand and annotate on the paper those concepts which require more work. Using this method you can review a large quantity of material in a short time and reduce the apparent workload to a manageable amount. Now go back to your time schedule and allocate the remaining time according to the needs of the subject under consideration. Return to the material that requires work and review it or study it until you are satisfied that you can solve problems covering this material. When you have finished the review, you are ready to solve problems.

Solving problems requires practice. To use the problems in this book effectively, you should cover the solution and try to solve the problem on your own. If this is not possible, map out a strategy and then check to see if you have the correct procedure. Remember that most problems that are not solved correctly were never started correctly. Merely reviewing the solutions will not help you to start the problem when you see it again at a later date. Read the problems carefully and in parts. Many people teach that reading the whole problem gives the best overview of what is to come; however, solutions are developed from small clues that are in parts of a sentence. For example: "An engine operates on an Otto Cycle," tells a great deal about the thermodynamic processes, the maximum temperature, the compression ratio, and the theoretical thermal efficiency. **Read the problem and break it into manageable parts.** Next, **try to avoid numbers until the problem is well formulated.** Too often numbers are substituted into equations early and become show stoppers. You will need numbers, just use them after the algebra has been completed. **Be mechanical,** list the knowns, the requirements of the problem, and check off those bits of knowledge you have as they appear. Checking off the intermediate answers and information you know is a positive attitude builder. Continue to solve problems until you are confident or you exceed the time allowed for a subject area.

As soon as you complete one subject, move to the next. Retain all of your notes as you complete each section. You will need these for your final overall review right before the exam. After you have completed the entire review, you may want to take a practice exam. Taking a practice exam will test your understanding of all the engineering subject areas and will help you identify sections that need additional study. With the test and the notes that you retained from the section reviews, you can determine weak areas requiring some additional work.

You should be ready for the exam if you follow these guidelines:

- Program your time wisely.
- Maintain a positive attitude.
- Develop good study habits.
- Review the material smartly to maximize the learning process.
- Do practice problems and a practice test.
- Review again to finalize your preparation.

GOOD LUCK!

CHAPTER THREE

Mathematics

Chapter 3

MATHEMATICS

When reviewing for the mathematics portion of the exam, it is strongly recommended that you use the mathematics section of your engineering handbook in addition to this chapter. Your handbook has formulas for areas, surfaces, and volumes. It has tables for derivatives, integrals, statistical functions, and Laplace Transforms. It has definitions and applications. This chapter reviews the fundamental concepts that are emphasized on the exam. Use this chapter along with your handbook to reinforce your mathematical skills.

The supplied reference handbook you will be given for use during the exam contains limited mathematical tables and formulas. Be sure you understand how to use all entries in the handbook, but do not be concerned with memorizing longer tables or additional formulas. The exam seeks to determine how well you understand and can apply the concept of integration, for example, rather than how quickly you can find the correct entry in a Table of Integrals.

ALGEBRA

Algebra defines the rules to allow us to rearrange, expand, and simplify mathematical relationships.

Commutative law for addition and multiplication:

$$a + b = b + a \qquad a \cdot b = b \cdot a$$

Associative law for addition and multiplication:

$$a + (b + c) = (a + b) + c \quad a \cdot (b \cdot c) = (a \cdot b) \cdot c$$

Distributive law:

$$a \cdot (b + c) = a \cdot b + a \cdot c$$

Partial Fraction Expansion

A ratio of polynomials

$$f(x) = \frac{g(x)}{h(x)}$$

can be written as a sum of terms each of which has a root of $h(x)$ in the denominator.

–Each single real root of $h(x)$ will generate a term of the form:

$$\frac{A}{x-r}$$

–A real root appearing n times will generate a set of terms:

$$\frac{A_1}{(x-r)}+\frac{A_2}{(x-r)^2}+\frac{A_3}{(x-r)^3}+.....+\frac{A_n}{(x-r)^n}$$

–A single pair of complex conjugate roots will generate a term of the form:

$$\frac{Ax+B}{x^2+rx+s}$$

Terms for complex conjugate roots $a+bi$ and $a-bi$ can also be written as

$$\frac{C_1}{x-(a+bi)}+\frac{C_2}{x-(a-bi)}$$

where the constants C_1 and C_2 are a pair of complex conjugates.

–Pairs of complex conjugate roots appearing n times will generate a set of terms:

$$\frac{A_1x+B_1}{x^2+rx+s}+\frac{A_2x+B_2}{\left(x^2+rx+s\right)^2}+.....+\frac{A_nx+B_n}{\left(x^2+rx+s\right)^n}$$

The function

$$f(x)=\frac{x^2+2x-2}{x^3(x+2)\left(x^2+x+9\right)^2}$$

can be expanded as

$$\frac{A_1}{x}+\frac{A_2}{x^2}+\frac{A_3}{x^3}+\frac{B}{x+2}+\frac{C_1x+D_1}{x^2+x+9}+\frac{C_2x+D_2}{\left(x^2+x+9\right)^2}.$$

EXAMPLE: Expand the following function into partial fractions and solve for the constants

$$f(x)=\frac{x^2+1}{(x-2)(x-1)(2x+1)}=\frac{A}{x-2}+\frac{B}{x-1}+\frac{C}{2x+1}$$

SOLUTION: To solve for the unknown constants A, B, and C, multiply the equation by $(x-2)(x-1)(2x+1)$.

$x^2+1=A(x-1)(2x+1)+B(x-2)(2x+1)+C(x-2)(x-1)$

Set $x=2$ $5=A(1)(5)+0+0$ $A=1$

Set $x=1$ $2=0+B(-1)(3)+0$ $B=-\dfrac{2}{3}$

Set $x=-\dfrac{1}{2}$ $\dfrac{5}{4}=0+0+C\left(-\dfrac{5}{4}\right)\left(-\dfrac{3}{2}\right)$ $C=\dfrac{1}{3}$

TRIGONOMETRY

A right triangle contains one 90° angle.

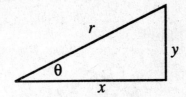

Figure 1. A right triangle

The following definitions and relations apply:

$$x^2 + y^2 = r^2$$

$$\sin\theta = \frac{y}{r}$$

$$\cos\theta = \frac{x}{r}$$

$$\tan\theta = \frac{y}{x}$$

$$\csc\theta = \frac{r}{y} = \frac{1}{\sin\theta}$$

$$\sec\theta = \frac{r}{x} = \frac{1}{\cos\theta}$$

$$\cot\theta = \frac{x}{y} = \frac{1}{\tan\theta}$$

$$\sin^2\theta + \cos^2\theta = 1$$
$$1 + \tan^2\theta = \sec^2\theta$$
$$1 + \cot^2\theta = \csc^2\theta$$

$$\sin 2\theta = 2(\sin\theta)(\cos\theta)$$
$$\cos 2\theta = \cos^2\theta - \sin^2\theta$$
$$= 1 - 2\sin^2\theta$$

$$\sin\theta = 2\left[\sin\left(\frac{1}{2}\theta\right)\cos\left(\frac{1}{2}\theta\right)\right]$$

$$\sin\left(\frac{1}{2}\theta\right) = \pm\sqrt{\frac{1}{2}(1 - \cos\theta)}$$

$$\sin(\theta + \phi) = [\sin\theta][\cos\phi] + [\cos\theta][\sin\phi]$$

$$\sin(\theta - \phi) = [\sin\theta][\cos\phi] - [\cos\theta][\sin\phi]$$
$$\cos(\theta + \phi) = [\cos\theta][\cos\phi] - [\sin\theta][\sin\phi]$$
$$\cos(\theta - \phi) = [\cos\theta][\cos\phi] + [\sin\theta][\sin\phi]$$

Laws for a General Triangle

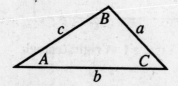

Figure 2. A general triangle

Law of Sines: $\dfrac{\sin A}{a} = \dfrac{\sin B}{b} = \dfrac{\sin C}{c}$

Law of Cosines: $a^2 = b^2 + c^2 - 2bc(\cos A)$

Area $= \dfrac{1}{2}ab(\sin C)$

Angles, Quadrants, and Signs

Angles are measured from the positive horizontal axis, and the positive direction is counterclockwise.

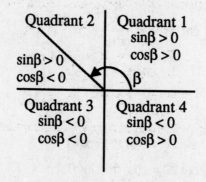

One complete circle is 360° or 2π radians. A right angle of 90° is equivalent to $\dfrac{\pi}{2}$ radians. One radian is equivalent to 57.3°.

LINEAR ALGEBRA

Finding the solution of a set of n linear algebraic equations in n unknowns is a primary application of matrix manipulation. Before this application is illustrated, some definitions are reviewed.

A **scalar** is a single number which has magnitude only, and requires no subscript.

$$C = 4$$

A **vector** is a set of numbers which have both magnitude and direction. They are written with one subscript.

$$\mathbf{a} = \begin{bmatrix} A_1 \\ \cdot \\ \cdot \\ A_n \end{bmatrix}$$

A **matrix** is an array of vectors requiring two subscripts. The first subscript identifies the **row** position; the second identifies the **column**.

$$\mathbf{A} = \begin{bmatrix} A_{11} & A_{12} & .... & A_{1n} \\ A_{21} & A_{22} & .... & A_{2n} \\ \cdot & & & \\ \cdot & & & \\ A_{m1} & A_{m2} & .... & A_{mn} \end{bmatrix}$$

To add two vectors or two matrices of the same size, add the corresponding terms.

$$\mathbf{A} + \mathbf{B} = \mathbf{C} \qquad A_{ij} + B_{ij} = C_{ij}$$

$$\begin{bmatrix} 1 & 4 \\ 3 & 9 \\ 2 & 0 \end{bmatrix} + \begin{bmatrix} 4 & 0 \\ 1 & 0 \\ 2 & 3 \end{bmatrix} = \begin{bmatrix} 5 & 4 \\ 4 & 9 \\ 4 & 3 \end{bmatrix}$$

The transpose of a matrix $\mathbf{A}$ is denoted by $\mathbf{A}^T$ and is formed by interchanging rows and columns.

$$\mathbf{A} = \begin{bmatrix} 1 & 2 & 3 \\ 4 & 5 & 6 \end{bmatrix}$$

$$\mathbf{A}^T = \begin{bmatrix} 1 & 4 \\ 2 & 5 \\ 3 & 6 \end{bmatrix}$$

Multiplication of Matrices

$$AB = C$$

Each element of C is given by $C_{ij} = \sum_{k=1}^{n} A_{ik} B_{kj}$.

Note that multiplication is valid only if the *number of rows in A is the same as the number of columns in B*. Matrix multiplication is not commutative; AB is not necessarily the same as BA.

Let
$$A = \begin{bmatrix} 1 & 2 & 3 \\ 4 & 1 & 2 \end{bmatrix} \quad B = \begin{bmatrix} 1 & 2 \\ 4 & 1 \\ 3 & 2 \end{bmatrix}$$

$$AB = \begin{bmatrix} 1 & 2 & 3 \\ 4 & 1 & 2 \end{bmatrix} \begin{bmatrix} 1 & 2 \\ 4 & 1 \\ 3 & 2 \end{bmatrix}$$

$$= \begin{bmatrix} 1 + 8 + 9 & 2 + 2 + 6 \\ 4 + 4 + 6 & 8 + 1 + 4 \end{bmatrix}$$

Thus,

$$AB = \begin{bmatrix} 18 & 10 \\ 16 & 13 \end{bmatrix}$$

Determinant

The **determinant** $|A|$ is a scalar quantity associated with matrix A. The determinant of a two-by-two matrix is

$$|A| = \begin{vmatrix} A_{11} & A_{12} \\ A_{21} & A_{22} \end{vmatrix} = A_{11}A_{22} - A_{12}A_{21}$$

The determinant of a three-by-three matrix is

$$|B| = \begin{vmatrix} B_{11} & B_{12} & B_{13} \\ B_{21} & B_{22} & B_{23} \\ B_{31} & B_{32} & B_{33} \end{vmatrix}$$

$$= B_{11}B_{22}B_{33} + B_{12}B_{23}B_{31} + B_{13}B_{21}B_{32}$$
$$- B_{13}B_{22}B_{31} - B_{12}B_{21}B_{33} - B_{11}B_{23}B_{32}$$

EXAMPLE: Find the determinant of an arbitrary three-by-three matrix, and the $|A|$ where:

$$A = \begin{bmatrix} -5 & 0 & 2 \\ 6 & 1 & 2 \\ 2 & 3 & 1 \end{bmatrix}$$

SOLUTION:

Let
$$A = \begin{bmatrix} b_{11} & b_{12} & b_{13} \\ b_{21} & b_{22} & b_{23} \\ b_{31} & b_{32} & b_{33} \end{bmatrix}$$

The two-by-two matrix inside the dotted box (- - -) is called a minor.

$$\begin{bmatrix} b_{11} & b_{12} & b_{13} \\ b_{31} & b_{32} & b_{33} \end{bmatrix}$$

Expand the above determinant by minors, using the first column.

$$|A| = +b_{11}\begin{vmatrix} b_{22} & b_{23} \\ b_{32} & b_{33} \end{vmatrix} - b_{21}\begin{vmatrix} b_{12} & b_{13} \\ b_{32} & b_{33} \end{vmatrix}$$
$$+ b_{31}\begin{vmatrix} b_{12} & b_{13} \\ b_{22} & b_{23} \end{vmatrix}$$

$$|A| = b_{11}(b_{22}b_{33} - b_{32}b_{23}) - b_{21}(b_{12}b_{33} - b_{32}b_{13}) + b_{31}(b_{12}b_{23} - b_{22}b_{13})$$

$$|A| = -b_{21}(b_{12}b_{33} - b_{32}b_{13}) + b_{22}(b_{11}b_{33} - b_{31}b_{13}) - b_{23}(b_{11}b_{32} - b_{31}b_{12})$$
$$= b_{22}b_{11}b_{33} - b_{22}b_{31}b_{13} - b_{23}b_{11}b_{32} + b_{23}b_{31}b_{12} - b_{21}(b_{12}b_{33} - b_{32}b_{13})$$
$$= b_{11}(b_{22}b_{33} - b_{32}b_{23}) - b_{21}(b_{12}b_{33} - b_{32}b_{13}) + b_{31}(b_{12}b_{23} - b_{22}b_{13})$$

Clearly, this is the same as the first answer. Note, also, that $|A|$ can be rearranged algebraically until it can be written as:

$$|A| = b_{11}b_{22}b_{33} + b_{12}b_{23}b_{31} + b_{13}b_{32}b_{21}$$
$$- (b_{13}b_{22}b_{31} + b_{23}b_{32}b_{11} + b_{33}b_{21}b_{12})$$

Expand the determinant by minors, using the first column.

$$|A| = -5\begin{vmatrix} 1 & 3 \\ 3 & 1 \end{vmatrix} - 6\begin{vmatrix} 0 & 2 \\ 3 & 1 \end{vmatrix} + 2\begin{vmatrix} 0 & 2 \\ 1 & 2 \end{vmatrix}$$
$$= -5(1-6) - 6(0-6) + 2(0-2) = 25 + 36 - 4 = 57$$

Now expand the determinant by minors, using the second row.

$$|A| = -b_{21}\begin{vmatrix} b_{12} & b_{13} \\ b_{32} & b_{33} \end{vmatrix} + b_{22}\begin{vmatrix} b_{11} & b_{13} \\ b_{31} & b_{33} \end{vmatrix}$$

$$-b_{23}\begin{vmatrix} b_{11} & b_{12} \\ b_{31} & b_{32} \end{vmatrix}$$

Inverse

A **nonsingular** matrix A possesses a unique **inverse** A^{-1} such that

$$A^{-1}A = A\,A^{-1} = I$$

where I is the **unit** matrix, which has 1's on the diagonal and 0's in all other positions.

$$I = \begin{bmatrix} 1 & 0 & 0 & \ldots \\ 0 & 1 & 0 & \ldots \\ & \ldots & & \end{bmatrix}$$

If A is a two-by-two matrix

$$A = \begin{bmatrix} A_{11} & A_{12} \\ A_{21} & A_{22} \end{bmatrix}$$

its inverse is $A^{-1} = \dfrac{1}{|A|}\begin{bmatrix} A_{22} & -A_{12} \\ -A_{21} & A_{11} \end{bmatrix}$.

If A is a general nonsingular square matrix, its inverse is

$$A^{-1} = \frac{\text{Adj A}}{|A|}$$

where Adj A is the **adjoint of A**, which is formed by replacing each element by its **cofactor** and then interchanging rows and columns.

A **singular** matrix has a determinant of 0.

The **rank** of a matrix is the order of the largest square array in that matrix (formed by deleting rows and columns) whose determinant does not vanish. The rank indicates the number of independent relations.

Let
$$A = \begin{bmatrix} 3 & -1 & -2 \\ 0 & 2 & -1 \\ 4 & -10 & 2 \end{bmatrix}$$

The **minor** M_{12} is found by deleting the first row and second column

$$\begin{bmatrix} 3 & -1 & -2 \\ 0 & 2 & -1 \\ 4 & -10 & 2 \end{bmatrix}$$

and calculating the determinant of the resulting two-by-two matrix

$$M_{12} = \begin{vmatrix} 0 & -1 \\ 4 & 2 \end{vmatrix} = 0 - (-4) = 4$$

The **cofactor of A_{12}** is

$$C_{12} = (-1)^{1+2} M_{12} = -4.$$

One way to find |A| is to add the products of the elements of the top row with their cofactors.

$$|A| = A_{11} C_{11} + A_{12} C_{12} + A_{13} C_{13}$$
$$= 3(-1)^{1+1}(4 - 10) + (-1)(-1)^{1+2}(0 + 4) + (-2)(-1)^{1+3}(0 - 8)$$
$$= -18 + 4 + 16 = 2$$

A Set of Linear Equations

The equations

$$A_{11}x_1 + A_{12}x_2 + A_{13}x_3 + \ldots\ldots + A_{1n}x_n = b_1$$
$$A_{21}x_1 + A_{22}x_2 + A_{23}x_3 + \ldots\ldots + A_{2n}x_n = b_2$$
$$A_{31}x_1 + A_{32}x_2 + A_{33}x_3 + \ldots\ldots + A_{3n}x_n = b_3$$
$$\ldots\ldots\ldots\ldots\ldots\ldots\ldots$$
$$A_{n1}x_1 + A_{n2}x_2 + A_{n3}x_3 + \ldots\ldots + A_{nn}x_n = b_n$$

can be written as $\mathbf{Ax = b}$

where

    A is the coefficient matrix

    x is the vector of unknowns

    b is the vector of constants

$$\mathbf{A} = \begin{bmatrix} A_{11} & A_{12} & A_{13} & A_{14} & \ldots\ldots & A_{1n} \\ A_{21} & A_{22} & A_{23} & A_{24} & \ldots\ldots & A_{2n} \\ A_{31} & A_{32} & A_{33} & A_{34} & \ldots\ldots & A_{3n} \\ \ldots\ldots & & & & & \\ A_{n1} & A_{n2} & A_{n3} & A_{n4} & \ldots\ldots & A_{nn} \end{bmatrix}$$

$$\mathbf{x} = \begin{bmatrix} x_1 \\ x_2 \\ x_3 \\ .. \\ x_n \end{bmatrix} \qquad \mathbf{b} = \begin{bmatrix} b_1 \\ b_2 \\ b_3 \\ .. \\ b_n \end{bmatrix}$$

Solution via the Inverse of A

One solution technique is to multiply the equation by $\mathbf{A}^{-1}$.

$$\mathbf{Ax} = \mathbf{b}$$

$$\mathbf{A}^{-1}\mathbf{Ax} = \mathbf{A}^{-1}\mathbf{b}$$

$$\mathbf{Ix} = \mathbf{x} = \mathbf{A}^{-1}\mathbf{b}$$

Solution via Cramer's Rule

Each unknown x_i can be found by

$$x_i = \frac{|\Delta_i|}{|\mathbf{A}|}$$

where $|\mathbf{A}|$ is the determinant of the coefficient matrix

$|\Delta_i|$ is the determinant of a matrix formed by replacing the ith column of the coefficient matrix $\mathbf{A}$ by the vector of constants $\mathbf{b}$.

Eigenvalues

Eigenvalues, characteristic values or latent roots are scalars and are roots of

$$|\mathbf{A} - \lambda\mathbf{I}| = 0$$

$$\mathbf{A} = \begin{bmatrix} 5 & 4 \\ 1 & 2 \end{bmatrix}$$

$$\mathbf{A} - \lambda\mathbf{I} = \begin{bmatrix} 5 & 4 \\ 1 & 2 \end{bmatrix} - \lambda \begin{bmatrix} 1 & 0 \\ 0 & 1 \end{bmatrix}$$

$$= \begin{bmatrix} 5-\lambda & 4 \\ 1 & 2-\lambda \end{bmatrix}$$

$$\begin{vmatrix} 5-\lambda & 4 \\ 1 & 2-\lambda \end{vmatrix} = (5-\lambda)(2-\lambda) - 4 = 0$$

$$\lambda^2 - 7\lambda + 6 = (\lambda - 6)(\lambda - 1) = 0$$

The Eigenvalues are 1, 6.

VECTOR ANALYSIS

A vector has a direction and a magnitude. In three-dimensional space, vectors can be written as

$$A = a_x i + a_y j + a_z k \qquad B = b_x i + b_y j + b_z k$$

where i, j, and k are unit vectors pointing in the positive x, y, and z directions, respectively.

The magnitude of $A = \sqrt{a_x^2 + a_y^2 + a_z^2}$.

The **scalar product**, or **dot product**, is a scalar defined by

$$A \cdot B = |A|\,|B| \cos \beta$$

where $|A|$ and $|B|$ are the magnitudes of A and B, and β is the angle between them. Written in terms of the three components of each vector, the dot product is

$$A \cdot B = a_x b_x + a_y b_y + a_z b_z.$$

The dot product of perpendicular vectors is 0. The dot product is used to define the projection of vector A in the direction of B.

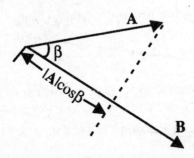

Figure 3. Projection of vector A in the direction of B

The unit vector that points in the direction of B is $\dfrac{B}{|B|}$. The magnitude of the projection of vector A in the direction of B is $|A| \cos \beta$.

Proj of A on $B = |A|\cos\beta \dfrac{B}{|B|}$.

The **vector product**, or **cross product**, is written as $C = A \times B$. The magnitude of C is

$$|C| = |A|\,|B| \sin \beta.$$

The magnitude of **C** is the area of the parallelogram with sides **A** and **B**. The direction of **C** is perpendicular to the plane of **A** and **B** and follows the right-hand rule. If the curled fingers of the right hand are wrapped in the direction from vector **A** to vector **B**, the thumb then points in the direction of **C**. Written in terms of the three components of each vector, the cross product is

$$\mathbf{A} \times \mathbf{B} = (a_y b_x - a_z b_y)i + (a_z b_x - a_x b_z)j + (a_x b_y - a_y b_x)k.$$

This is the determinant of

$$\mathbf{A} \times \mathbf{B} = \begin{vmatrix} a_x & a_y & a_z \\ b_x & b_y & b_z \\ i & j & k \end{vmatrix}$$

The **scalar triple product** is defined as

$$\mathbf{A} \times \mathbf{B} \cdot \mathbf{C} = \begin{vmatrix} a_x & a_y & a_z \\ b_x & b_y & b_z \\ c_x & c_y & c_z \end{vmatrix}$$

$$= a_x(b_y c_z - b_z c_y) - a_y(b_x c_z - b_z c_x) + a_z(b_x c_y - b_y c_x)$$

The scalar triple product is the volume of a parallelepiped with side **A**, **B**, and **C**.

It is convenient to define ∇, a **vector differential operator** as

$$\nabla = \frac{\partial}{\partial x}i + \frac{\partial}{\partial y}j + \frac{\partial}{\partial z}k.$$

The **gradient** of a scalar function $F(x, y, z)$ is a vector

$$\nabla F = \frac{\partial F}{\partial x}i + \frac{\partial F}{\partial y}j + \frac{\partial F}{\partial z}k.$$

The direction of the gradient vector is perpendicular to the surface $F(x, y, z) =$ constant.

The **divergence** of a vector function V is a scalar, which is the dot product of the differential operator and **A**.

$$\nabla \cdot \mathbf{A} = \frac{\partial v_x}{\partial x} + \frac{\partial v_y}{\partial y} + \frac{\partial v_z}{\partial z}$$

The **curl** of a vector function **V** is a vector, which is the cross product of the differential operator and **V**.

$$\nabla \times \mathbf{V} = \left(\frac{\partial v_z}{\partial y} - \frac{\partial v_y}{\partial z}\right)i + \left(\frac{\partial v_x}{\partial z} - \frac{\partial v_z}{\partial x}\right)j + \left(\frac{\partial v_y}{\partial x} - \frac{\partial v_x}{\partial y}\right)k$$

A vector function **V** is conservative or irrotational if it is the gradient of a scalar function $F(x, y, z)$. The curl of a conservative vector function is 0.

Let $\mathbf{A} = 3i + 2j + k$ and $\mathbf{B} = i - 4j + 3k$.

The magnitude of A is $|\mathbf{A}| = \sqrt{3^2 + 2^2 + 1^2} = \sqrt{14}$.

The magnitude of B is $|\mathbf{B}| = \sqrt{1^2 + (-4)^2 + 3^2} = \sqrt{26}$.

The dot product of A and B is
$$\mathbf{A} \cdot \mathbf{B} = 3(1) + 2(-4) + 1(3) = 7.$$

The angle between A and B is given by
$$\cos\beta = \frac{\mathbf{A} \cdot \mathbf{B}}{|\mathbf{A}||\mathbf{B}|} = \frac{7}{\sqrt{14 \cdot 26}} = 0.33669 \qquad \beta = 68.5°$$

The cross product between A and B is
$$\mathbf{A} \times \mathbf{B} = (2(1) - 1(-4))i + (1(1) - 3(3))j + (3(-4) - 2(1))k$$
$$= 6i - 8j - 14k$$

The projection of A in the direction of B is
$$|\mathbf{A}|\cos\beta\frac{\mathbf{B}}{|\mathbf{B}|} = 0.2471(i - 4j + 3k)$$
$$= 0.2471i - 0.9883j + 0.7412k$$

EXAMPLE: Let $F(x, y, z) = x^2 + 5xy - 8xyz^3 - 5$.

SOLUTION: The gradient of this function is
$$\nabla F = (2x + 5y - 8yz^3)i + (5x - 8xz^3)j + (-24xyz^2)k.$$

Let $\mathbf{V} = \nabla F$ and find the curl of V.
$$\nabla \times \mathbf{V} = (-24xz^2 - (-24xz^2))i + (-24yz^2 - (-24yz^2))j + ((5 - 8z^3) - (5 - 8z^3))k = 0$$

The curl is 0 as expected because V is the gradient of $F(x, y, z)$.

Complex Numbers

A complex number z has a real part and an imaginary part.

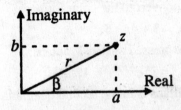

Figure 4. Components of a complex number

From the diagram, $r = \sqrt{a^2 + b^2}$ and $\beta = \arctan\left(\frac{b}{a}\right)$.

$$a = r \cos \beta \text{ and } b = r \sin \beta$$

The number z can be expressed in rectangular form,

$$z = a + ib \text{ where } i = \sqrt{-1},$$

or polar form,

$$z = re^{i\beta}.$$

These forms are completely equivalent because of **Euler's Equations**:

$$e^{i\beta} = \cos\beta + i\sin\beta$$

$$e^{-i\beta} = \cos\beta - i\sin\beta$$

$$\sin\beta = \frac{e^{i\beta} - e^{-i\beta}}{2i}$$

$$\cos\beta = \frac{e^{i\beta} + e^{-i\beta}}{2}$$

Complex conjugates in the rectangular form have imaginary parts of opposite signs. The complex conjugate of $7 + 2i$ is $7 - 2i$. Complex conjugates in the polar form are symmetrical with respect to the horizontal axis. When adding and subtracting, it is convenient to use the rectangular form. For finding powers and roots, the polar form is preferred.

Let $z_1 = 3 + 4i$ and $z_2 = 1 - i$.

The sum of z_1 and z_2 is

$$z_1 + z_2 = 3 + 4i + 1 - i = 4 - 3i.$$

The product of z_1 and z_2 is

$$z_1 z_2 = (3 + 4i)(1 - i) = 3 - 3i + 4 - 4i^2 = 3 + i - 4(-1) = 7 + i.$$

To divide z_1 by z_2, multiply the numerator and denominator by the conjugate of z_2.

$$\frac{z_1}{z_2} = \frac{3+4i}{1-i}\frac{1+i}{1+i} = \frac{3+3i+4i+4i^2}{1-i^2} = \frac{3+7+4(-1)}{1-(-1)^2}$$

To find the cube of z_1, convert to the polar form

$$z_1 = 3 + 4i = 5e^{i0.927}.$$

Note that the angle is expressed as 0.927 radians.

$$z_1^3 = (5e^{i0.927})^3 = 125e^{i2.781}$$

The angle $\beta = 2.781$ radians is in the second quadrant.

To convert to rectangular form

$$a = r\cos\beta = 125\cos(2.781) = 125(-0.936) = -116.0$$
$$b = r\sin\beta = 125\sin(2.781) = 125(0.3528) = 44.1$$

Thus, the cube of $3 + 4i$ is $-116.0 + 44.1i$.

ANALYTICAL GEOMETRY

Straight Lines

The equation of a straight line is $y = mx + b$, where m is the slope and b is the intercept. If two straight lines are perpendicular, their slopes are negative reciprocals.

Conic Sections

The general second degree equation has the form

$$Ax^2 + Bxy + Cy^2 + Dx + Ey + F = 0$$

and describes one of three conic sections:

If $B^2 - 4AC < 0$,   the equation describes an **ellipse**.
                       If $B = 0$ and if $A = C$ this is a **circle**.

If $B^2 - 4AC = 0$,   the equation describes a **parabola**.

If $B^2 - 4AC > 0$,   the equation describes a **hyperbola**.

Circle

The general form for a circle is

$$(x - a)^2 + (y - b)^2 = r^2$$

where a and b is the center, and r is the radius.

Similarly, the general form of a sphere is

$$(x - a)^2 + (y - b)^2 + (z - c)^2 = r^2.$$

Ellipse

The sum of the distances from the two foci F to any point on an ellipse is a constant. For an ellipse centered at the origin

$$\frac{x^2}{a^2} + \frac{y^2}{b^2} = 1$$

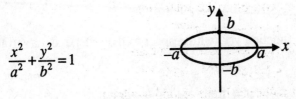

Figure 5. Ellipse at origin

where a and b are the semimajor and semiminor axes. The foci are at $+c, 0$ and $-c, 0$ where $c^2 = a^2 - b^2$.

Parabola

All points on a parabola are equidistant from the focus F and a line called the directrix. If the **vertex** is at the origin and the parabola opens to the right, the equation is

$$y^2 = 4px$$

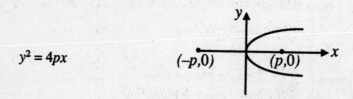

Figure 6. Parabola

where the **focus** is at $(p, 0)$ and the **directrix** is at $x = -p$.

Hyperbola

The difference of the distances from the foci F to any point on the curve is a constant. If the hyperbola is centered at the origin and opens left and right, the standard form is

$$\frac{x^2}{a^2} + \frac{y^2}{b^2} = 1$$

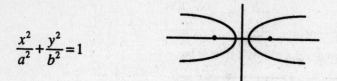

Figure 7. Hyperbola

DIFFERENTIAL CALCULUS

Derivatives

The **derivative** of a function $y(x)$ with respect to the independent variable x is defined as

$$\frac{dy}{dx} = y' = \lim_{\Delta x \to 0} \frac{y(x + \Delta x) - y(x)}{\Delta x}$$

and represents the slope, or how much y changes for a very small change in x. Derivatives of standard forms are given in Table 1.

```
┌─────────────────────────────────────────┐
│              TABLE 1                      │
│           DERIVATIVES                     │
├───────────────────────────────────────────┤
```

$$d(au) = a\ du$$

$$d(u + v - w) = du + dv - dw$$

$$d(uv) = u\ dv + v\ du$$

$$d\left(\frac{u}{v}\right) = \frac{v\,du - u\,dv}{v^2}$$

$$d(u^n) = nu^{n-1}\ du$$

$$d(u^v) = vu^{v-1}\ du + u^v(\log_e u)\ dv$$

$$d(e^u) = e^u\ du$$

$$d(e^{au}) = ae^{au}\ du$$

$$d(a^u) = a^u(\log_e a)\ du$$

$$d(\log_e u) = u^{-1}du$$

$$d(\log_a u) = u^{-1}(\log_a e)\ du$$

$$d(u^u) = u^u(1 + \log_e u)\ du$$

$$d\sin u = \cos u\ du$$

$$d\cos u = -\sin u\ du$$

$$d\tan u = \sec^2 u\ du$$

$$d\cot u = -\csc^2 u\ du$$

$$d\sec u = \tan u \sec u\ du$$

$$d\csc u = -\cot u \csc u\ du$$

Maxima and Minima

For a function $y(x)$, it is necessary that $\dfrac{dy}{dx} = 0$ at points of **maximum** and **minimum** values of y. Such points of 0 slope are called **critical points**.

At a critical point,

if $\dfrac{d^2y}{dx^2} < 0$, the point is a **local maximum**;

if $\dfrac{d^2y}{dx^2} > 0$, the point is a **local minimum**;

if $\dfrac{d^2y}{dx^2} = 0$, the test fails, no conclusion can be made, and additional analysis is necessary.

Consider the function

$$y = x(x - 1)^3.$$

The first derivative is

$$\frac{dy}{dx} = 3x(x-1)^2 + (x-1)^3(1) = (x-1)^2(4x-1)$$

and the second derivative is

$$\frac{d^2y}{dx^2} = (x-1)^2 4 + (4x-1)2(x-1)$$
$$= (x-1)(4x-4+8x-2)$$
$$= (x-1)(12x-6) = 6(x-1)(2x-1)$$

Critical points exist where

$$\frac{dy}{dx} = 0 \text{ or } x = 1 \text{ and } x = \frac{1}{4}.$$

For $x = 1$, $\frac{d^2y}{dx^2} = 0$. This point could be a local maximum, a local minimum, or neither.

For $x = \frac{1}{4}$, $\frac{d^2y}{dx^2} = 6\left(-\frac{3}{4}\right)\left(-\frac{1}{2}\right) > 0$ and the point is a local minimum.

Test for Increasing or Decreasing Functions

Let y be a function that is differentiable on the interval (a, b).

1. If $\frac{dy}{dx} > 0$ for all x in (a, b), then y is increasing on (a, b).

2. If $\frac{dy}{dx} < 0$ for all x in (a, b), then y is decreasing on (a, b).

3. If $\frac{dy}{dx} = 0$ for all x in (a, b), then y is constant on (a, b).

Test for Concavity

Let y be a function whose second derivative exists on an open interval I.

1. If $\frac{d^2y}{dx^2} > 0$ for all x in I, then the graph of y is concave upward.

2. If $\frac{d^2y}{dx^2} < 0$ for all x in I, then the graph of y is concave downward.

Points of Inflection

If $(c, y, (c))$ is a point of inflection of the graph of y, then either $\dfrac{d^2y(c)}{dx^2} = 0$ or $\dfrac{d^2y}{dx^2}$ is undefined at $x = c$.

Second Derivative Test

Let y be a function such that $\dfrac{d^2y(c)}{dx^2} = 0$ and the second derivative of y exists on an open interval containing c.

1. If $\dfrac{d^2y(c)}{dx^2} > 0$, then $y(c)$ is a relative minimum.

2. If $\dfrac{d^2y(c)}{dx^2} < 0$, then $y(c)$ is a relative maximum.

3. If $\dfrac{d^2y(c)}{dx^2} = 0$, then the test fails.

EXAMPLE: Determine the relative maxima, relative minima, and points of inflection of the function:

$$f(x) = \frac{1}{4}x^4 - \frac{3}{2}x^2.$$

SOLUTION: The derivatives are

$$f'(x) = x^3 - 3x \text{ and } f''(x) = 3x^2 - 3.$$

The critical points are solutions of $x^3 - 3x = 0$. We obtain $x = 0, \sqrt{3}, -\sqrt{3}$. The Second Derivative Test tells us that

$$x = 0 \text{ is a relative maximum;}$$

$$x = \sqrt{3}, -\sqrt{3} \text{ are relative minima.}$$

The possible points of inflection are solutions of $3x^2 - 3 = 0$; that is, $x = +1, -1$.

Partial Derivatives

The **partial derivative** of a function $F(x, y, z)$ with respect to x is written as $\dfrac{\partial F}{\partial x}$ and is obtained by considering y and z to be held constant.

The **total differential** of $F(x, y, z)$ is

$$dF = \frac{\partial F}{\partial x}dx + \frac{\partial F}{\partial y}dy + \frac{\partial F}{\partial z}dz.$$

To define the derivative $\dfrac{dy}{dx}$ for an equation in the implicit form

$$F(x, y) = 0$$

first write the total differential, which has a value of 0, because F has a constant value of 0.

$$dF = \frac{\partial F}{\partial x}dx + \frac{\partial F}{\partial y}dy = 0$$

Solve for the desired derivative

$$\frac{dy}{dx} = -\frac{\frac{\partial F}{\partial x}}{\frac{\partial F}{\partial y}}$$

Let $F(x, y) = x^3 + y^2 - 2xy = 0$ then $\frac{\partial F}{\partial x} = 3x^2 - 2y$ and $\frac{\partial F}{\partial y} = 2y - 2x$.

The total differential is

$$dF = (3x^2 - 2y)dx + (2y - 2x)dy = 0$$

and the derivative of y with respect to x in this implicit function is

$$\frac{dy}{dx} = \frac{3x^2 - 2y}{2y - 2x}.$$

Limits and L'Hopital's Rule

When seeking a limit, an indeterminate form sometimes results

$$\lim_{x \to a} y(x) = \frac{0}{0} \quad \text{or} \quad \lim_{x \to a} y(x) = \frac{\infty}{\infty}$$

in which case the limit might be found by

$$\lim_{x \to a} = \frac{F(x)}{G(x)} = \lim_{x \to a} \frac{F'(x)}{G'(x)}$$

where $F'(x) = \frac{dF}{dx}$ and $G'(x) = \frac{dG}{dx}$.

L'Hopital's Rule is valid only if the right-hand side exists. The rule can be applied several times in succession if necessary.

INTEGRATION

The **indefinite integral** of $f(x)$ generates a function, which, when differentiated, results in the original $f(x)$. The indefinite integral contains an arbitrary constant of integration.

$$\int \left(x^2 + \frac{1}{x} \right) dx = \frac{x^3}{3} + \ln x + c$$

Table 2 contains integrals of common functions. For simplicity the constant of integration has not been printed, but is understood to be present in every integral.

The **definite integral** of $f(x)$ between $x = a$ and $x = b$ defines the change in the value of the integral as x changes from a to b.

$$\int_a^b (1-x^2)dx = \left[x - \frac{x^3}{3} \right]_a^b$$

$$= b - \frac{b^3}{3} - \left(a - \frac{a^3}{3} \right)$$

TABLE 2
INTEGRALS

$$\int a\, dx = ax$$

$$\int a \bullet f(x)dx = a\int f(x)dx$$

$$\int (u+v)dx = \int u\, dx + \int dx$$

$$\int x^n dx = \frac{x^{n+1}}{n+1}, \ n \neq 1$$

$$\int \frac{dx}{x} = \ln x$$

$$\int \frac{dx}{a+bx} = \frac{1}{b}\ln(a+bx)$$

$$\int \frac{dx}{(a+bx)^2} = -\frac{1}{b(a+bx)}$$

$$\int \frac{dx}{(a+bx)} = -\frac{1}{2b(a+bx)^2}$$

$$\int \sin x\, dx = -\cos x$$

$$\int \cos x\, dx = \sin x$$

$$\int \tan x\, dx = -\ln \cos x$$

$$\int \cot x\, dx = \ln \sin x$$

$$\int \sec x\, dx = \ln(\sec x + \tan x) = \ln \tan\left(\frac{\pi}{4} + \frac{x}{2} \right)$$

$$\int \csc x\, dx = \ln(\csc x - \cot x) = \ln \tan \frac{x}{2}$$

$$\int \sin^2 x\, dx = -\frac{1}{2}\cos x \sin x + \frac{1}{2}x = \frac{1}{2}x - \frac{1}{4}\sin 2x$$

$$\int \sin^3 x\, dx = -\frac{1}{3}\cos x (\sin^2 x + 2)$$

$$\int \sin^n x\, dx = -\frac{\sin^{n-1} x \cos x}{n} + \frac{n-1}{n}\int \sin^{n-2} x\, dx$$

$$\int e^x dx = e^x$$

$$\int e^{ax} dx = \frac{e^{ax}}{a}$$

$$\int b^{ax} dx = \frac{b^{ax}}{a\ln b}$$

$$\int \ln x\, dx = x\ln x - x$$

$$\int a^x \ln a\, dx = a^x$$

Integration by Parts

Integrate the formula for the derivative of a product

$$d(uv) = u\,dv + v\,du$$

to obtain

$$\int d(uv) = \int u\,dv + \int v\,du$$

Rearrange to obtain the formula for integration by parts.

$$\int u\,dv = uv - \int v\,du$$

EXAMPLE: Find $\int x \sin 2x\,dx$.

SOLUTION: Let $u = x \quad dv = \sin 2x\,dx$

$$du = dx \qquad v = \int \sin 2x\,dx = -\frac{1}{2}\cos 2x$$

With these definitions for u and v,

$$\int x \sin 2x\,dx = -\frac{1}{2}x\cos 2x + \frac{1}{2}\int \cos 2x\,dx$$

$$= \frac{1}{2}x\cos 2x + \frac{1}{4}\sin 2x + C$$

Integration and Area

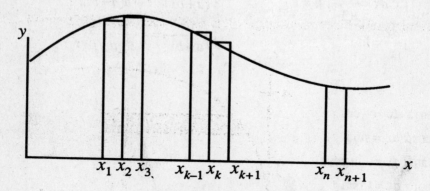

Figure 8. Integration of area under a curve

Let $a = x_1$ and $b = x_{n+1}$. The area bounded vertically by the horizontal axis and the continuous function $y(x)$, and bounded horizontally by the ordinates $x = a$ and $x = b$, can be approximated by the sum of the areas of n rectangles. The approximation improves as the number of rectangles increases.

$$\text{Area} = \lim_{\max \Delta x_k \to 0} \sum_{k=1}^{n} f(x_k)\Delta x_k = \int_a^b f(x)\,dx$$

This is the Fundamental Theorem of Integral Calculus and demonstrates that the definite integral of a continuous function corresponds geometrically to an area.

The computation of areas is easily viewed geometrically by defining a rectangle with an infinitesimal thickness within the region and then performing the integration.

Find the area above the horizontal axis bounded by the parabola $y^2 = 4x$, the x-axis, and the line $x = 4$.

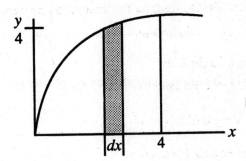

Figure 9. Area under a parabola with respect to x

The area of the vertical rectangle is $y\,dx$, which can be integrated between $x = 0$ and $x = 4$.

$$\int_0^4 y\,dx = \int_0^4 2x^{\frac{1}{2}}dx = \left[2\left(\frac{2}{3}\right)x^{\frac{3}{2}}\right]_0^4 = \frac{32}{3}$$

A horizontal differential rectangle could also be used. Rectangle area $= (4 - x)\,dy$.

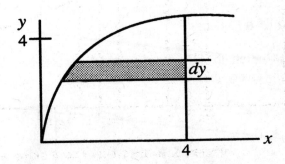

Figure 10. Area under a parabola with respect to y

$$\int_0^4 (4 - x)\,dy = \int_0^4 \left(4 - \frac{y^2}{4}\right)dy$$

$$= \left[4y - \frac{y^3}{12}\right]_0^4 = 16 - \frac{16}{3} = \frac{32}{3}$$

Revolve the same area around the line $x = a$ and find the volume generated.

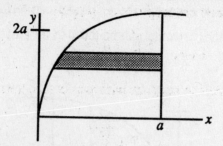

Figure 11. Volume generated by revolving curve about $x = a$

$$\text{Volume} = \pi \int_0^{2a} (a-x)^2 \, dy$$

$$= \pi \int_0^{2a} \left(a - \frac{y^2}{4a}\right)^2 dy$$

$$= \pi \int_0^{2a} \left(a^2 - \frac{y^2}{2} + \frac{y^4}{16a^2}\right) dy$$

$$= \pi \left[a^2 y + \frac{y^3}{2\cdot 3} + \frac{y^5}{5\cdot 16a^2}\right]_0^{2a}$$

$$= \frac{16}{15}\pi a^3$$

Length of a Curve

To find the length of the arc of a plane curve between two given points, we begin by drawing a very short segment of the curve.

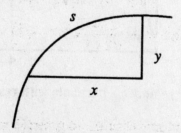

Figure 12. Length of a curve

$$\Delta s = \sqrt{\Delta x^2 + \Delta y^2} = \sqrt{1 + \left(\frac{\Delta y}{\Delta x}\right)^2}\, \Delta x$$

In the limit as Δx becomes infinitesimally small, Δs will follow the curvature of the function, and the total **arc length** is an integral.

$$\text{length} = \int_a^b \sqrt{1+\left(\frac{dy}{dx}\right)^2} \, dx$$

If it is more convenient to integrate with respect to y, the integral can be written as

$$\text{length} = \int_c^d \sqrt{\left(\frac{dx}{dy}\right)^2 + 1} \, dy \, .$$

If x and y are given in terms of a parameter t, the integral is

$$\text{length} = \int_{t_1}^{t_2} \sqrt{\left(\frac{dx}{dt}\right)^2 + \left(\frac{dy}{dt}\right)^2} \, dt \, .$$

Find the length of the curve $y = \ln \sin x$ from $x = \frac{\pi}{4}$ to $\frac{\pi}{2}$.

$$\frac{dy}{dx} = \frac{\cos x}{\sin x} = \cot x$$

$$\text{length} = \int_{\frac{1}{4}\pi}^{\frac{1}{2}\pi} \sqrt{1 + \cot^2 x} \, dx$$

$$= \int_{\frac{1}{4}\pi}^{\frac{1}{2}\pi} \csc x \, dx = \left[\ln(\csc x - \cot x)\right]_{\frac{1}{4}\pi}^{\frac{1}{2}\pi}$$

$$= \ln(1-0) - \ln\left(\sqrt{2}-1\right) = -\ln\left(\sqrt{2}-1\right)$$

$$\text{length} = -\ln\frac{2-1}{\sqrt{2}+1} = \ln\left(1+\sqrt{2}\right)$$

Solids of Revolution

If a given curve is revolved around an axis, it generates a **surface of revolution**. The area of the surface of revolution may be found from a summation of small (infinitesimal) cylinders. The cylinder has a radius x, and hence a circumference of $2\pi x$. The height ds of the cylinder is defined from

$$ds = \sqrt{dy^2 + dx^2} = \sqrt{1+\left(\frac{dy}{dx}\right)^2} \, dx \, .$$

The surface of revolution, therefore, has an area equal to $\int 2\pi x \, ds$, or

$$2\pi \int_a^b x \sqrt{1 + \left(\frac{dy}{dx}\right)^2} \, dx$$

where the limits a and b are the values of x at the beginning and end of the portion of the curve to be considered.

EXAMPLE: Find the volume of $y = f(x) = 2x$, when rotated about the x-axis and bounded by $x = 2$.

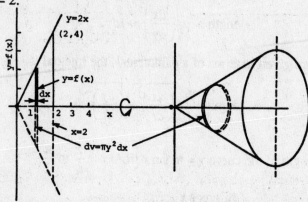

a) **Two-dimensional representation of $f(x)$.**   b) **Schematic of disk volume generated by rotating $f(x)$ about the x-axis.**

Figure 13. Volume of a curve

SOLUTION: Solution #1, Disk Method.

The shaded strip, as shown in Figure 13(a), when rotated about the x-axis, sweeps a volume expressed by the disk as in Figure 13(b). The radius of this disk is y. Hence, its volume is $\pi y^2 \, dx$, where dx is the thickness of the disk. The sum of the volumes of all such disks for $0 \le x \le 2$ and passing to the limit gives:

$$V = \int_0^2 \left(\pi y^2\right) dx = \pi \int_0^2 (2x)^2 \, dx$$

$$= 4\pi \int_0^2 x^2 \, dx = 4\pi \frac{x^3}{3} \bigg]_0^2 = \frac{32\pi}{3}$$

Since an increment of the volume is a disk, this is called the **disk method**.

Solution #2, Shell Method

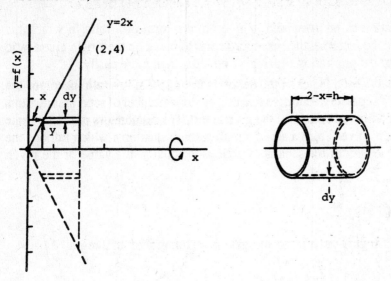

a) Two-dimensional representation of *f(x)*   **b) Element of volume is shell**
Figure 14. Shell Method

This volume might be conceived of as a constantly expanding cylindrical shell, the radius *y* of which increases from 0 to 4. The height (*x* value) varies from 2 down to 0. The element of volume is obtained by multiplying circumference by height and by differential of thickness:

$$dV = 2\pi r\, h\, dy$$
$$=2\pi y(2 - x)dy$$

The height of the element is really the length on the coordinate axis between the outer boundary and the equation of the line *y* – 2*x*. Since *x* is the distance to the function, 2 – *x* equals the height of the shell. All the unknowns in the *dV* expression must be in terms of *y*, since the differential term is *dy*. Therefore, we subsititute $\frac{y}{2}$ for *x* (from the original equation):

$$dV = 2\pi y\left(2 - \frac{y}{2}\right)dy$$

$$\int dV = 2\pi \int_0^4 \left(2y - \frac{y^2}{2}\right)dy$$

$$V = 2\pi\left(y^2 - \frac{y^3}{6}\right)_0^4 = \frac{32\pi}{3}$$

Integration by Partial Fractions

When a function to be integrated is given in the form of a ratio in which the denominator can be factored, the best approach is to break up the single given ratio into a number of simpler ratios which may be integrated more easily.

In this approach, each factor of the denominator of the given ratio becomes the denominator of a separate fraction, so that the resulting number of separate fractions is equal to the number of factors of the given ratio. The numerators of the separate fractions are then solved from a set of simultaneous equations, which impose the condition that the sum of the separate fractions is equal to the value of the given function.

Double Integral

The **double integral** or **iterated integral** is an integral of an integral

$$\int_a^b \int_{y_1}^{y_2} f(x,y)\, dy\, dx$$

The limits of the inner integral y_1 and y_2 are usually functions of x, while the limits of the outer integral a and b are constants.

EXAMPLE: Use double integration to find the area enclosed by $y = x^2$ and $x + y - 2 = 0$.

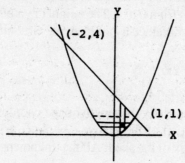

Figure 15. Area of a parabola

SOLUTION: The formula for area in Cartesian coordinates, using double integrals, is:

$$A = \iint dy\, dx \qquad dx = \int_a^b dx \int_{f(x)}^{F(x)} dy.$$

The limits a and b of the integral with respect to x are the x-coordinates of the points of intersection of the two curves. To find the points of intersection, we set $y = x^2$ equal to $y = 2 - x$ and solve for x.

$$x^2 = 2 - x \qquad (x + 2)(x - 1) = 0$$
$$x = -2 \quad x = 1$$

The limits of the integral with respect to y are the two functions. The lower limit is the lower function, the parabola $y = x^2$. The upper limit is the upper function, the line $y = 2 - x$. Therefore,

$$A = \int_{-2}^{1} dx \int_{x^2}^{2-x} dy$$

$$= \int_{-2}^{1} dx [y]_{x^2}^{2-x}$$

$$= \int_{-2}^{1} [2 - x - x^2] dx = \left[2x - \frac{x^2}{2} - \frac{x^3}{3} \right]_{-2}^{1}$$

$$= \frac{7}{6} + \frac{10}{3} = \frac{27}{6}$$

$$= \frac{9}{2} \text{ sq. units.}$$

Computations with Series

Series are very useful for numerical computations of such constants as c, π etc., and for the computations of terms such as $\log x$, $\sin x$, etc.

Hence, if, by one of a number of possible procedures, we can find a series for a function, then that series can be used for computational purposes, but only IF IT CONVERGES. That is, the only kind of series that can be used for computation is a convergent series. The reason for this is that, if a series were not convergent but divergent, more and more terms are added, a continually different result would be obtained. With a convergent series, however, the series approaches a limit and gives an ever more accurate result, the more terms we add. Adding more terms does not change the result substantially after the first serveral terms, it only makes the result more accurate.

Assuming, then, that a series converges, it can be differentiated, integrated, added to other series, subtracted from other series, and mutltiplied by a constant, for example. A series, therefore, has a great deal of utility. If, for example, we know the series for $\sin x$, we can find the series for $\cos x$ by differentiating that series term by term, since we know that $(d/dx)(\sin x) = \cos x$.

In a computation, once convergence is established, the number of terms to be used in the computation depends only on the accuracy desired.

Series

The sum of the first n numbers:

$$\Sigma(n) = 1 + 2 + 3 + 4 + 5 \ldots + n = \frac{n(n+1)}{2}$$

The sum of the squares of the first n numbers:

$$\Sigma(n^2) = 1^2 + 2^2 + 3^2 + 4^2 + 5^2 \ldots + n^2 = \frac{n(n+1)(2n+1)}{6}$$

The sum of the cubes of the first n numbers:

$$\Sigma(n^3) = 1^3 + 2^3 + 3^3 + 4^3 + 5^3 \ldots + n^3 = \frac{n^2(n+1)^2}{4}$$

Arithmetic Series

Let C_1 be the first term and C_i be the ith term. Each term differs from the previous term by a constant, d.

$$C_{i+1} = C_i + d$$

The nth term is $C_1 + (n-1)d$, and the sum of n terms is $\frac{n}{2}(2a + (n-1)d)$.

An arithmetic series with an infinite number of terms always diverges. The sum of its terms is not defined.

Geometric Series

Let C_1 be the first term and C_i be the ith term. Each term is defined by multiplying the previous term by a constant, r.

$$C_{i+1} = rC_i$$

The nth term is $C_1 r^{n-1}$.

The sum of a **geometric series** with a finite number of terms, n, is

$$S_n = \frac{C_1(1 - r^n)}{1 - r}.$$

The sum of a series with an infinite number of terms is

$$S_\infty = \frac{C_1}{1 - r}.$$

The infinite geometric series converges for $-1 < r < 1$ and diverges otherwise.

Harmonic Series

The general term of this series is

$$C_n = \frac{1}{a+(n-1)d}$$

where a and d are constants.

The sum of n terms where n is finite is

$$S_n = \frac{2}{n(2a+(n-1)d)}.$$

The infinite series always diverges.

The *p*-Series

The general term is $C_n = \frac{1}{n^p}$.

The sum of n terms is

$$\sum_{i=1}^{n} = \frac{1}{1^p} + \frac{1}{2^p} + \frac{1}{3^p} + \dots + \frac{1}{i^p} + \dots + \frac{1}{n^p}.$$

The infinite series converges if $p > 1$ and diverges otherwise.

Convergence of Infinite Series

If $\lim\limits_{n \to \infty} S_n$ exists, then the series converges. S_n is the sum of n terms.

If $\lim\limits_{n \to \infty} C_n \neq 0$, then the series diverges.

If $\lim\limits_{n \to \infty} C_n = 0$, no conclusions can be made. Additional tests are necessary.

Ratio Test for Convergence

Calculate $\lim\limits_{n \to \infty} \frac{C_{n+1}}{C_n} = h$.

The series converges if $h < 1$.

The series diverges if $h > 1$.

The ratio test fails if $h = 1$.

The ratio test will determine convergence for series with terms of alternating signs if the absolute value of the ratio is used.

Definition of Power Series

If x is a variable, then an infinite series of the form

$$\sum_{n=0}^{\infty} a_n x^n = a_0 + a_1 x + a_1 x^2 + a_3 x^3 + \ldots + a_n x^n + \ldots$$

is called a **power series**. More generally, we call a series of the form

$$\sum_{n=0}^{\infty} a_n (x-c)^n = a_0 + a_1(x-c) + a_2(x-c)^2 + \ldots + a_n(x-c)^n + \ldots$$

a **power series centered at** c, where c is a constant.

Definition of Taylor and Maclaurin Series

If a function f has derivatives of all orders at $x = c$, then the series

$$\sum_{n=0}^{\infty} \frac{f^{(n)}(c)}{n!}(x-c)^n = f(c) + f'(c)(x-c) + \ldots + \frac{f^{(n)}(c)}{n!}(x-c)^n + \ldots$$

is called the **Taylor series for** $f(x)$ at c. Moreover, if $c = 0$, then this series is called the **Maclaurin series for** f.

Comparison Test for Convergence

If G_n and H_n are both series and $g_i < h_i$ for all i, then H_n diverges if G_n diverges, and G_n converges if H_n converges.

Does the following infinite series converge?

$$1 - \frac{1}{2} + \frac{2}{2^2} - \frac{3}{2^3} + \ldots + \frac{(-1)^n(-n)}{2^n} + \frac{(-1)^{n+1}(n+1)}{2^{n+1}} + \ldots$$

Use the ratio test. Because this is an alternating sign series, we will use the absolute value of the ratio.

$$\lim_{n\to\infty}\left|\frac{Cn+1}{Cn}\right| = \lim_{n\to\infty}\frac{\frac{n+1}{2^{n+1}}}{\frac{n}{2^n}} = \lim_{n\to\infty}\frac{n+1}{2n}$$

$$= \lim_{n\to\infty}\left(\frac{1}{2} + \frac{1}{2n}\right) = \frac{1}{2} < 1$$

Thus, the series converges.

DIFFERENTIAL EQUATIONS

A differential equation displays the relationship among derivatives of a dependent variable

$$a_2 \frac{d^2 y}{dx^2} + a_1 \frac{dy}{dx} + a_0 y = f(x)$$

where x is the **independent variable** and y is the **dependent variable**.

A differential equation is linear if no term contains the dependent variable or any of its derivatives to other than the first power. The example equation is linear if coefficients a_0, a_1, a_2 are constants or functions of x.

The equation is **homogeneous** if all terms contain the dependent variable or its derivatives. The example would be homogeneous if $f(x) = 0$.

The order of the equation is the order of its highest derivative. The example is second order. The general solution includes a number of arbitrary constants equal to the order. Numerical value of these constants are determined from initial and/or boundary conditions.

First Order

A linear, first order equation can be manipulated into the form

$$\frac{dy}{dx} + h(x)y = g(x).$$

We multiply the equation by an integrating factor

$$u(x) = e^{\int h(x)dx}.$$

The left-hand side forms the derivative of a product.

$$d(y \cdot u(x)) = u(x) \cdot g(x)dx$$

After integration and rearrangement, we get the solution

$$y(x) = \frac{1}{u(x)} \int u(x) \cdot g(x)dx + \frac{C}{u(x)}.$$

EXAMPLE: Find the solution of

$$\frac{dy}{dx} - \frac{1}{x}y = x^2 + 3x - 2.$$

SOLUTION: The integrating factor $u(x) = e^{-\int \frac{dx}{x}} = e^{-\ln x} = \frac{1}{x}$.

Multiply the entire equation by the integrating factor.

$$\frac{1}{x}dy - \frac{y}{x^2}dx = \left(x + 3 - \frac{2}{x}\right)dx$$

$$d\left(\frac{y}{x}\right) = \left(x + 3 - \frac{2}{x}\right)dx$$

Integrate both sides.

$$\frac{y}{x} = \frac{1}{2}x^2 + 3x - 2\ln x + C$$

Finally, solve for y.

$$y = \frac{1}{2}x^3 + 3x^2 - 2x\ln x + Cx$$

Linear Differential Equations with Constant Coefficients

$$a_n\frac{d^n y}{dx^n} + a_{n-1}\frac{d^{n-1}y}{dx^{n-1}} + a_{n-2}\frac{d^{n-2}y}{dx^{n-2}} + \ldots + a_2\frac{d^2 y}{dx^2} + a_1\frac{dy}{dx} + a_0 y = f(x)$$

$a_n, a_{n-1}, a_{n-2}, \ldots, a_2, a_1, a_0$ are constant.

Homogeneous solution y_h:

The general solution is $y = y_h + y_p$, where y_h is the solution to the homogeneous equation corresponding to $f(x) = 0$ and y_p is the particular solution.

Homogeneous solution y_h:

The characteristic equation corresponding to the homogeneous form of the differential equation is found by replacing the kth derivative with m_k:

$$a_n m^n + a_{n-1}m^{n-1} + a_{n-2}m^{n-2} + \ldots a_2 m^2 + a_1 m + a_0 = 0$$

Each root of this equation, r_k, appears in y_h in the form $e^{r}k^x$.

–Each distinct real root r_1, will generate $C_j e^{r_1 x}$.

–Repeated real roots, r_2, will generate

$$C_2 e^{r_2 x} + C_3 x e^{r_2 x}.$$

–Complex roots $r_5 = a + ib$ and $r_6 = a - ib$ can be written in the standard form

$$C_5 e^{r_5 x} + C_6 e^{r_6 x}$$

where the constant coefficients C_5 and C_6 are complex conjugates.

It is convenient to write solution terms corresponding to complex roots with trigonometric functions.

$$e^{ax}(C_7\sin bx + C_8 \cos bx)$$

Here C_7 and C_8 are real constants. Note that the real part of the complex root, a, appears in the exponent, and the coefficient of the imaginary part, b, appears in the sin and cos arguments.

If the **deferential** equation is m^{th} order, there will be m terms in the homogeneous solution with m constants, the numerical values of which must be determined from m specified data values.

Particular Solution y_p

The particular solution is formed from the sum of $f(x)$ and its derivatives, each term of which has an unknown constant coefficient to be determined by substituting the particular solution into the differential equation. Table 3 displays forms of particular solutions.

TABLE 3 PARTICULAR SOLUTIONS	
$f(x)$	y_p
a	A
$ax + b$	$Ax + B$
$ax^n + bx^{n-1} + cx^{n-2} + ...$	$Ax^n + Bx^{n-1} + Cx^{n-2} + ... + Cx + H$
$a \sin cx,\ a \cos cx,$ or $a \sin cx + b \cos cx$	$A \sin cx + B \cos cx$
e^{ax}	Ae^{ax}

Note: If $f(x) = e^{ax}$, and the term e^{ax} also appears in the homogeneous solution, the particular solution has the form Axe^{ax}.

If $f(x) = a \sin cx$ or $a \cos cx$, and these terms also appear in the homogeneous solution, the particular solution has the form $A x \sin cx + B x \cos cx$.

EXAMPLE: Solve

$$2\frac{dy}{dx} + y = \sin 5x$$

where $y = 10$ at $x = 0$.

SOLUTION: To find the homogeneous solution, write the characteristic equation

$$2m + 1 = 0 \text{ which has a root } m = -\frac{1}{2}.$$

The homogeneous solution is $y_h = Ae^{-\frac{x}{2}}$, where A is a constant to be determined later.

To find the particular solution, we use Table 3 to determine the form is

$$y_p = C \sin 5x + D \cos 5x.$$

To determine the constants C and D, substitute y_p into the differential equation.

$$2(5C \cos 5x - 5D \sin 5x) + (C \sin 5x + D \cos 5x) = \sin 5x$$

Equate coefficients of the sin and cos terms.

$$-10D + C = 1 \qquad 10C + D = 0$$

Solving for C and D gives $C = \dfrac{1}{100}$ and $D = -\dfrac{10}{101}$.

The total solution is $y = y_h + y_p$.

$$y = Ae^{-\frac{x}{2}} + \frac{1}{100} \sin 5x - \frac{10}{101} \cos 5x$$

To find A, use $y = 10$ when $x = 0$.

$$10 = A(1) + 0 - \frac{10}{101}(1)$$

Thus, $A = 10 + \dfrac{10}{101} = 1.099$ and the solution is

$$y = 10.099e^{-\frac{x}{2}} + \frac{1}{100} \sin 5x - \frac{10}{101} \cos 5x.$$

Linear, Second Order with Constant Coefficients

$$a\frac{d^2y}{dx^2} + b\frac{dy}{dx} + cy = f(x)$$

(This equation, with $b > 0$ and $c > 0$, plays an important role in dynamics.)
The homogeneous part of the solution is based on the characteristic equation

$$am^2 + bm + c = 0.$$

This is a quadratic equation and the roots are given by

$$r_1, r_2 = \frac{-b \pm \sqrt{b^2 - 4ac}}{2}.$$

If $b^2 - 4ac > 0$, there are two distinct real roots, the solution is overdamped, and the homogeneous solution has the form

$$C_1 e^{r_1 x} + C_2 e^{r_2 x}.$$

If $b^2 - 4ac = 0$, there are two repeated roots, the solution is critically damped, and the homogeneous solution has the form

$$C_1 e^{r_1 x} + C_2 x e^{r_1 x}.$$

If $b^2 - 4ac < 0$, the roots are complex conjugates

$$r_1 = \lambda - i\mu \quad r_2 = \lambda - i\mu.$$

The solution is underdamped, and the homogeneous solution has the form

$$e^{\lambda x}(C_1 \cos \mu x + C_2 \sin \mu x).$$

EXAMPLE: Find the solution to

$$\frac{d^2y}{dx^2} + 3\frac{dy}{dx} + 2x = 1 \text{ at } x = 0, y = 0, \text{ and } \frac{dy}{dx} = 0$$

SOLUTION: To find the homogeneous part, write and factor the characteristic equation.

$$m^2 + 3m + 2 = (m + 2)(m + 1) = 0$$

There are two distinct real roots, -2 and -1, and the system is overdamped.

$$y_h = C_1 e^{-2x} + C_2 e^{-x}$$

To find the particular solution, note that the nonhomogeneous term is a constant, and from Table 3, the form of the particular solution is a constant.

$$y_p = A$$

Substitute this particular solution into the original differential equation.

$$0 + 0 + 2A = 1 \text{ Thus, } A = \frac{1}{2}$$

The total solution is

$$y = y_h + y_p = C_1 e^{-2x} + C_2 e^{-x} + \frac{1}{2}.$$

Differences

If y is a function of x, and y is given at discrete values of x, this yields a table of values.

The first difference of $f(x)$ is obtained when we subtract $f(x_{i+1}) - f(x_i) i = 0,\ldots, n$ and is written as

$$f(x_{i+1}) - f(x_i) = \Delta f(x_i) \quad \text{or}$$
$$y_{i+1} - y_i = \Delta y_i$$

which is more commonly used.

The second difference is the difference of the first differences, indicated as

$$\Delta^2 y_i = \Delta y_{i+1} - \Delta y_i.$$

The nth differences are obtained in a similar manner. So we have for any n, an integer, the nth difference of $y = f(x)$ is given by

$$\Delta^n y_i = \Delta^{n-1} y_{i+1} - \Delta^{n-1} y_i.$$

Another way of obtaining the nth difference is by continuously substituting function values. For example,

$$\Delta^2 y_0 = \Delta y_1 - \Delta y_0 = (y_2 - y_1) - (y_1 - y_0) = y_2 - 2y_1 + y_0$$

$$\Delta^2 y_1 = \Delta y_2 - \Delta y_1 = (y_3 - y_2) - (y_2 - y_1) = y_3 - 2y_2 + y_1$$

$$\Delta^3 y_0 = \Delta^2 y_1 - \Delta^2 y_0 = (y_3 - 2y_2 + y_1) - (y_2 - 2y_1 + y_0)$$

$$= y_3 - 3y_2 + 3y_1 - y_0$$

We then have a general formula given by

$$\Delta^k y_0 = y_k - \binom{k}{1} y_{k-1} + \binom{k}{2} y_{k-2} - \dots + (-1)^k \left\{ \begin{matrix} k \\ k \end{matrix} \right\} y_0 = \sum_{i=0}^{k} (-1)^i \binom{k}{i} y_{k-1}$$

where $\binom{k}{i}$ is the binomial coefficient.

Considering the reverse, we may express a value of y in terms of the preceding values of y and the differences. We had

$$y_{i+1} - y_i = \Delta y_i$$

so

$$y_1 = y_0 + \Delta y_0$$

and

$$\Delta^n y_i = \Delta^{n-1} y_{i+1} - \Delta^{n-1} y_i$$

therefore

$$\Delta^3 y_0 = y_3 - 3y_2 + 3y_1 - y_0$$

$$y_3 = y_0 - 3y_1 + 3y_2 + \Delta^3 y_0$$

$$= y_0 + 3(y_2 - y_1) + \Delta^3 y_0$$

$$= y_0 + 3\Delta y_1 + \Delta^3 y_0$$

$$= y_0 + 3(\Delta^2 y_0 + \Delta y_0) + \Delta^3 y_0$$

$$= y_0 + 3\Delta y_0 + 3\Delta^2 y_0 + \Delta^3 y_0$$

So it appears that y_k can be expressed by the general formula

$$y_k = \sum_{i=0}^{k} (-1)^i \binom{k}{i} y_{k-1}$$

where $\binom{k}{i}$ is the binomial coefficient.

$$y_k = y_0 + \left[\begin{matrix} k \\ 1 \end{matrix} \right] \Delta y_0 + \left[\begin{matrix} k \\ 2 \end{matrix} \right] \Delta^2 y_0 + \dots + \Delta^k y_0$$

$$= \sum_{i=0}^{k} \left[\begin{matrix} k \\ i \end{matrix} \right] \Delta^i y_0$$

LAPLACE TRANSFORMS

The **Laplace Transform** is an integral transform. A function $f(t)$ can be transformed to $F(s)$ by the following:

$$F(s) = L\big(f(t)\big) = \int_0^\infty f(t)e^{-st}dt$$

Given $F(s)$, the corresponding function $f(t)$ can be found by

$$f(t) = \frac{1}{2\pi i}\int_{a-i\infty}^{a+i\infty} F(s)e^{st}ds.$$

In practice a table of transforms such as Table 4 is used. It is important to note that the Laplace Transform is linear.

$$L\big(cf(t)\big) = cL\big(f(t)\big)$$

$$L\big(f_1(t) + f_2(t)\big) = L\big(f_1(t)\big) + L\big(f_2(t)\big)$$

Also note that the Laplace Transform of a derivative is s multiplied into the transform of the function being differentiated. These transforms are used primarily to solve linear differential equations with constant coefficients.

TABLE 4 LAPLACE TRANSFORMS	
$f(t)$	$F(s)$
1	$\dfrac{1}{s}$
e^{-at}	$\dfrac{1}{s+a}$
t	$\dfrac{1}{s^2}$
$\sin at$	$\dfrac{a}{s^2+a^2}$
$\cos at$	$\dfrac{s}{s^2+a^2}$
$\dfrac{df(t)}{dt}$	$s\,F(s) - f(0)$
$\dfrac{d^2 f(t)}{dt^2}$	$s^2 F(s) - sf(0) - \dfrac{df}{dt}(0)$
$\int_0^t f(t)dt$	$\dfrac{F(s)}{s}$

Find the transform of $f(t) = e^{-at}$.

$$\mathcal{L}\left(e^{-at}\right) = \int_0^\infty e^{-at} e^{-st} dt = \int_0^\infty e^{-(a+s)t} dt$$

$$= -\frac{e^{-(a+s)t}}{a+s}\Bigg]_0^\infty = 0 - \left(-\frac{1}{s+a}\right) = \frac{1}{s+a}$$

Laplace Transform Properties

The step function

The transform of the step function is given by

$$\mathcal{L}[f(t)] = \int_0^x e^{-st} dt = \frac{1}{s}.$$

Thus $\qquad\qquad f(t) \leftrightarrow \dfrac{1}{s}.$

The ramp function

The ramp is $tf(t)$. Hence,

$$\mathcal{L}[tf(t)] = \int_0^\infty te^{-st} dt = \frac{1}{s^2}$$

and $\qquad\qquad tf(t) \leftrightarrow \dfrac{1}{s^2}.$

The impulse function

The Laplace Transform of an impulse existing at $t = 0$ is

$$\mathcal{L}[\delta(t)] = \int_0^\infty \delta(t) e^{-st} dt = 1.$$

The exponential function

For a causal exponential function

$$\mathcal{L}\left[e^{-\alpha t} f(t)\right] = \int_0^\infty e^{-\alpha t} e^{-st} dt = \frac{1}{s+\alpha}.$$

This transform exists even if α is negative—that is, the exponential is a growing one — because σ can always be chosen so that $e^{-(\alpha+\sigma)t} \to 0$ as $t \to \infty$.

Linearity

If $\qquad\qquad f_1(t) \leftrightarrow F_1(s)$ and $f_2(t) \leftrightarrow F_2(s),$

then $\qquad\qquad a_1 f_1(t) + a_2 f_2(t) \leftrightarrow a_1 F_1(s) + a_2 F_2(s).$

Scaling

A change in the time scale can bring about time expansion or compression of a signal depending on the magnitude of the scale change. In the following result, time reflection is disallowed.

If
$$f(t) \leftrightarrow F(s),$$

then
$$f(at) \leftrightarrow \frac{1}{a} F\left(\frac{s}{a}\right), \quad a > 0.$$

Integration

If
$$f(t) \leftrightarrow F(s),$$

$$g(t) = \int_{-\infty}^{t} f(\tau) d\tau,$$

then
$$\mathcal{L}[g(t)] = G(s) = \frac{1}{s} F(s) + \frac{1}{s} \int_{-\infty}^{0} f(\tau) d\tau.$$

This can also be written

$$G(s) = \frac{1}{s} F(s) + \frac{g(0)}{s}$$

because
$$g(0) \equiv \int_{-\infty}^{0} f(\tau) d\tau.$$

Differentiation

The differentiation theorem will be most important in the solution of differential equations using the Laplace Transform. It is analogous to the time-advance theorem of the z-transform.

If
$$f(t) \leftrightarrow F(s),$$

then
$$pf(t) \leftrightarrow sF(s) - f(0),$$

$$p^2 f(t) \leftrightarrow s^2 F(s) - sf(0) - pf(0),$$

$$\vdots$$

$$p^n f(t) \leftrightarrow s^n F(s) - s^{n-1} f(0) - s^{n-2} pf(0) - \dots - p^{n-1} f(0)$$

Periodic functions

Consider a causal function that is periodic for $t > 0$. We denote the part of $f(t)$ in the first period as $f_T(t)$. That is,

$$f_T(t) = \begin{cases} f(t), & 0 \le t < T, \\ 0, & \text{elsewhere} \end{cases}$$

Then, if

$$f_T(t) \leftrightarrow F_T(s),$$

$$f(t) \leftrightarrow F_T(s)\frac{1}{1+e^{-Ts}}.$$

STATISTICS

For a set of n values, x_i, $i = 1, 2, 3, ..., n$

−the **mean** or **average** is $\dfrac{\sum\limits_{i=1}^{n} x_i}{n}$.

−the **mode** is the value that occurs most often.

−the **median** is the middle value. The set must be ordered in increasing or decreasing order. If there are an odd number of values, the median is the middle value. If there are an even number of values, the median is the mean of the two values in the middle.

−the **range** is the maximum value−the minimum value.

−the **variance** for a sample is

$$\text{sample variance} = \frac{\sum\limits_{i=1}^{n}(x_i - \text{mean})^2}{n-1}$$

−the **standard deviation** for a sample is the square root of the variance.

The numerical values that are computed from the complete population of numbers are called population parameters. The numerical values that are computed from a random sample and are used to infer information about the complete population are called sample statistics. The equation for the sample variance is given above. The variance computed from complete population data has the formula

$$\text{population variance} = \frac{\sum\limits_{i=1}^{n}(x_i - \text{mean})^2}{n}$$

EXAMPLE: The following 10 values represent a random sample:

1, 3, 4, 6, 7, 9, 9, 9, 14, 18

SOLUTION: The mean is $\dfrac{1+3+4+6+7+(3\bullet 9)+14+18}{10} = 8.$

The mode is 9.

There are an even number of values and the median is

$$\frac{7+9}{2} = 8.$$

The range is $18 - 1 = 17$.

The variance is

$$\frac{(1-8)^2 + (3-8)^2 + (4-8)^2 + (6-8)^2 + (7-8)^2 + 3(9-8)^2 + (14-8)^2 + (18-8)^2}{10-1}$$

$$= \frac{234}{9} = 26$$

and the sample standard deviation is $\sqrt{26} = 5.10$.

Combinations and Permutations

The number of **combinations** of n things taken x at a time is

$$\binom{n}{x} = C(n,x) = \frac{n!}{x!(n-x)!}.$$

The number of **permutations** of n things taken x at a time is

$$P(n,x) = \frac{n!}{(n-x)!}.$$

In a permutation each order or arrangement of the x items is counted. There are more permutations than combinations.

Consider the set of four letters a, b, c, and d. For this set $n = 4$. We wish to consider pairs of letters from this set; $x = 2$. The number of combinations is

$$C(4,2) = \frac{4!}{2!2!} = \frac{4 \cdot 3 \cdot 2 \cdot 1}{2 \cdot 1 \cdot 2 \cdot 1} = 6.$$

These six combinations are ab, ac, ad, bc, bd, and cd.

The number of permutations is

$$P(n,x) = \frac{4!}{2!} = \frac{4 \cdot 3 \cdot 2 \cdot 1}{2 \cdot 1} = 12.$$

These 12 permutations are ab, ba, ac, ca, ad, da, bc, cb, bd, db, cd, and dc.

Rules of Probability

Let A and B represent independent events.

The **probability** of A occurring is

$$0 \le p(A) \le 1.$$

If $p(A) = 1$, then the occurrence is a certainty. If $p(A) = 0$, then it will certainly not occur. The probability of A not occurring is

$$p(\text{not } A) = 1 - p(A).$$

The probability of either A or B or both occurring is

$$p(A + B) = p(A) + p(B) - p(A)p(B).$$

The probability of either A or B but not both occurring is

$$p(A \text{ or } B) = p(A) + p(B) - 2\,p(A)p(B).$$

The probability of both A and B occurring is

$$p(A \text{ and } B) = p(A)p(B).$$

Consider a pair of standard six-sided dice.

We roll one die. The probability of rolling a 4 is $\dfrac{1}{6}$.

The probability of not rolling a 4 is $1 - \dfrac{1}{6} = \dfrac{5}{6}$.

The probability of rolling at least a 4 means the probability of rolling a 4, a 5, or a 6, which is

$$p(4) + p(5) + p(6) = \frac{1}{6} + \frac{1}{6} + \frac{1}{6} = \frac{1}{2}.$$

We roll both dice. The probability of rolling a 4 two times is

$$\left(\frac{1}{6}\right)\left(\frac{1}{6}\right) = \frac{1}{36}.$$

We roll both dice. The probability of rolling a 4 on one or the other die, or on both dice is

$$\left(\frac{1}{6}\right) + \left(\frac{1}{6}\right) - \left(\frac{1}{6}\right)\left(\frac{1}{6}\right) = \frac{11}{36}.$$

We roll both dice. The probability of rolling a 4 on one or the other die, but not on both dice (exclusive or), is

$$\left(\frac{1}{6}\right) + \left(\frac{1}{6}\right) - 2\left(\frac{1}{6}\right)\left(\frac{1}{6}\right) = \frac{10}{36}.$$

Normal Distribution

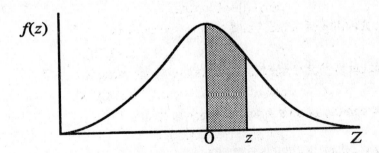

$f(z)$

0   z   Z

Figure 16. Normal distribution

The **normal, gaussian,** or **bell-shaped** distribution is described by its **probability density function**

$$f(Z) = \frac{1}{\sqrt{2\pi}} e^{-\frac{z^2}{2}}$$

where
$$Z = \frac{X - \text{mean}}{\text{standard deviation}}.$$

The probability of finding values between Z_1 and Z_2 is obtained by integrating the density function between these limits.

$$\int_{Z_1}^{Z_2} f(Z) dZ$$

In practice, these probabilities are read from the Appendix, which displays the integral from $Z = 0$ to $Z = z$. Note that the density function is symmetric. Probabilities are indicated by areas under the density function curve.

Probability Density Functions for Continuous Random Variables

In the case of X being a continuous random variable, its probability density function obeys the following conditions:

1) $f(x) \geq 0$ for all x

2) $\int_{-\infty}^{\infty} f(x) dx = 1$

3) $P(a \leq X \leq b) = \int_{a}^{b} f(x) dx$

Where here $P(a < X < b)$ is the probability of X having a value greater than a and less than b. (Notice "greater than" and "greater than or equal to" give the same value since X is a continuous random variable).

Consider the function given by

$$f(x) = \begin{cases} \dfrac{1}{x^2} & \text{for } x \geq 1 \\ 0 & \text{otherwise} \end{cases}$$

We show this is a density function. Since $f(x) > 0$ for all x, 1) is satisfied. We see that 2) is satisfied since

$$\int_{-\infty}^{\infty} f(x)dx = \int_{1}^{\infty} \frac{1}{x^2}dx = -\frac{1}{x}\bigg|_{1}^{\infty} = 1.$$

The probability function is given by

$$P(a \leq X \leq b) = \int_{a}^{b} f(x)dx = \int_{a}^{b} \frac{1}{x^2}dx.$$

(Here we assume that a and b are greater than 1).

For example,

$$P(3 \leq X \leq 5) = \int_{3}^{5} \frac{1}{x^2}dx = -\frac{1}{x}\bigg|_{3}^{5} = \frac{1}{3} - \frac{1}{5} = \frac{2}{15}.$$

EXAMPLE: Find the constant C that makes $f(x)$ given by

$$f(x) = \begin{cases} \dfrac{C}{x^4} & \text{for } x \geq 1 \\ 0 & \text{otherwise} \end{cases}$$

a density function.

SOLUTION: In order for $f(x)$ to be a probability density, it must satisfy 2). Integrating we have

$$\int_{-\infty}^{\infty} f(x)dx = \int_{1}^{\infty} \frac{C}{x^4}dx$$

$$= \frac{Cx^{-3}}{-3}\bigg|_{1}^{\infty} = \frac{C}{3}.$$

Hence,

$$\frac{C}{3} = 1$$

$$C = 3$$

Probability Density Functions for Discrete Random Variables

A probability density function tells how a distribution is weighted from $-\infty$ to ∞. In the case of X being a discrete random variable which has outcomes x_i with probability $p(x_i)$ for $i = 1, 2, 3, ...$ The probability density function $p(x)$ obeys the following conditions:

1) $p(x_i) \geq 0$ for all x

2) $\sum p(x_i) = 1$

3) $P(a \leq X \leq b) = \sum\limits_{x : \sum [a,b]} p(x_i)$

where here $P(a \leq X \leq b)$ is the probability of X having an outcome within the interval $[a, b]$.

Binomial Distributions

Let X be a binomial random variable with n repetitions. The probability distribution is given by

$$P(X = k) = \binom{n}{k} p^k (1-p)^{h-k} \qquad k = 0,1,2,...,n \text{ and } 0 < p < 1$$
$$0 \qquad \qquad \text{otherwise}$$

Note that this is a discrete distribution where the sum of the weights adds up to 1, i.e.,

$$\sum_{k=0}^{n} P(X = k) = \sum_{k=0}^{n} \binom{h}{k} p^k (1-p)^{h-k}$$

$$= [p + (1-p)]^n$$
$$= 1^n$$
$$= 1$$

Taking $p = \dfrac{1}{2}$ we see that $P(X = k)$ gives the probability of obtaining k heads when flipping a fair coin n times.

Poisson Distribution

Let X be a Poisson random variable. The probability distribution is given by

$$P(X = k) = \frac{e^{-\lambda} \lambda^k}{k!} \qquad k = 0,1,2,3...$$

Here λ is called the parameter of the Poisson distribution.

As with the binomial distribution this is a discrete distribution where the sum of the weights add up to 1, i.e.,

$$\sum_{k=0}^{\infty} P(X=k) = \sum_{k=0}^{\infty} \frac{e^{-\lambda}\lambda^k}{k!}$$

$$= e^{-\lambda}e^{\lambda}$$

$$= 1$$

The Poisson distribution is applied in many instances to physical phenomena.

Exponential Distribution

Let x be an exponential random variable. The probability density function is given by

$$f(x) = \begin{cases} \lambda e^{-\lambda x} & \text{for } x \geq 0 \\ 0 & \text{otherwise} \end{cases}$$

Integrating we obtain the probability distribution given by (here $x \geq 0$)

$$P(X \leq x) = \int_0^x \lambda e^{-\lambda x} dx = -e^{-\lambda x}\big|_0^x = 1 - e^{-\lambda x}.$$

An interesting property of the exponential distribution is the fact that it has "no memory," i.e.,

$$P(X > s+t \mid X > s) = P(X > t).$$

This distribution is used in many instances to model the failure rate of electrical components.

EXAMPLE: Let T denote the lifetime of a component. Define the reliability of a component by

$$R(t) = P(T > t) = 1 - P(t \leq T)$$

$$= e^{-\lambda T}$$

SOLUTION: Assume $\lambda = 2$. We calculate the number of operating hours given the reliability is specified at 80%. From the equation above we have

$$R(t) = .8 = e^{-2x}$$

hence $\qquad\qquad t = .112.$

This can be interpreted as follows: if 100 components are operating for .112 hours, then about 80 of them will not fail during that time.

Hypothesis Testing

The purpose of hypothesis testing is to choose between competing hypotheses about the value of a population parameter. The two competing mutually exclusive hypotheses are usually called the null hypothesis and the alternative hypothesis. For example, suppose the computer industry claims the price of computers has increased by \$100 over last year. You believe that the price has increased by more than \$100. Letting μ be the parameter which measures the price increase, we establish the null and alternative hypothesis by

$$H_0 : \mu = \$100 \text{ (null hypothesis)}$$

$$H_\mu : \mu > \$100 \text{ (alternative hypothesis)}$$

The above is called a "one-sided" test due to the fact that the alternative hypothesis specifies as completely above the value of 100. If one believes that the price of computers has either increased by more than \$100 or increased by less than \$100 we have a "two-sided" test where the null and alternative hypothesis are given by

$$H_0 : \mu = \$100 \text{ (null hypothesis)}$$

$$H_a : \mu \neq \$100 \text{ (alternative hypothesis)}$$

There are four possibilities that can occur when a decision is made in the tests above:

   1) accepting H_0 when H_0 is true,

   2) rejecting H_0 when H_0 is false,

   3) rejecting H_0 when H_0 is true,

   4) accepting H_0 when H_0 is false.

Note that 3) and 4) are incorrect conclusions. These are labeled Type 1 error and Type 2 error, respectively. We denote by the probability of Type 1 error. The confidence level is then defined by

$$\text{Confidence level} = 1 - P(\text{Type 1 error})$$

$$= 1 - \alpha$$

$$= P(\text{accept H | H is true}).$$

α is also called the level of significance.

In a similiar way we denote by β the probability of Type 2 error. The power is then defined by

$$\text{Power} = 1 - \beta = P(\text{reject H | H is false}).$$

Note that α and β need not add up to 1.

EXAMPLE: A firm producing light bulbs wants to know if it can claim that its light bulbs last 1,000 burning hours. To answer this question, the firm takes a random sample of 100 bulbs from those it has produced and finds that the average lifetime for this sample is 970 burning hours. The firm knows that the standard deviation of the lifetime of the bulbs it produces is 80 hours. Can the firm claim that the average lifetime of its bulbs is 1,000 hours at the 5% level of significance?

   SOLUTION: Since the firm is claiming that the average lifetime of its bulbs is 1,000 hours, we have

$$H_0 : \mu = 1,000, H_1 : \mu \neq 1,000.$$

Figure 17 depicts the data from this problem.

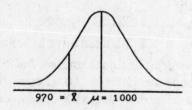

Figure 17. Normal distribution

The statistic $\dfrac{(\overline{X}-\mu)}{\sigma}$ has a standard normal distribution with a mean of 0 and a standard deviation of 1. We calculate this value, which is called z,

$$z = \frac{\overline{X}-\mu}{\sigma_{\overline{X}}}, \text{ where } \sigma_{\overline{X}} = \frac{\sigma}{\sqrt{n}}$$

and compare the value of z to a critical value. If z lies beyond this critical value, we will reject H_0. For this problem, where we have $\alpha = 5\%$ and a two-tailed test, our critical value is 1.96, since for the standard normal distribution, 2.5% of scores will have a z-value above 1.96 and 2.5% of scores will have a value below −1.96. Therefore, we use the following decision rule: reject H_0 if $z > 1.96$ or $z < -1.96$. Accept H_0 if $-1.96 < z \le 1.96$.

For the data of this problem

$$\sigma_{\overline{X}} = \frac{80}{\sqrt{100}} = 8 \text{ and } z = \frac{970-1{,}000}{8} = \frac{-30}{8} = -3.75.$$

Since $-3.75 < -1.96$, we reject H_0 and conclude that the average lifetime of the firm's bulbs is not 1,000 hours.

REVIEW PROBLEMS
PROBLEM 1

Find the inverse of

$$A = \begin{bmatrix} 3 & -1 & -2 \\ 0 & 2 & -1 \\ 4 & -10 & 2 \end{bmatrix}$$

SOLUTION: In the previous example, the determinant was found to be $|A| = 2$. In order to find the adjoint of A the cofactor of each of the nine elements is needed.

$$C_{11} = (+1)(4-10) \qquad C_{12} = (-1)(0+4) \qquad C_{13} = (+1)(0-8)$$
$$C_{21} = (-1)(-2-20) \qquad C_{22} = (+1)(6+8) \qquad C_{23} = (-1)(-30+4)$$
$$C_{31} = (+1)(1+4) \qquad C_{32} = (-1)(-3+0) \qquad C_{33} = (+1)(6-0)$$

$$\text{Adj } A = \begin{bmatrix} -6 & -4 & -8 \\ 22 & 14 & 26 \\ 5 & 3 & 6 \end{bmatrix}^T$$

$$= \begin{bmatrix} -6 & 22 & 5 \\ -4 & 14 & 3 \\ -8 & 26 & 6 \end{bmatrix}$$

The inverse of **A** is

$$A^{-1} = \frac{\text{Adj } A}{|A|} = \frac{\text{Adj } A}{2}$$

$$= \begin{bmatrix} -3 & 11 & 2.5 \\ -2 & 7 & 1.5 \\ -4 & 13 & 3 \end{bmatrix}$$

We can verify the inverse A^{-1} by multiplying it by A.

$$A^{-1}A = \begin{bmatrix} -3 & 11 & 2.5 \\ -2 & 7 & 1.5 \\ -4 & 13 & 3 \end{bmatrix}\begin{bmatrix} 3 & -1 & -2 \\ 0 & 2 & -1 \\ 4 & -10 & 2 \end{bmatrix}$$

$$= \begin{bmatrix} -9+0+10 & 3+22-25 & 6-11+5 \\ -6+0+6 & 2+14-15 & 4-7+3 \\ -12+0+12 & 4+26-30 & 8-13+6 \end{bmatrix}$$

$$= \begin{bmatrix} 1 & 0 & 0 \\ 0 & 1 & 0 \\ 0 & 0 & 1 \end{bmatrix} = I$$

PROBLEM 2

Solve the following set of equations by a) using the inverse of the coefficient matrix and b) using Cramer's Rule.

$$3x_1 - x_2 - 2x_3 = 4$$
$$2x_2 - x_3 = 2$$
$$4x_1 - 10x_2 + 2x_3 = 1$$

SOLUTION: These equations can be written as $\mathbf{Ax = b}$, where

$$\mathbf{A} = \begin{bmatrix} 3 & -1 & -2 \\ 0 & 2 & -1 \\ 4 & -10 & 2 \end{bmatrix} \qquad \mathbf{x} = \begin{bmatrix} x_1 \\ x_2 \\ x_3 \end{bmatrix} \qquad \mathbf{b} = \begin{bmatrix} 4 \\ 2 \\ 1 \end{bmatrix}$$

a) Solution via the inverse of $\mathbf{A}$. The coefficient matrix $\mathbf{A}$ is the same as the matrix in Practice Problem 1 used to illustrate the inverse of a general nonsingular matrix. From that problem, the inverse is

$$\mathbf{A}^{-1} = \begin{bmatrix} -3 & 11 & 2.5 \\ -2 & 7 & 1.5 \\ -4 & 13 & 3 \end{bmatrix}$$

The $\mathbf{x}$ vector is found by

$$\mathbf{x} = \mathbf{A}^{-1}\mathbf{b} = \begin{bmatrix} -3 & 11 & 2.5 \\ -2 & 7 & 1.5 \\ -4 & 13 & 3 \end{bmatrix} \begin{bmatrix} 4 \\ 2 \\ 1 \end{bmatrix}$$

$$= \begin{bmatrix} -12 + 22 + 2.5 \\ -8 + 14 + 1.5 \\ -16 + 26 + 3 \end{bmatrix} = \begin{bmatrix} 12.5 \\ 7.5 \\ 13 \end{bmatrix}$$

The solution is

$$x_1 = 12.5$$
$$x_2 = 7.5$$
$$x_3 = 13$$

b) Solution via Cramer's Rule.

The determinant of the coefficient matrix $\mathbf{A}$ is

$$|\mathbf{A}| = \begin{vmatrix} 3 & -1 & -2 \\ 0 & 2 & -1 \\ 4 & -10 & 2 \end{vmatrix} = 2$$

The three determinants $|\Delta_1|$, $|\Delta_2|$, and $|\Delta_3|$ are from matrices formed by replacing the first, second, and third columns, respectively, of the **A** matrix by the *b* vector.

$$|\Delta_1| = \begin{vmatrix} 4 & -1 & -2 \\ 2 & 2 & -1 \\ 1 & -10 & 2 \end{vmatrix} = 25$$

$$|\Delta_2| = \begin{vmatrix} 3 & 4 & -2 \\ 0 & 2 & -1 \\ 4 & 1 & 2 \end{vmatrix} = 15$$

$$|\Delta_3| = \begin{vmatrix} 3 & -1 & 4 \\ 0 & 2 & 2 \\ 4 & -10 & 1 \end{vmatrix} = 26$$

The elements of the **x** vector are given by

$$x_1 = \frac{|\Delta_1|}{|A|} = \frac{25}{2} = 12.5$$

$$x_2 = \frac{|\Delta_2|}{|A|} = \frac{15}{2} = 7.5$$

$$x_3 = \frac{|\Delta_3|}{|A|} = \frac{26}{2} = 13$$

PROBLEM 3

Find the equation of a plane perpendicular to the vector $\mathbf{B} = i - 4j + 3k$.

SOLUTION: Let $\mathbf{G} = xi + yj + zk$ be a vector in the plane. The vectors **G** and **B** should be perpendicular. Their dot product is zero,

$$\mathbf{G} \cdot \mathbf{B} = 0 = x(1) + y(-4) + z(3)$$

This is the equation of a plane that is perpendicular to **B**. But an infinite number of parallel planes are perpendicular to **B**. An additive arbitrary constant can be included.

$$x - 4y + 3z = \text{constant}$$

PROBLEM 4

Let $F(x, y) = x^3 + y^2 - 2xy = 0$, find $\dfrac{dy}{dx}$.

SOLUTION: $\dfrac{\partial F}{\partial x} = 3x^2 - 2y$ and $\dfrac{\partial F}{\partial y} = 2y - 2x$.

The total differential is

$$dF = (3x^2 - 2y)dx + (2y - 2x)dy = 0$$

and the derivative of y with respect to x in this implicit function is

$$\frac{dy}{dx} = \frac{3x^2 - 2y}{2y - 2x}.$$

PROBLEM 5

Integrate the expression: $\displaystyle\int \frac{x}{\left(4 - x^2\right)}dx$.

SOLUTION: Let $u = 4 - x^2$ from which $du = -2xdx$.

Using the rule $\displaystyle\int \frac{du}{u} = \ln|u| + C$, we obtain

$$\int \frac{(-2x)}{4 - x^2}dx = \ln\left|4 - x^2\right| + C.$$

But, to make this result applicable to the original problem, we require a (-2) in the numerator to obtain the form $\dfrac{du}{u}$. Because this is a constant, it is permissable to multiply the numerator under the integral sign by (-2), as long as we multiply the integral by $\left(-\dfrac{1}{2}\right)$ outside of the integral sign, in order to leave the resultant value unchanged.

Hence,

$$\int \frac{x}{4 - x^2}dx = -\frac{1}{2}\int \frac{(-2x)}{4 - x^2}dx.$$

Now we can use

$$-\frac{1}{2}\int\frac{du}{u} = -\frac{1}{2}\ln|u| + C$$

$$= \frac{1}{2}\ln\left|4 - x^2\right| + C$$

PROBLEM 6

Find the volume of the solid generated by revolving about the Y-axis the region bounded by the parabola: $y = -x^2 + 6x - 8$, and the X-axis.

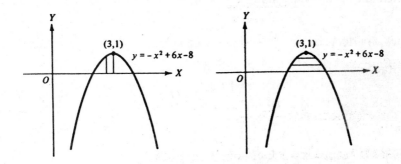

a) **Shell method**   b) **Disk method**

Figure 18. Volume of a solid with repect to Y

SOLUTION: Method 1. We use the method of cylindrical shells. The curve
$$y = -x^2 + 6x - 8$$
cuts the X-axis at $x = 2$ and $x = 4$.

The cylindrical shells are generated by the strip formed by the two lines parallel to the Y-axis, at distances x and $x + \Delta x$ from the Y-axis, $2 \le x \le 4$, as shown in Figure 18 (a). When this strip is revolved about the Y-axis, it generates a cylindrical shell of average height y^*, $y \le y^* \le y + \Delta y$, thickness Δx, and average radius x^*, $x \le x^* \le x + \Delta x$. The volume of this element is

$$\Delta V = 2\pi x^* y^* \Delta x$$

where $2\pi x^* y^*$ is the surface area. Expressing y in terms of x and passing to the limits, the sum of the volumes of all such cylindrical shells is the integral:

$$V = 2\pi\int_2^4 x\left(-x^2 + 6x - 8\right)dx$$

$$= 2\pi\int_2^4\left(-x^3 + 6x^2 - 8x\right)dx$$

$$= 2\pi \left(-\frac{x^4}{4} + 2x^3 - 4x^2 \right) \Big|_2^4$$

$$= 2\pi \left((-64 + 128 - 64) - (-4 + 16 - 16) \right)$$

$$= 8\pi$$

Method 2. This can also be thought of as the volume comprising a series of concentric disks with variable outer and inner radii, as sectionally shown in Figure 18 b. The variable radii are as follows: Since

$$y = -x^2 + 6x - 8,$$

we solve for x.

To complete the square, we require a 9, so that

$$x^2 - 6x + 9$$

consitutues a perfect square. Rewriting the equation,

$$x^2 - 6x + 9 - 9 + 8 = -y.$$

$$x^2 - 6x + 9 = 1 - y.$$

$$(x - 3)^2 = 1 - y.$$

Therefore,

$$x = 3 \pm \sqrt{1 - y}$$

which shows the disks, y units from the x-axis, have an

inner radius $x_{in} = 3 - \sqrt{1-y}$ and an

outer radius $x_o = 3 + \sqrt{1-y}$

(The particular one on the x-axis has $x_{in} = 2$ and $x_o = 4$.)

The volume of this disk with thickness dy is:

$$dV = \pi \left(x_o^2 - x_{in}^2 \right) dy,$$

or

$$dV = \pi \left[(x_0 + x_{in})(x_0 - x_{in}) \right] dy.$$

Subsituting the values for x_o and x_{in},

$$dV = \left(\left(\left(3 + \sqrt{1-y} \right) + \left(3 - \sqrt{1-y} \right) \right) \bullet \left(\left(3 + \sqrt{1-y} \right) \left(3 + \sqrt{1-y} \right) \right) \right) dy$$

$$= \pi \left(12\sqrt{1-y} \right) dy$$

Since y varies from 0 to 1, the desired volume is:

$$V = 12\pi \int_0^1 (1-y)^{\frac{1}{2}} dy$$

$$= 12\pi \left(-\frac{2}{3}(1-y)^{\frac{3}{2}} \Big|_0^1 \right) = 8\pi.$$

PROBLEM 7

Integrate: $\int \frac{2-x}{x^2+x}\,dx.$

SOLUTION: To integrate this expression, we use partial fractions. We first find the factors of the denominator. They are: (x) and $(x+1)$. Now we find two numbers, A and B, such that

$$\frac{2-x}{x^2+x} = \frac{A}{x} + \frac{B}{x+1}.$$

Multiplying both sides of this equation by the common denominator, $x(x+1)$, we obtain: $2 - x = A(x+1) + B(x)$. When $x(x+1) = 0$, $x = 0$ or $x = -1$. When $x = 0$, $2 - 0 = A(0+1) + B(0)$, and $A = 2$. When $x = -1$, $2 - (-1) = A(0) + B(-1)$ and $B = -3$. We can now write the integral as follows:

$$\int \frac{2-x}{x^2+x}\,dx = \int \left[\frac{2}{x} - \frac{3}{x+1}\right]dx$$

$$= 2\int \frac{1}{x}\,dx - 3\int \frac{1}{x+1}\,dx$$

$$= 2\ln x - 3\ln(x+1) + C.$$

PROBLEM 8

Solve the differential equation

$$\frac{d^2x}{dt^2} + 3\frac{dx}{dt} + 2x = 1 \quad \text{when } t = 0, x = 0, \text{ and } \frac{dx}{dt} = 0.$$

SOLUTION: We use Table 4 to find the transform of each term.

$$\left(s^2 X(s) - s0 - 0\right) + 3\left(sX(s) - 0\right) + 2X(s) = \frac{1}{s}$$

$$\left(s^2 + 3s + 2\right)X(s) = \frac{1}{s}$$

Solve for $X(s)$.

$$X(s) = \frac{1}{s\left(s^2 + 3s + 2\right)} = \frac{1}{s(s+1)(s+2)}$$

Expand into partial fractions and solve.

$$X(s) = \frac{\frac{1}{2}}{s} + \frac{\frac{1}{2}}{s+2} - \frac{1}{s+1}$$

Use Table 4 to transform to time functions.

$$x(t) = \frac{1}{2} + \frac{1}{2}e^{-2t} - e^{-t}$$

PROBLEM 9

The tensile strength of pins produced with a new process is distributed normally with a mean of 35 and a standard deviation of 2.20.

What fraction of pins have a strength exceeding 37?

$$Z = \frac{37 - 35}{2.20} = 0.91$$

SOLUTION: From the Appendix, the probability of finding a strength between $Z = 0$ and $Z = 0.91$ is 0.3186. The probability of finding a strength for $Z > 0$ is 0.5; thus the probability of finding a strength corresponding to $Z > 0.91$ is

$$0.5 - 0.3186 = 0.1814.$$

The probability of finding a strength less than 37 or $Z < 0.91$ is

$$0.5 + .3186 = 0.8186.$$

What is the probability of the strength of a pin being between 32 and 36?

$$Z_1 = \frac{32 - 35}{2.20} = -1.36$$

$$Z_2 = \frac{36 - 35}{2.20} = 0.45$$

From the Appendix, the probability is $0.1736 + 0.4131 = 0.5867$.

PROBLEM 10

Consider a fair coin tossed three times. If the random variable x denotes the total number of heads attained in the three tosses, and the density function is given by:

$$p(x) \begin{cases} \dfrac{1}{8} & \text{for } x = 0 \\[2mm] \dfrac{3}{8} & \text{for } x = 1 \\[2mm] \dfrac{3}{8} & \text{for } x = 2 \\[2mm] \dfrac{1}{8} & \text{for } x = 3 \end{cases}$$

What is the probability of getting two or more heads?

SOLUTION:

$$P(X \geq 2) = \sum_{x_i \in [2,3]} p(x_i)$$
$$= p(2) + p(3)$$
$$= \frac{3}{8} + \frac{1}{8}$$
$$= \frac{1}{4}.$$

CHAPTER FOUR

Electrical Circuits

Chapter 4

ELECTRICAL CIRCUITS

DC CIRCUITS

Current and Voltage

Current i is the net charge crossing a cross section of a conductor per unit time. If a net charge q crosses in a t time interval then:

$$i = \frac{q}{t}.$$

Unit of current = ampere (A), 1A = 1 coulomb of net charge crossing in 1 second

Instantaneous current (i) = time rate of change of charge = $\dfrac{dq}{dt}$

CURRENT FLOW

In metallic conductors, the net movement of charge is due to the displacement of electrons. The flow of current is opposite to the direction of the movement of electrons by convention.

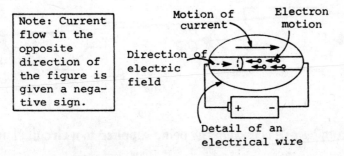

Note: Current flow in the opposite direction of the figure is given a negative sign.

Direction of electric field

Motion of current

Electron motion

Detail of an electrical wire

Figure 1. Current flow convention (Note: Current flow in the opposite direction of the figure is given a negative sign.)

Direct Current (DC) is defined as a current which is constant due to a steady, unchanging, unidirectional flow of charge.

Figure 2. Direct current (DC)

The **voltage** (V or v), or the potential difference between two points, is the measure of the work required to move a unit charge from one point to another. The unit of voltage is the volt (V or v), and is equal to one joule per coulomb. The voltage sign convention is illustrated in Figure 3, where it is assumed that a positive current is supplied by an external source entering terminal 1.

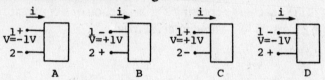

Figure 3. Voltage sign convention

In C and D of Figure 3, terminal 1 is 1 volt positive with respect to terminal 2. In A and B of Figure 3, terminal 2 is 1 volt positive with respect to terminal 1.

In Figures C and D the current enters the positive terminal of the circuit in the box. This means that electrical energy is being transformed (absorbed by the circuit) into another form of energy. In a resistor the electric energy is transformed into heat, in an electric motor into mechanical energy.

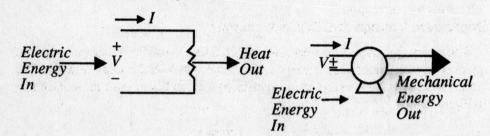

Figure 4. Examples of electric energy being supplied to a circuit element

If the current flows out of the positive terminal, then electric energy is generated in the circuit, as in Figure 5.

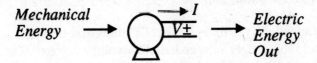

Figure 5. Example of electric energy generation

Circuit Elements Found in DC Circuits

Independent Voltage Source

As shown in Figure 6, the same amount of voltage output is supplied continuously by an **independent DC voltage source** regardless of the amount of current drawn from it. The circuit symbols used to represent independent DC voltage sources are shown in Figure 7.

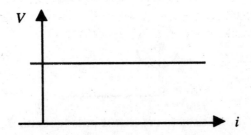

Figure 6. Voltage-current relationship for independent DC voltage sources

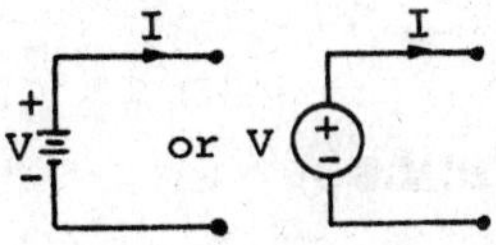

Figure 7. Circuit symbols used to represent independent DC voltage sources

Dependent Voltage and Current Sources

The source quantity of a **dependent source** is determined by a voltage or current existing at some other location in the electrical system under consideration. The circuit symbols used to represent dependent voltage and current sources are shown in Figure 8.

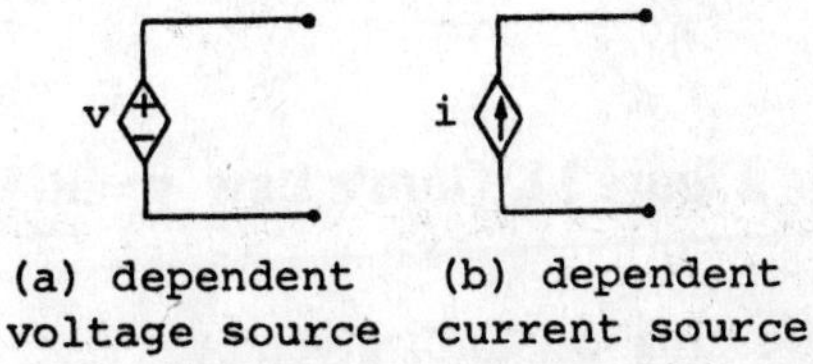

(a) dependent   (b) dependent
voltage source   current source

**Figure 8. Circuit symbols used to represent dependent DC voltage
and current sources**

Independent Current Source

The current supplied by an **independent DC current source** is constant, as shown in Figure 9, and is completely independent of the voltage across it. The circuit symbol used to represent an independent DC current source is shown in Figure 10.

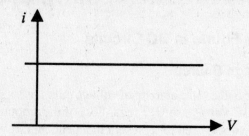

Figure 9. Voltage-current relationship for independent DC current sources

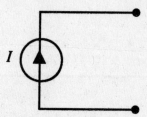

Figure 10. Circuit symbol used to represent independent DC current sources

Both independent current and voltage sources are approximations for a physical element.

Resistance and Conductance

Ohm's Law

The voltage across a conducting material is directly proportional to the current through the material, i.e., $v = Ri$, where R(resistance) is the proportionality constant. This is **Ohm's Law** as illustrated in Figure 11.

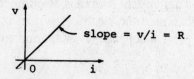

Figure 11. Ohm's Law $v = Ri$

Hence, absorbed power in a resistor is given by:

$$p = vi = i^2R = \frac{v^2}{R}.$$

This power is in the form of heat because a resistor is a passive element; it neither delivers power nor stores energy.

Resistance

Resistance (R) is the measure of the tendency of a material to impede the flow of electric charges through it. Resistance is therefore low in good conductors but high in poor conductors (insulators). The unit of resistance is the ohm (Ω), which is volts per ampere. The circuit symbol used to represent a resistor is shown in Figure 12.

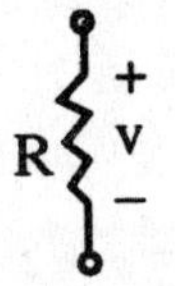

$$R \lessgtr \begin{matrix} + \\ v \\ - \end{matrix}$$

Figure 12. Circuit symbol used to represent a resistor

Two particular cases of resistance are the open and short circuit, shown in Figure 13. If the resistance between two points of a circuit is zero, it is said that a short circuit exists between the two points. If the resistance between two points is infinite, an open circuit is said to exist between these two points.

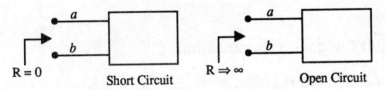

R = 0   Short Circuit $R \Rightarrow \infty$ Open Circuit

Figure 13. Short circuit and open circuit

Conductance

Conductance (G) is the reciprocal of resistance, or the ratio of current to voltage:

$$G = \frac{1}{R} = \frac{i}{v}.$$

The unit of conductance is the mho($\mho$), or siemens (S).

EXAMPLE: With reference to Figure 14, find v if;
  (a) $G = 10^{-2}$ $\mho$ and $i = 2.5A$
  (b) $R = 40\Omega$ and the resistor absorbs 250W.
  (c) $i = 2.5A$ and the resistor absorbs 500W.

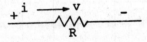

Figure 14. Resistor

SOLUTION: (a) Apply Ohm's Law.

$$v = iR\frac{i}{G}$$

$$v = \frac{-2.5}{10^{-2}} = -250\text{V}.$$

(b) We know that $p = vi$ since $v = iR$;

$$p = i^2R = \frac{v^2}{R} \tag{1}$$

Solving for v

$$v = \sqrt{pR}$$

$v = \sqrt{250(40)} = \pm 100\text{V}$. Since R absorbs power, $v = +100\text{V}$.

(c) Solving for R in equation (1)

$$R = \frac{p}{i^2} = \frac{500}{(2.5)^2} = 80 \text{ ohms}.$$

Using Ohm's Law:

$$v = iR = 2.5(80) = 200\text{V}.$$

Resistor and Conductor Combinations

For a series combination of n resistors:

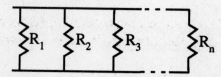

R_1 R_2 R_3 R_4 R_n

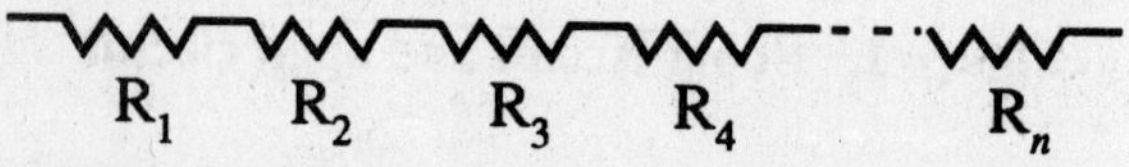

Figure 15. Series combination of *n* resistors

$$R_{eq} = R_1 + R_2 + ... + R_n$$

and
$$\frac{1}{G_{eq}} = \frac{1}{G_1} + \frac{1}{G_2} + ... + \frac{1}{G_n}$$

For a parallel combination of n resistors:

R_1 R_2 R_3 R_n

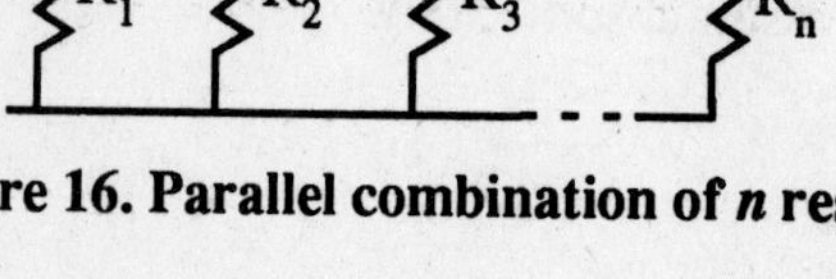

Figure 16. Parallel combination of *n* resistors

$$\frac{1}{R_{eq}} = \frac{1}{R_1} + \frac{1}{R_2} + \dots + \frac{1}{R_n}$$

and
$$G_{eq} = G_1 + G_2 + \dots + G_n.$$

Wheatstone Bridge

The **Wheatstone Bridge** is used to determine an unknown resistance. Referring to Figure 17, the bridge is balanced by adjusting a variable resistor until no current flows through the ammeter. When the current flow through the ammeter is zero, the current through R_1 will equal the current through R_3, and the current through R_2 will equal the current through R_4. The following equation can then be used to solve for the unknown resistance:

$$\frac{R_1}{R_2} = \frac{R_3}{R_4}.$$

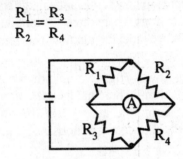

Figure 17. Wheatstone Bridge

Kirchhoff's Laws

Kirchhoff's Current Law (KCL) states that the algebraic sum of all currents entering a node equals the algebraic sum of all currents leaving it, i.e., for a given node Σ currents entering = Σ currents leaving, or:

$$\sum_{n=1}^{N} i_n = 0.$$

Kirchhoff's Voltage Law (KVL) states that the algebraic sum of all voltages around a closed loop (or path) is zero, i.e., for a closed loop, Σ potential rises = Σ potential drops.

EXAMPLE: Analyze the circuit in Figure 18 using Kirchoff's Laws. Find the magnitude and direction of I_2, the current in the upper 2Ω resistor.

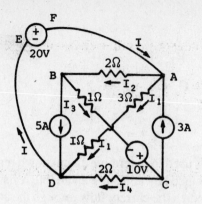

Figure 18. Circuit

SOLUTION: The circuit contains two constant voltage sources and two constant current sources. It has four junctions where three or more wires join, labelled A, B, C, and D, as shown. To begin, we label the unknown currents, I, I_1, I_2, I_3, and I_4, as indicated in Figure 18.

This problem has five unknowns (the five unknown currents) and, therefore, requires five independent equations. Three independent equations are obtained by applying Kirchoff's Current Sum rule at any three of the four junctions. For convenience, we ignore units and choose junctions A, B, and C:

$$\text{at A, } I + 3 = I_2 + 5; \tag{1}$$
$$\text{at B, } I_2 = I_3 + 5; \tag{2}$$
$$\text{at C, } I_3 = I_4 + 3. \tag{3}$$

The two additional independent equations required for the solution are obtained by applying Kirchoff's Voltage Sum rule about closed loops in the circuit. As it is difficult to represent the potential drops across the constant current sources, we choose the loops ADEF and ABCDEF:

$$\text{loop ADEF, } 20 = 3I_1 + I_1, \tag{4}$$
$$\text{loop ABCDEF, } 20 = 2I_2 + I_3 - 10 + 2I_4. \tag{5}$$

From (4), by inspection, $I_1 = 5$ amperes. Multiplying (2) by minus two gives

$$-2I_2 + 2I_3 = -10,$$

and, rewriting (5),

$$2I_2 + I_3 + 2I_4 = 30.$$

The sum of these two equations is

$$3I_3 + 2I_4 = 20. \tag{6}$$

Multiplying (3) by two gives

$$2I_3 - 2I_4 = 6$$

which, when summed with (6), results in

$$5I_3 = 26$$

or, $I_3 = \dfrac{26}{5}$ amperes. Substituting this result back into (6) gives

$$3\left(\frac{26}{5}\right) + 2I_4 - 20,$$

or $I_4 = \dfrac{11}{15}$ amperes.

The remaining two currents are found by substituting these results, first into (2) and then into (1):

$$I_2 = \left(\frac{26}{5}\right) + 5,$$
$$I = I_2 + 2.$$

Therefore, the current in the upper 2Ω resistor has the magnitude $I_2 = 51/5$ amperes, and flows in the direction of the arrow shown in the figure. The total current flowing in the circuit is $I = 61/5$ ampere.

DC Circuit Analysis Techniques

Two common **DC circuit analysis techniques** are mesh analysis and nodal analysis. The general approach to these techniques is listed below. It should be noted that mesh analysis (sometimes called loop analysis) is only applicable to a planar network. By definition, a planar network is a circuit whose diagram can be drawn on a plane surface so that no branch passes over or under any other branch.

Nodal Analysis – General Approach

(1) Convert all voltage sources to current sources.
(2) In each network, determine the number of nodes.
(3) Choose a reference node and assign voltages to the remaining node.
(4) Apply KCL at each node, except at the reference node.
(5) Solve the unknown equations for nodal voltages.

Mesh Analysis — General Approach

(1) Assign closed loops of current called mesh currents, clockwise, to each loop of the circuit.
(2) Apply KVL around each closed loop.
(3) Solve resulting equations for the assumed loop currents.

Thevenin and Norton Equivalent Circuits

A **Thevenin equivalent circuit** consists of a voltage source in series with a resistor, as shown in Figure 19. This equivalent circuit is used to represent a linear, two-terminal network, which may contain both independent and dependent current and voltage sources. The voltage and current sources are set to zero by shorting the voltage sources and treating current sources as open circuits. The Thevenin equivalent resistance, R_{TH}, is the resistance across terminals denoted A and B in the figure, and the open circuit voltage, V_{TH}, (also called V_{OC}), is the Thevenin equivalent voltage.

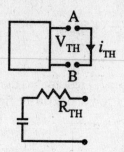

Figure 19. The Thevenin equivalent circuit

ELECTRIC AND MAGNETIC FIELDS

Coulomb's Law

By definition, the force between two point charges of arbitrary positive or negative strengths is given by **Coulomb's Law** as follows:

$$F = k\frac{Q_1 Q_2}{d^2}$$

where Q_1 and Q_2 = Positive or negative charges on either object in coulombs (C)

d = Distance separating the two point charges

k = The constant of proportionality

= $(4\pi\varepsilon_0)^{-1} = 8.987 \times 10^9$ newton-meter²/coulomb² (Nm²/C²)

ε_0 = Permittivity in free space = 8.854×10^{-12} Farad/meter (F/m)

Note that $\varepsilon = \varepsilon_0 \varepsilon_r$ is used for media other than free space, where ε_r is the relative permittivity of the media.

These equations are expressed for free space, where permittivity (ε) is equal to the permittivity in free space (ε_0).

The force F can be expressed in a vector form to indicate its direction as follows:

$$F = \frac{Q_1 Q_2}{4\pi\varepsilon_0 d^2} a_d$$

where unit vector a_d is in the direction of d and $a_d = \frac{d}{|d|} = \frac{d}{d}$.

EXAMPLE: A negative point charge of 10^{-6} coulomb is situated in air at the origin of a rectangular coordinate system. A second negative point charge of 10^{-4} coulomb is situated on the positive x axis at a distance of 50 cm from the origin. What is the force on the second charge?

SOLUTION: By Coulomb's law the force

$$F = i\frac{\left(-10^{-6}\right)\left(-10^{-4}\right)}{4\pi \times 0.5^2 \times \dfrac{10^{-9}}{36\pi}}$$

$$= +i3.6 \text{ newtons}$$

That is, there is a force of 3.6 newtons (0.8 lb) in the positive x direction on the second charge.

Electric Field Intensity

Assuming a fixed position point charge Q, then by Coulomb's law, the force on a test charge Q_t due to Q is:

$$F_t = \frac{QQ_t}{4\pi\varepsilon_0 d_t^2} a_{d_t}.$$

Now, by definition, the **electric field intensity** E due to Q equals the force per unit charge in V/m or N/C.

Hence,

$$E = \frac{F_t}{Q_t} = \frac{Q}{4\pi\varepsilon_0 d_t^2} a_{d_t}.$$

Thus, in general, the electric field intensity E is:

$$E = \frac{Q}{4\pi\varepsilon_0 d^2} a_d,$$

where d is the magnitude of the vector **d**.

In general, if Q is located at $d' = aa_x + ba_y + ca_z$ and the field is at $d = xa_x + ya_y + za_z$, then:

$$E = \frac{Q(d - d')}{4\pi\varepsilon_0 |d - d'|^3}.$$

EXAMPLE: The figure shows eight point charges situated at the corners of a cube. Find the electric field intensity at each point charge, due to the remaining seven point charges.

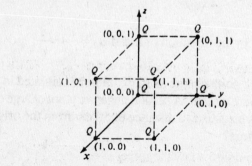

Figure 20. A cubical arrangement of point charges

SOLUTION: First note that the electric field intensity at a point $B(x_2, y_2, z_2)$ due to a point charge Q at point A (x_1, y_1, z_1) is given by

$$E_B = \frac{Q}{4\pi\varepsilon_0(AB)^2}i_{AB} = \frac{Q}{4\pi\varepsilon_0(AB)^2}\frac{AB}{(AB)} = \frac{Q(AB)}{4\pi\varepsilon_0(AB)^3}$$

$$= \frac{Q}{4\pi\varepsilon_0}\frac{(x_2-x_1)i_x + (y_2-y_1)i_y + (z_2-z_1)i_z}{\left[(x_2-x_1)^2 + (y_2-y_1)^2 + (z_2-z_1)^2\right]^{3/2}}$$

(1)

Now consider the point $(1,1,1)$. Applying (1) to each of the charges at the seven other points and using superposition, the electric field intensity at the point $(1,1,1)$ is

$$E_{(1,1,1)} = \frac{Q}{4\pi\varepsilon_0}\left[\frac{i_x}{(1)^{3/2}} + \frac{i_y}{(1)^{3/2}} + \frac{i_z}{(1)^{3/2}}\right.$$

$$\left. + \frac{i_y+i_z}{(2)^{3/2}} + \frac{i_z+i_x}{(2)^{3/2}} + \frac{i_x+i_y}{(2)^{3/2}} + \frac{i_x+i_y+i_z}{(3)^{3/2}}\right]$$

$$= \frac{Q}{4\pi\varepsilon_0}\left(1 + \frac{1}{\sqrt{2}} + \frac{1}{3\sqrt{3}}\right)(i_x+i_y+i_z)$$

$$= \frac{3.29Q}{4\pi\varepsilon_0}\left(\frac{i_x+i_y+i_z}{\sqrt{3}}\right)$$

Noting that $\dfrac{i_x+i_y+i_z}{\sqrt{3}}$ is the unit vector directed from $(0,0,0)$ to $(1,1,1)$, the electric field intensity at $(1,1,1)$ is directed diagonally away from $(0,0,0)$ with a magnitude equal to $\dfrac{3.29Q}{4\pi\varepsilon_0}$ N/C. From symmetry considerations, it then follows that the electric field intensity at each point charge, due to the remaining seven point charges, has a magnitude $\dfrac{3.29Q}{4\pi\varepsilon_0}$ N/C, and it is directed away from the corner opposite to that charge.

Electric Field Intensity Due to Point Charges, Volume Charge Distribution, Line of Charge, and Sheet of Charge

The electric field intensity due to two **point charges** as illustrated in Figure 21 is:

$$E = E_1 + E_2 = \frac{Q_1}{4\pi\varepsilon_0 d_1^2}a_{d_1} + \frac{Q_2}{4\pi\varepsilon_0 d_2^2}a_{d_2}.$$

Thus, the differential electric field dE due to dQ in Figure 22 is:

$$dE = \frac{dQ}{4\pi\varepsilon_0 d^2}a_d = \frac{\rho_v dV}{4\pi\varepsilon_0 d^2}a_d,$$

and

$$E = \int_v \frac{\rho_v dV}{4\pi\varepsilon_0 d^2}a_d = \text{total electric field at point } P \text{ in the figure.}$$

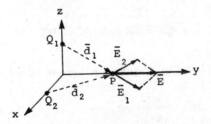

Figure 21. Electric fields produced by point charges

In general, for k point charges, the electric field intensity is:

$$E = \sum_{j=1}^{k} \frac{Q_j}{4\pi\varepsilon_0 d_j^2}a_{d_j}.$$

The field due to a **volume charge distribution** is determined by Gauss's Law. Gauss's law for electric fields in free space is defined as:

$$\oint_s (\varepsilon_0 E)\cdot ds = \int_v \rho_v du \equiv Q$$

where Q is the total charge within a finite volume:

$$Q = \int_V \rho_V dV = \int_V dQ,$$

and ρ_v is the volume charge density:

$$\rho_V = \frac{dQ}{dV} \text{ in C/m}^3.$$

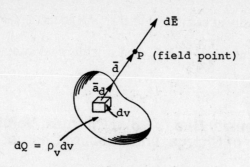

Figure 22. Volume charge

The electric field intensity due to a **line of charge** is determined as follows, using Figure 23. The line charge density is denoted by ρ_ℓ.

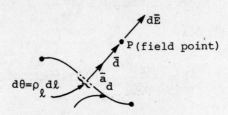

Figure 23. Line of charge

The differential electric field dE due to the line of charge is:

$$dE = \frac{dQ}{4\pi\varepsilon_0 d^2}\, a_d.$$

Hence,

$$dE = \frac{\rho_\ell d\ell}{4\pi\varepsilon_0 d^2}\, a_d,$$

$$E = \int_\ell \frac{\rho_\ell}{4\pi\varepsilon_0 d^2}\, a_d d\ell.$$

For an infinitely long wire, the electric field intensity at a distance d from the line charge is:

$$E = \frac{\rho_\ell}{2\pi\varepsilon_0 d}.$$

The electric field intensity due to a **sheet of charge** is determined using Figure 24, where ρ_s is the surface charge density.

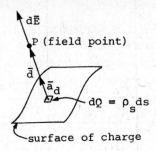

Figure 24. Sheet of charge

The differential electric field *dE* due to the sheet of charge is:

$$dE = \frac{\rho_s ds}{4\pi\varepsilon_0 d^2} a_d,$$

where

$$dQ = \rho_s ds.$$

Hence,

$$E = \int_s \frac{\rho_s ds}{4\pi\varepsilon_0 d^2} a_d.$$

Work in an Electric Field

In Figure 25 shown below, a small test charge Q is presented in an electric field *E*.

Figure 25. Forces on test charge in electric field *E*

The work done, *d*W, or the energy needed to move a point charge through a distance *dL*, is expressed as

$$d\mathrm{W} = \mathrm{F}' \bullet d\mathrm{L}$$
$$= Q\mathrm{E} \bullet d\mathrm{L}$$

where *d*W is the differential work done by moving the point charge through a differential distance *dL*.

Or

$$W = -Q\int_A^B E \bullet dL$$

where A and B specify the starting and ending positions of the travelling point charge. Hence, work done is independent of the path taken. Also note, by convention, the work done by the electric field is a negative quantity.

The following table gives the equation for dL in different coordinate systems:

Coordinate systems	dL
Cartesian	$a_x dx + a_y dy + a_z dz$
Cylindrical	$a_r dr + a_\phi r d\phi + a_z dz$
Spherical	$a_r dr + a_\phi r d\phi + a_\phi r \sin\theta d\phi$

Biot-Savart Law

The **Biot-Savart Law** states that the differential magnetic field strength dH at any point P, produced by a differential element $d\ell$ carrying the current I, is proportional to $Id\ell \times a_R$, where a_R is a unit vector leading from $d\ell$ to the point P, as shown in Figure 26. The differential magnetic field strength dH is also inversely proportional to the square of the distance from the differential element to the point P.

In mathematical form:

$$dH = \frac{Id\ell \times a_R}{4\pi R^2} \quad \text{(Ampere/meter, i.e., A/m).}$$

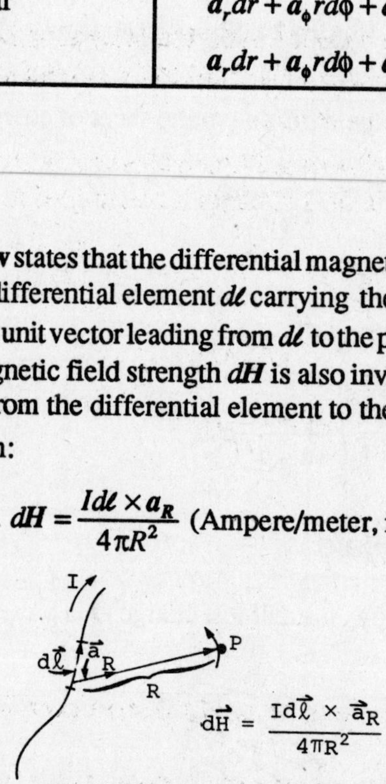

$$d\vec{H} = \frac{Id\vec{\ell} \times \vec{a}_R}{4\pi R^2}$$

Figure 26. Illustration of Biot-Savart Law

The integral form of the Biot-Savart Law is:

$$H = \int_s \frac{K \times a_R}{4\pi R^2} ds,$$

where K is the surface current density.

Other forms of the Biot-Savart Law include:

$$H = \oint \frac{Id\ell \times a_R}{4\pi R^2},$$

and

$$H = \int_V \frac{J \times a_R}{4\pi R^2} dV$$

where J is the current density.

The Poynting vector, S, is defined as:

$$S = E \times H,$$

and is the instantaneous power density of an electromagnetic wave, with units of webers per square meter (W/m²).

Ampere's Circuital Law

Ampere's Circuital Law states that the line integral of the tangential component of H about any closed path is exactly equal to the current enclosed by that path:

$$\oint H \cdot d\ell = I.$$

For an application such as in Figure 27, the path is a circle of radius r. Using Ampere's Circuital Law, the magnetic field strength is determined as follows:

$$\oint H \cdot d\ell = \int_0^{2\pi} H_\phi r d\phi = H_\phi r \int_0^{2\pi} d\phi$$

$$= H_\phi \cdot 2\pi r = I$$

$$H = \frac{I}{2\pi r} a_\phi.$$

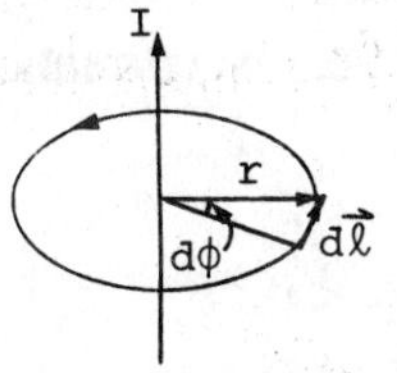

Figure 27. An infinitely long straight filament carrying a direct current *I*

Magnetic Flux and Magnetic Flux Density

Magnetic flux density, *B* is defined as:

$$B = \mu_o H$$

where μ_o is the permeability of free space:

$$\mu_o = 4\pi \times 10^{-7} \text{ Henry/meter (H/m)}.$$

The use of μ_o assumes free space, for other media:

$$\mu = \mu_o \mu_r$$

where μ_r is the relative permeability of the media. The unit of magnetic flux density is the tesla (T), or webers per square meter (Wb/m²).

Magnetic flux, ϕ, has units of weber, and is defined in terms of magnetic flux density as follows:

$$B = \frac{\phi}{A},$$

where A is the area prependicular to the flux.

The following example illustrates the use of Ampere's Circuital Law to determine magnetic flux and magnetic flux density.

EXAMPLE: Find the flux between the conductors of the coaxial line shown in the figure. The conductor extends a length L.

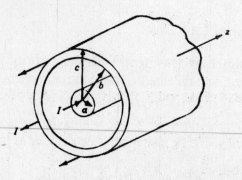

Figure 28. Cross section of a coaxial cable

SOLUTION: By applying Ampere's Law to the region between the conductors,

$$\oint H \bullet d\ell = I_{\text{enclosed}}$$

In the region between the conductors, the current enclosed is I, therefore

$$H = 2\pi r = I$$

$$H = \frac{I}{2\pi r} \quad (a < r < b)$$

and therefore

$$B = \mu_0 H = \frac{\mu_0 I}{2\pi r} a_\phi$$

The magnetic flux contained between the conductors in a length L is the flux crossing any radial plane extending from $r = a$ to $r = b$ and from, say, $z = 0$ to $z = L$

$$\phi = \int_s B \bullet dS = \int_0^L \int_a^b \frac{\mu_0 I}{2\pi r} a_\phi \bullet dr\, dz a_\phi$$

or

$$\phi = \frac{\mu_0 I L}{2\pi} \ln \frac{b}{a}.$$

CAPACITANCE AND INDUCTANCE

The Parallel Plate Capacitor

A **parallel plate capacitor** is shown in Figure 29. **Capacitance** (C) is defined as:

$$C = K\varepsilon_0 \frac{A}{d}$$

$$= \frac{q}{v}.$$

ε_0 is equal to 8.854 pF/m, and K, the relative dielectric constant, is:

$$K = \frac{\varepsilon}{\varepsilon_0}.$$

For the time variant case:

$$C = \frac{q(t)}{v(t)}$$

Note that for direct current, the capacitor acts as an open circuit. The unit of capacitance is the farad (F), which is coulombs per volt. The circuit symbol is used to represent capacitance are shown in Figure 30.

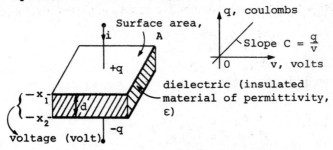

Figure 29. Parallel plate capacitor

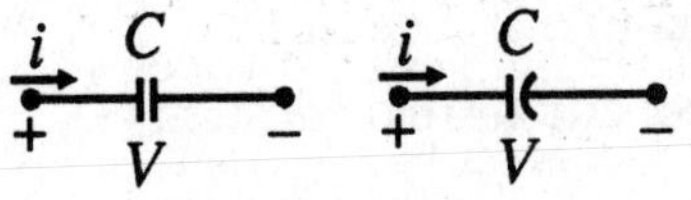

Figure 30. Circuit symbols used to represent capacitors

$$v(t) = \frac{1}{C} \int_{-\infty}^{t} i(\tau)d(\tau)$$

$$i(t) = \frac{dq(t)}{dt} = C\frac{dv(t)}{dt}$$

$$p = Cv\left(\frac{dv}{dt}\right)$$

and

$$W = \frac{1}{2}Cv^2 = \text{stored energy [joules]}$$

or

$$W = \frac{Q^2}{2C}.$$

EXAMPLE: Consider a capacitor with capacitance $C = 10^{-6}$ farad. Assume that initial voltage across this capacitor is $v_c(0) = 1$ volt. Find the voltage $v_c(t)$ at time $t \geq 0$ on this capacitor if the current through it is $i_c(t) = \cos(10^6 t)$.

  SOLUTION: We use the definition,

$$i = C\frac{dv}{dt}$$

solving for v

$$\frac{1}{C} \int_{-\infty}^{t} i(\tau)d\tau = v(t).$$

If we have an initial voltage at time t_o, $-\infty < t_o < t$, we may state that

$$\frac{1}{C} \int_{-\infty}^{t_o} i(\tau)d\tau + \frac{1}{C} \int_{t_o}^{t} i(\tau)d\tau = v(t)$$

$$v(t_o) + \frac{1}{C} \int_{t_o}^{t} i(\tau)d\tau = v(t). \tag{1}$$

  In this problem we are given $v(t_o) = v_c(0) = 1$ volt, $C = 10^{-6}$ farad, and $i_c(t) = \cos(10^6 t)$. We are asked to find $v_c(t)$ at time $t \geq 0$.

  Substituting the above conditions into equation (1)

$$v_c(t) = 1 + \frac{1}{10^{-6}} \int_{o}^{t} \cos(10^6 \tau)d\tau$$

$$v_c(t) = 1 + \frac{1}{10^6 \cdot 10^{-6}} \left[\sin(10^6 \tau)\right]_{o}^{t}$$

$$v_c(t) = 1 + \sin(10^6 \tau).$$

The Iron-Core Inductor

An **iron-core inductor** is shown in Figure 31. Inductance (L) is defined as:

$$L = \frac{\mu N^2 A}{\ell};$$

where

      N = Number of turns of coil

      μ = Permeability of the core

      A = Cross-sectional area of core

      ℓ = Mean length of core.

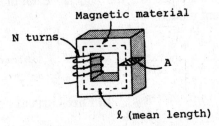

Figure 31. Iron-core inductor

The magentic flux (ϕ), magnetic field intensity (H), and magnetic flux density (B) relationships of an inductor are:

$$\phi(t) = \left(\frac{\mu N^2 A}{\ell} \right) i(t)$$

$$= Li(t)$$

$$H = \frac{Ni}{\ell}$$

$$B = \mu H$$

$$= \frac{\phi}{A}.$$

The *B-H* curve is used to relate magnetic field intensity and magnetic flux density, and is shown in Figure 32 for the iron-core inductor.

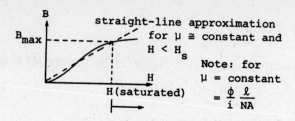

Figure 32. *B-H* curve

The concept of self-inductance is illustrated in Figure 33, which shows the production of magnetic flux by a current. Note that the slope of the graph relating magnetic flux and current is equal to *L*, the inductance. Figure 33 includes the circuit symbol used to represent inductors.

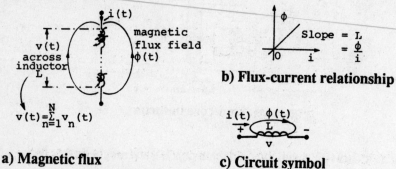

a) **Magnetic flux**

b) **Flux-current relationship**

c) **Circuit symbol**

Figure 33. The concept of self-inductance

Inductance can also be expressed as:

$$L = \frac{N d\phi(i)}{di}.$$

Voltage, current, energy, and power relationships for inductors are:

$$\text{Inductance } (L) = \frac{N d\phi(i)}{di} \text{ [henry (H) or volt-second/ampere]}$$

$$\text{Voltage } v(t) = N(\text{no. of turns of coil}) \times \frac{d\phi(t)}{dt}$$

(rate of change of ϕ with respect to time), or

$$v(t) = \frac{N d\phi(i)}{di} \quad \frac{di}{dt} = L \frac{di(t)}{dt}$$

$$i(t) = \frac{1}{L} \int_{-\infty}^{t} v \, dt$$

$$W = \frac{1}{2}Li^2$$

$$p = vi = Li\frac{di}{dt}[W]$$

In direct current sicuits, an inductor acts a short circuit.

Mutual Inductance

If the magnetic flux produced by the current in due coils links with a second coil, and if this flux changes due to changes in the current, then the change in the current in one coil induces a back voltage in the second. It is said that a mutual inductance exists between the two coils.

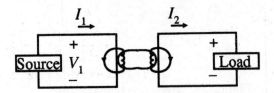

Figure 34. Mutual inductance

The following equations and definitions refer to the circuit in Figure 34:

$\phi_{1\ell}$ = Flux linking only coil 1 ϕ_m = Mutual flux (linking both coils)

$\phi_{2\ell}$ = Flux linking only coil 2 ϕ_1 = $\phi_{1\ell} + \phi_m$

$\phi_{2\ell}$ = $\phi_{2\ell} + \phi_m$ ϕ_1 = Total flux linking coil 1

ϕ_2 = Total flux linking coil 2

In linear magnetic materials

ϕ_1 = $L_{1\ell}i_1$ $\phi_{2\ell}$ = $L_{2\ell}i_2$

ϕ_m = $L_{m1}(i_1) + L_{m2}(i_2)$

The voltage across L_1 in Figure 34 is:

$$\vartheta_1 = L_1\frac{di_1}{dt} + M_{12}\frac{di_2}{dt},$$

where M_{12} denotes that a voltage response is produced at L_1 due to a current source at L_2, and the voltage across L_2 is:

$$\vartheta_2 = L_2\frac{di_2}{dt} + M_{21}\frac{di_1}{dt},$$

where M_{21} denotes a voltage responsed at L_2 due to a current source at L_1. It is not necessary to use these subscripts for the mutual inductance, however, as:

$$M_{12} = M_{21} + M,$$

where M is the proportionality constant between two coils (units of henry):

$$M = K\sqrt{L_1 L_2},$$

where the coupling coefficient K is:

$$K = \frac{\phi_m}{\phi_1}.$$

The Iron-Core Transformer

An **iron-core transformer** is shown in Figure 35.

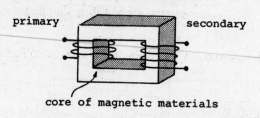

primary   secondary

core of magnetic materials

Figure 35. Iron-core transformer

The ratio of primary to secondary windings is the turns ratio:

$$\text{Turns ratio} = \frac{\text{no. of turns on the primary}(N_p)}{\text{no. of turns on the secondary}(N_s)},$$

or

$$\frac{V_p}{V_s} = \frac{N_p}{N_s} \text{ and } \frac{I_p}{I_s} = \frac{N_s}{N_p},$$

where V_p is the voltage across the primary coil and V_s is the voltage across the secondary coil of the transformer. The same notation is used for the current. The number of turns on the primary and secondary coils are used to differentiate between step-up and step-down transformers, as shown in Figure 36.

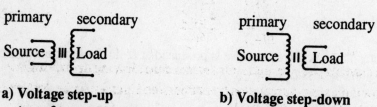

primary   secondary   primary   secondary

Source }|||{ Load   Source }||{ Load

a) Voltage step-up   b) Voltage step-down
    transformer $N_p < N_s$ transformer $N_p > N_s$

Figure 36. Voltage step-up and step-down transformers

Dot Notation

The sign of the voltage across mutually coupled coils may be determined using **dot notation**. The procedure of assigning the dots on a pair of mutually coupled coils is given below, using Figure 37. The numbers in the procedure correspond to the circled numbers in the figure.

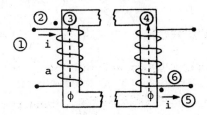

Figure 37. Assigning the dots on a pair of coupled coils

Procedure:

(1) Select a current direction in one of the coils.
(2) Assign a dot where the current enters the winding. (Note: This is the positive terminal with respect to point *a*.)
(3) Use the right-hand rule to assign flux direction.
(4) Assign opposite flux direction for the second coil.
(5) Use right-hand rule to find the current direction in the second coil which produces flux in the direction found in 4.
(6) Assign a dot where the current leaves the winding.
(7) Obtain simplified diagram as shown:

(8) Assign the sign to M using this convention:

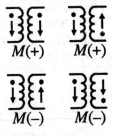

M = + (Both currents pass through coils and are either leaving or entering dots.)

M = − (If arrow indicating current direction through coil is entering the dot for one coil and leaving the dot for another.)

TRANSIENTS

Simple RL and RC Circuits

A source-free RL circuit is shown in Figure 38.

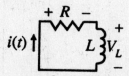

Figure 38. Source-free RL circuit

The following equation is produced by summing the voltages in the circuit:

$$v_R + v_L = Ri + L\frac{di}{dt} = 0.$$

The current $i(t)$ is:

$$i(t) = I_0 e^{\frac{-Rt}{L}} = I_0 e^{\frac{-t}{\tau}},$$

where I_0 is defined as the current $i(t)$ at time $t = 0$, and τ is the time constant:

$$\tau = \frac{L}{R},$$

as shown in Figure 39. The power dissipated in the resistor, p_R, is:

$$p_R = i^2 R = I_0^2 \operatorname{Re}^{\frac{-2Rt}{L}},$$

and the total energy (in terms of heat) in the resistor is:

$$W_R = \frac{LI_0^2}{2}.$$

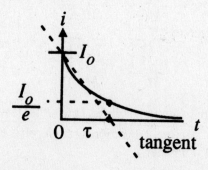

Figure 39. Time constant τ for a source-free RL circuit

EXAMPLE: A 30-mH inductor is in series with a 400Ω resistor. If the energy stored in the coil at $t = 0$ is 0.96µJ, find the magnitude of the current at (a) $t = 0$; (b) $t = 100$µs; (c) $t = 300$µs.

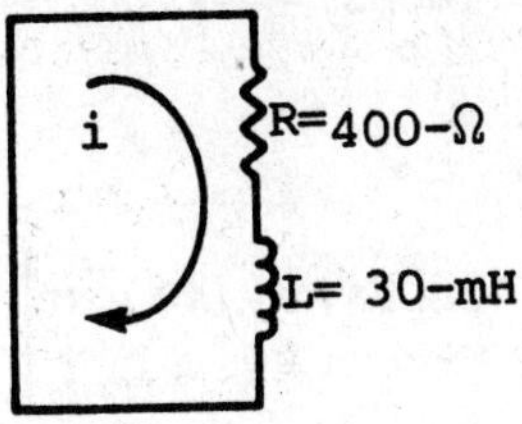

Figure 40. Inductor and resistor

SOLUTION: (a) Find the initial current ($i(0)$) by making use of the energy relationship for an inductor

$$W = \frac{1}{2}Li^2.$$

Since we are given W and asked to find i,

$$i = \sqrt{\frac{2W}{L}}$$

$$i = \sqrt{\frac{2(0.96 \times 10^{-6})}{0.03}}$$

$$i = \sqrt{\frac{1.92 \times 10^{-6}}{3 \times 10^{-2}}} = \sqrt{64 \times 10^{-6}}$$

$$i = 8 \times 10^{-3} = 8\text{mA}$$

(b) After $t = 0$ the current through the inductor is governed by the response of the series RL circuit.

$$i(t) = I_o e^{\frac{-Rt}{L}}.$$

To find i at 100µs,

$$i(100\mu s) = (.008)\exp\left[\frac{-400(100 \times 10^{-6})}{0.03}\right]$$

$$i(100\mu s) = (.008)(2.64) = 2.11\text{mA}.$$

(c) To find i at 300µs,

$$i(300\mu s) = (.008)\exp\left[\frac{-400(300 \times 10^{-6})}{0.03}\right]$$

$$i(300\mu s) = (.Cv(t)e^{-4} = (.008)(.018) = 0.15\text{mA}.$$

A **source-free RC circuit** is shown in Figure 41.

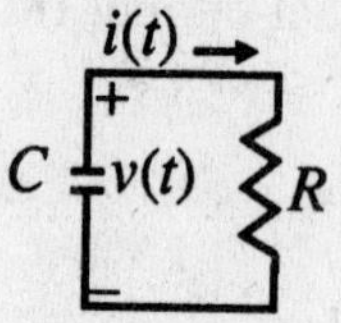

Figure 41. Source-free RC circuit

The following equation is produced by summing the current through the resistor and capacitor:

$$C\frac{dv}{dt} + \frac{v}{R} = 0.$$

The voltage $v(t)$ is:

$$v(t) = v(0)e^{\frac{-t}{RC}} = V_0 e^{\frac{-t}{RC}},$$

where V_o is the voltage at time $t =$ zero, and the time constant τ is:

$$\tau = RC,$$

as shown in Figure 42. The current $i(t)$ may be expressed as follows, where I_o is the current at time $t =$ zero:

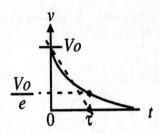

Figure 42. Source-free RC circuit

$$i(t) = i(0)e^{\frac{-t}{RC}} = I_o e^{\frac{-t}{RC}}.$$

Natural and Forced Response of RL and RC Circuits

The complete response of RL and RC circuits to the application of energy is composed of two responses, the **natural response** and the **forced response**. The natural response is the complementary solution of a linear differential equation, and has the form:

$$i_n = Ae^{-Pt}.$$

The forced response is the particular solution of a alinear differential equation, and has the form:

$$i_f = \frac{Q}{P}$$

for a sudden application of a DC source. P is a general function of time and Q is the forcing function (Q is constant for a suddenly applied DC source) in the question above. The complete current response is:

$$i(t) = \frac{Q}{P} + Ae^{-Pt},$$

The complete response is always:

Complete response = natural response + forced response

Total solution = complementary solution + particular solution.

The procedure to fnd the complete response $f(t)$ of RL and RC circuits with DC sources is shown in table form on the next page.

TABLE 1 **COMPLETE RESPONSE OF RL AND RC CIRCUITS** **WITH DC SOURCES**		
	RL	***RC***
1) Simplify the circuit by setting all independent sources to zero and determine:	R_{eq}, L_{eq} $$\left(\tau = \frac{L_{eq}}{R_{eq}}\right)$$	R_{eq}, C_{eq} $$\left(\tau = R_{eq}C_{eq}\right)$$
2) Consider: $\Rightarrow$ and use dc-analysis to find:	$L_{eq} \approx$ short circuit $i_L(0^-)$	$C_{eq} \approx$ open circuit $v_c(0^-)$
3) Repeat procedure 2 to find the forced response:	i.e., $f(t)$ as $t \rightarrow \infty$ $f(\infty)$	
4) Obtain the total response as the sum of the natural and forced responses:	$f(t) = Ae^{\frac{-t}{\tau}} + f(\infty)$	
5) Determine $f(0^+)$ by considering the conditions:	$i_L(0^+) = i_L(0^-)$	$v_c(0^+) = v_c(0^-)$
6) Then:	$f(0^+) = A + f(\infty)$ and $$f(t) = \left[f(0^+) - f(\infty)\right]e^{\frac{-t}{\tau}} + f(\infty)$$	

The RLC Circuits

A parallel, **source-free RLC circuit** is shown in Figure 43.

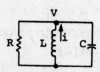

Figure 43. Parallel, source-free RLC circuit

Kirchoff's Current Law produces the follwing equation for the parallel RLC circuit:

$$\frac{v}{R} + \frac{1}{L}\int_{t_0}^{t} v\,dt - i(t_0) + C\frac{dv}{dt} = 0$$

and the corresponding linear, second-order homogeneous differential equation is:

$$C\frac{d^2v}{dt^2}+\frac{1}{R}\frac{dv}{dt}+\frac{v}{L}=0.$$

The natural response has the following form (as this is a source-free circuit, the forced response must be zero):

$$V = A_1 e^{s_1 t} + A_2 e^{s_2 t}$$

where

$$s_1 = \frac{-1}{2RC}+\sqrt{\left(\frac{1}{2RC}\right)^2 - \frac{1}{LC}}$$

$$s_2 = \frac{-1}{2RC}-\sqrt{\left(\frac{1}{2RC}\right)^2 - \frac{1}{LC}}$$

or

$$s_1 = -\alpha+\sqrt{\alpha^2 - \omega_0^2}$$

$$s_2 = -\alpha-\sqrt{\alpha^2 - \omega_0^2}$$

where the exponential damping coefficient, or neper frequency α is:

$$\alpha = \frac{1}{2RC},$$

and the resonant frequency ω_0 is:

$$\omega_0 == \frac{1}{\sqrt{LC}}.$$

Three special responses of parallel RLC circuits are the cases of a) overdamped response, b) critically damped response, and c) underdamped response, as shown in Table 2. The response curves are shown in Figure 44.

TABLE 2		
SPECIAL RESPONSE OF A PARALLEL RLC CIRCUIT		
a) Overdamped	b) Critical damping	c) Underdamped
1) $\alpha > \omega_0$ or if $LC > 4R^2C^2$	$\alpha = \omega_0$ or $LC = 4R^2C^2$ or $LC = 4R^2C$	$\alpha < \omega_0$
2) s_1 and $s_2 =$ negative real numbers, i.e., $\sqrt{\alpha^2 - \omega_0^2} < \alpha$	$s_1 = s_2 = \alpha$	s and s_2 composed of real and complex quantities.
or $\left(-\alpha - \sqrt{\alpha^2 - \omega_0^2}\right)$ $< \left(-\alpha + \sqrt{\alpha^2 - \omega_0^2}\right)$ < 0		
3) $v(t) \to A_1 e^{s_1 t} \to 0$, as $t \to \infty$	$v(t) = A_1 e^{s_1 t} + A_2 e^{s_2 t}$	$v(t) = e^{-\alpha t}\left(A_1 e^{j\omega_d t} + A_2 e^{j\omega_d t}\right)$ where $\omega_\alpha = \sqrt{\omega_0^2 - \alpha^2} =$ Natural Resonant Frequency or $v(t) = e^{-\alpha t}\left\{(A_1 + A_2)\left[\dfrac{e^{j\omega_d t} + e^{-j\omega_d t}}{2}\right] + j(A_1 - A_2)\left[\dfrac{e^{j\omega_d t} - e^{-j\omega_d t}}{2j}\right]\right\}$ or $v(t) = e^{-\alpha t}\left[(A_1 + A_2)\cos\omega_d t + j(A_1 - A_2)\sin\omega_d t\right]$

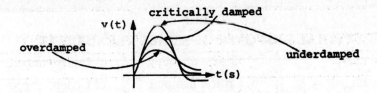

Figure 44. Three response curves for source-free parallel RLC circuits

A series, source-free RLC circuit is shown in Figure 45.

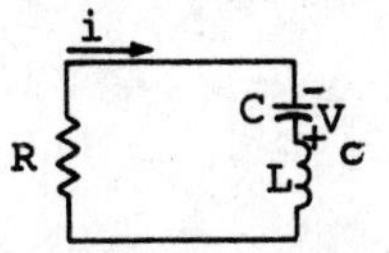

Figure 45. Series, source-free RLC circuit

Kirchoff's Voltage Law produces the following equation:

$$Ri + \frac{1}{C}\int_{t_0}^{t} i\,dt + L\frac{di}{dt} - v_c(t_0) = 0,$$

and the corresponding second-order differential equation in terms of i:

$$L\frac{d^2i}{dt^2} + R\frac{di}{dt} + \frac{i}{C} = 0$$

or in terms of v_c:

$$LC\frac{d^2v_c}{dt^2} + RC\frac{dv_c}{dt} - v_c = 0.$$

Cases of a) overdamped response, b) critically damped responses, and c) underdamped response are summarized in Table 3.

TABLE 3
SPECIAL RESPONSE OF SERIES RLC CIRCUIT

a) Overdamped	b) Critical damping	c) Underdamped
1) $\alpha > \omega_0$	$\alpha = \omega_0$	$\alpha < \omega_0$
2) $s_1, s_2 = \dfrac{-R}{2L} \pm$ $\sqrt{\left(\dfrac{R}{2L}\right)^2 - \dfrac{1}{LC}}$ or $= -\alpha \pm \sqrt{\alpha^2 - \omega_0^2}$ where $\alpha = \dfrac{R}{2L}$, $\omega_0 = \dfrac{1}{\sqrt{LC}}$	$s_1 = s_2 = \alpha$	$s_{1,2} = -\alpha \pm j\omega_d$ $\omega_d = \sqrt{\omega_0^2 - \alpha^2}$
3) $i(t) = A_1 e^{s_1 t} + A_2 e^{s_2 t}$	$i(t) = e^{-\alpha t}(A_1 t + A_2)$	$i(t) = e^{-\alpha t}(B_1 \cos\omega_d t + B_2 \sin\omega_d t)$

The general equation of a complete response of a second-order system in terms of voltage for an RLC circuit is given by:

$$v(t) = \underbrace{V_f} + \underbrace{Ae^{s_1 t} + Be^{s_2 t}}$$

forced response + natural response

where V_f is a constant (DC excitation).

AC CIRCUITS

Sinusoidal Current, Voltage, and Phase Angle

The sinusoidal forcing function in its general form is as follows:
$$v(t) = V_m \cos(\omega t + \theta)$$
where

V_m = Maximum value

ω = Angular frequency = $2\pi f = \dfrac{2\pi}{T}$ in $\dfrac{\text{radians}}{\text{sec}}$

$$f = \text{Frequency} = \frac{1}{T} \text{ in } \frac{\text{cycles}}{\text{sec}} \text{ or hertz (Hz)}$$

$$T = \text{Period (time duration of 1 cycle = sec)}$$

$$\theta = \text{Phase angle in degrees or radians.}$$

The voltage across resistance (R), inductance (L), or capacitance (C) is given below for the specified current.

		TABLE 4	
	RESISTANCE, INDUCTANCE, OR CAPACITANCE		
Element	voltage	$i = I_m \sin \omega t$	$i = I_m \cos \omega t$
R	V_R =	$RI_m \sin \omega t$	$V_R = RI_m \cos \omega t$
L	V_L =	$\omega L I_m \cos \omega t$	$V_L = \omega L I_m (-\sin \omega t)$
C	V_C =	$\dfrac{I_m}{\omega C}(-\cos \omega t)$	$V_C = \dfrac{I_m}{\omega C}\sin \omega t$

The current in R, L, or C for the specified voltage is given below in table form.

		TABLE 5	
	CURRENT, R, L, OR C		
Element	current	$v = V_m \sin \omega t$	$v = V_m \cos \omega t$
R	i_R =	$\dfrac{V_m}{R}\sin \omega t$	$i_R = \dfrac{V_m}{R}\cos \omega t$
L	i_L =	$\dfrac{V_m}{\omega L}(-\cos \omega t)$	$i_L = \dfrac{V_m}{\omega L}\sin \omega t$
C	i_C =	$\omega C V_m \cos \omega t$	$i_C = \omega C V_m (-\sin \omega t)$

Voltage and current were expressed in earlier sections of this chapter in integral and differential form for the resistor, capacitor and inductor, as follows:

$$i_R(t) = \frac{V_R(t)}{R}, \quad V_R(t) = i_R(t)R$$

$$i_L(t) = \frac{1}{L}\int_{-\infty}^{t} V_L(\tau)d\tau, \quad V_L(t) = L\frac{di_L(t)}{dt}$$

$$i_C(t) = C\frac{dV_c(t)}{dt}, \quad V_C(t) = \frac{1}{C}\int_{-\infty}^{t} i_C(\tau)d\tau$$

The characteristics of phase angle are summarized in Table 6.

TABLE 6 CHARACTERISTICS OF PHASE ANGLE IN A PURE ELEMENT			
Element	Current and voltage phase angle relationship	Impedance magnitude	Diagram
R	Current and voltage in phase	R	FIG
L	Current lags the voltage by $90°$ or $\dfrac{\pi}{2}$ rad.	ωL	
C	Current leads the voltage by $90°$ or $\dfrac{\pi}{2}$ rad.	$\dfrac{1}{\omega C}$	
Series RL	Current lags the voltage by $\tan^{-1}\left(\dfrac{\omega L}{R}\right)$.	$\sqrt{R^2 + (\omega L)^2}$	
Series RC	Current leads the voltage by $\tan^{-1}\left(\dfrac{1}{\omega CR}\right)$.	$\sqrt{R^2 + \left(\dfrac{1}{\omega C}\right)^2}$	

The Concept of Phasor

In general, the **phasor** form of a sinusoidal voltage or current is:

$$V = V_m \angle \theta° \text{ and } I = I_m \angle \theta°.$$

Thus, for the voltage source $v(t) = V_m \cos \omega t$, the corresponding phasor form is $V_m \angle \theta°$. For the current response $i(t) = I_m \cos(\omega t + \theta)$, the corresponding phasor form is $I_m \angle \theta°$. Most often, instead of using the maximum value of voltage or current, the rms value is used (see AC Circuit Analysis).

The following steps outline the procedure for transformation to and from the time domain and frequency domain. Current and voltage relationships in the time and frequency domain are shown in table form below, and illustrated in Figure 46.

Time domain → frequency domain:

$$i(t) = I_m \cos(\omega t + \theta) \rightarrow I = I_m \angle \theta$$

(1) Assume a sinusoidal function $i(t)$ in the time domain is given. Express $i(t)$ as a cosine wave with a phase angle.

(2) Using Euler's identity—$e^{j\theta} = \cos\theta + j\sin\theta$—express the cosine wave as the real part of a complex quantity.

(3) Drop the Re and the term $e^{j\omega t}$ to obtain the final phasor form (the frequency domain form).

Frequency domain → time domain:

(1) Given a phasor current or voltage in polar form in the frequency domain, express the complex expression in exponential form.

(2) Multiply the factor $e^{j\omega t}$ by the obtained exponential form.

(3) Apply Euler's identity and take the real part of the complex expression to obtain the time domain representation.

TABLE 7
TIME DOMAIN AND FREQUENCY DOMAIN RELATIONSHIPS OF VOLTAGE AND CURRENT FOR R, L, AND C

Element	Voltage & Current Relationship		Phasors
	Time Domain	Frequency Domain	
current R + voltage −	$v = Ri$	$V = RI$	
current L + voltage −	$v = \dfrac{Ldi}{dt}$	$V = (j\omega L)I$	
current C + voltage −	$v = \dfrac{1}{C}\int i\,dt$	$V = \left(\dfrac{1}{j\omega C}\right)I$	

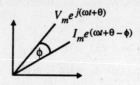 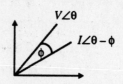

a) Time domain plot   **b) Frequency domain plot**

Figure 46. Time and frequency domain plot.

V_m and I_m are $\dfrac{1}{\sqrt{2}}$ times V and I

Impedance is a complex quantity with units of ohms, and is defined as:

$$\text{impedance } (Z) = \frac{\text{phasor voltage}}{\text{phasor current}} [\text{ohms}],$$

which in polar form is:

$$Z = Z \angle \theta°,$$

and in rectangular form:

$$Z = R + jX,$$

where X is the reacitive component of the impedance. In a similar fashion, admittance is defined as:

$$\text{Admittance } (Y) = \frac{1}{Z} = \frac{\text{phasor current}}{\text{phasor voltage}},$$

with the following polar form:

$$Y = Y \angle \theta°,$$

and the expected rectangular form:

$$Y = G + jB.$$

The imaginary part of the admittance Y is the susceptance B. The units of admittance is mhos.

AC Circuit Analysis

Procedures similar to DC analysis and theorems are used for **AC analysis** except that they are in terms of phasor voltage and current (V and I) and impedance (Z) or admittance (Y). The following examples illustrate the application of familar theorems to AC circuits. In the case of source conversions, the general format is as shown in Figure 47.

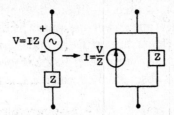

Figure 47. Source conversions

The rms, or effective, value of current and voltage are:

$$I_{eff} = I_{rms} = \sqrt{\frac{1}{T}\int_0^T [i(t)]^2\, dt}\,,$$

and

$$V_{eff} = V_{rms} = \sqrt{\frac{1}{T}\int_0^T [v(t)]^2\, dt}\,.$$

(1) Effective value of $a\sin\omega t$ and $a\cos\omega t = \dfrac{a}{\sqrt{2}}$

(2) I_{eff} for sinusoidal current $i(t)$ equals $I_m\cos(\omega t - \theta)$

with $T = \dfrac{2\pi}{\omega} = \dfrac{I_m}{\sqrt{2}} = 0.707 I_m$

The effective value of $a\sin(\omega t)$ and $a\cos(\omega t)$ is:

$$\frac{a}{\sqrt{2}},$$

therefore, for a sinusoidal current as follows:

$$i(t) = I_m\cos(\omega t - \theta)$$

the effective current, I_{eff} is:

$$I_{eff} = \frac{I_m}{\sqrt{2}} = 0.707 I_m,$$

where the period T is:

$$T = \frac{2\pi}{\omega}.$$

POWER

Instantaneous power is defined as follows:

$$p = vi.$$

This equation describes the power absorbed by the element as shown in Figure 48.

The passive sign convention is used to differentiate between power generation (absorption of negative power) and power absorption by the element. This convention states that if the current enters the positive terminal then the sign of the product of the current and voltage may be used to describe the absorbed power. If the product is negative, negative power is absorbed by the element, or in other words, power is generated by the element.

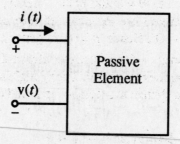

Figure 48. Power absorbed by an element

Instantaneous power in a resistive circuit is:

$$p = i^2 R = \frac{v^2}{R},$$

and for an inductive circuit:

$$p = Li\frac{di}{dt} = \frac{1}{L}\int_{-\infty}^{t} v\,dt,$$

and for a capacitive circuit:

$$p = Cv\frac{dv}{dt} = \frac{1}{C}\int_{-\infty}^{t} i\,dt.$$

Average power is defined mathematically as:

$$P = \frac{1}{t_2 - t_1}\int_{t_1}^{t_2} p(t)\,dt,$$

and for the sinusoidal steady state:

$$\text{Average power } (P) = \frac{1}{2}V_m I_m \cos\theta$$

$$= \underbrace{V_{rms}I_{rms}}_{\text{apparent power}} \cos\theta$$

where

$$V_{rms} = \frac{V_m}{\sqrt{2}}$$

$$I_{rms} = \frac{I_m}{\sqrt{2}}.$$

Power factor (*pf*) is defined as a ratio of average power to apparent power:

$$\text{Power factor } (pf) = \frac{\text{average power}(P)}{\text{apparent power}(V_{rms}I_{rms})} = \cos\theta,$$

where apparent power is equal to the product of *rms* voltage and *rms* current as shown above. The angle (θ) is referred to as the *pf* angle. For a purely resistive load, voltage and current are in phase, i.e., $\theta = 0$ and *pf* = 1. Hence, apparent power = average power. For a purely reactive load, the phase difference between the voltage and current is either +90° or –90°. Hence, *pf* = 0. In general networks, $0 < pf < 1$.

Power factor correction is used to reduce electrical utility charges by changing the *pf* angle without changing the real power. The *pf* angle may be changed by altering the circuit reactance. The change in reactance required to change the *pf* angle from θ_1 to θ_2 is:

$$\Delta\theta = P(\tan\theta_1 - \tan\theta_2)$$

where Q, the reactive power, is:

$$\underbrace{V_{rms}I_{rms}}_{\text{apparent power}} \sin\theta$$

in units of VAR, or volt-ampere reactance. The power factor can be corrected by adding inductance to a capacitive circuit, or adding capacitance to an inductive circuit. The capacitance in farads needed to correct the power factor is found by the application of the following equation:

$$C = \frac{\Delta Q}{2\pi f V^2},$$

where f is frequency.

Complex power, *S*, is the sum of the real average power and the reactive power

$$S = VI e^{j(\theta_v - \theta_i)}$$
$$= VI\cos\theta - j\,VI\sin\theta$$
$$= \underset{\text{real average power}}{P} - \underset{\text{reactive power}}{jQ,}$$

where

$$P = Re(VI^*)$$
$$Q = Im\,(VI^*).$$

The current and voltage are effective values, and I^* is the complex conjugated of I. The magnitude of the complex power S is the apparent power:

$$S = |VI^*|.$$

EXAMPLE: A circuit draws $4A$ at $25V_{rms}$, and dissipates $50W$. Find: (a) apparent power; (b) reactive power; (c) power factor and phase angle; and (d) impedance in both polar and rectangular forms.

   SOLUTION: (a) The apparent power is

$$|S| = V_{eff}I_{eff} = (4)(25) = 100VA.$$

   (b) Since

$$\vec{S} = P + jQ$$

and

$$|S| = \sqrt{P^2 + Q^2} = 100VA$$

we can find Q because we are given P (dissipated power) $= 50W$.

   Hence

$$Q = \sqrt{|S|^2 - P^2}$$

$$Q = \sqrt{100^2 - 50^2} = \sqrt{7500} = 86.6VA.$$

   (c) The power factor is defined as the ratio of dissipated power to apparent power. Hence,

$$pf = \frac{P}{|S|} = \frac{50}{100} = 0.5.$$

   The phase angle is the $\cos^{-1}pf$, thus $\phi = \cos^{-1}0.5 = 60°$.

   (d) The magnitude of the impedance can be found from

$$|z| = \frac{V_{eff}}{I_{eff}} = \frac{25}{4} = 6.25\Omega.$$

We can therefore write z in polar form:

$$\vec{z} = |z|\angle\phi$$

$$\vec{z} = 6.25\angle60°.$$

In rectangular form we must write

$$\vec{z} = 3.125 \pm j5.41\Omega$$

since we are not given any information to determine the polarity of the phase angle (that is, whether V lags or leads I). Thus the j term could be either plus or minus.

THREE-PHASE CIRCUITS

Three-Phase Systems

Figure 49 illustrates both positive, or ABC, sequence and negative, or CBA, sequence of voltages generated in a **three-phase system**. In the ABC sequence, $V_{A'A}$ reaches its peak before $V_{B'B}$, and $V_{B'B}$ reaches its peak before $V_{C'C}$. Reversing the rotation of the field magnet produces the CBA sequence in which $V_{C'C}$ reaches its peak before $V_{B'B}$, which in turn peaks before $V_{A'A}$.

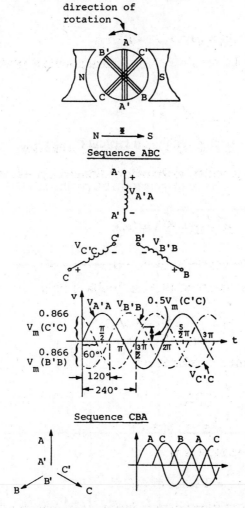

Figure 49. ABC and CBA sequences in three-phase system

At any instant of time, the summation of all three phase voltages is zero:

$$\Sigma V_{A'A} + V_{B'B} + V_{C'C} = 0.$$

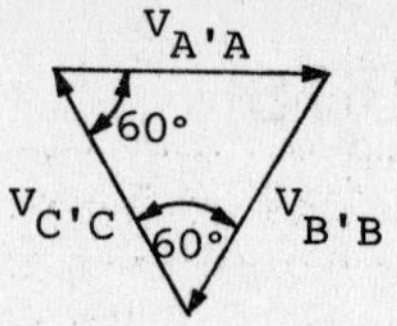

$V_{A'A}$

60°

$V_{C'C}$ 60° $V_{B'B}$

Figure 50. Summation of three phase voltages

In a three-phase system, the three coils on the rotor are placed 120° apart as in Figure 51. (Assume each coil has an equal number of turns.)

$V_{C'C}$

120°

120° $V_{A'A}$

120°

$V_{B'B}$

Figure 51. Spacing of coils on typical three phase system

Three phase voltages are listed in table form for positive and negative sequences.

TABLE 8	
THREE PHASE SYSTEM VOLTAGES	
Sequence (positive phase sequence)	(Note: V_L = line voltage)
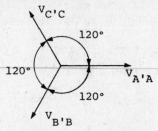	$V_{AN} = \left(\dfrac{V_L}{\sqrt{3}}\right)\angle 90°$ $V_{BN} = \left(\dfrac{V_L}{\sqrt{3}}\right)\angle -30°$ $V_{CN} = \left(\dfrac{V_L}{\sqrt{3}}\right)\angle -150°$ $V_{AB} = V_L\angle 120°$ $V_{BC} = V_L\angle 0°$ $V_{CA} = V_L\angle 240°$

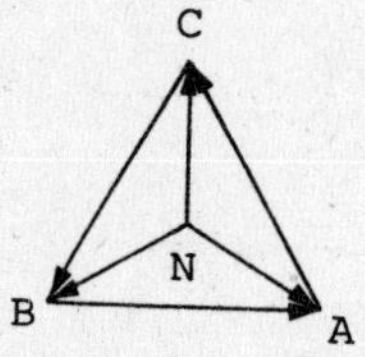

TABLE 8 cont	
Sequence (negative phase sequence)	(Note: V_L = line voltage)
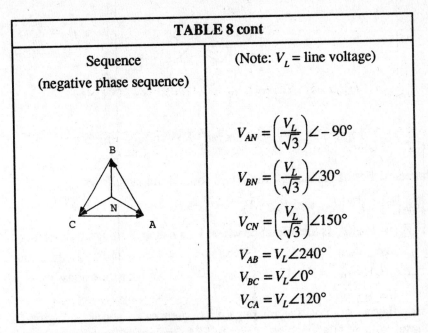	$V_{AN} = \left(\dfrac{V_L}{\sqrt{3}}\right) \angle -90°$ $V_{BN} = \left(\dfrac{V_L}{\sqrt{3}}\right) \angle 30°$ $V_{CN} = \left(\dfrac{V_L}{\sqrt{3}}\right) \angle 150°$ $V_{AB} = V_L \angle 240°$ $V_{BC} = V_L \angle 0°$ $V_{CA} = V_L \angle 120°$

(**Note:** Phase reference line voltage V_{AB}.)

The Wye and Delta Connections

Two common configurations found in three phase circuits are the **wye (Y)** and **delta (Δ) connections**, as shown in Figure 52. Figure 53 illustrates the connection of ideal voltage sources in wye configuration.

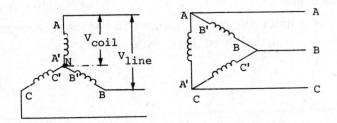

Y-Alternator Δ-Alternator

Figure 52. Wye and delta connections

$$I_{coil} = I_{line} \qquad\qquad I_{coil} = \frac{1}{\sqrt{3}} I_{line}$$

$$V_{line} = \sqrt{3}\, V_{coil} \qquad\qquad V_{line} = V_{coil}$$

I_{coil} is more commonly referred to as I_{phase}.

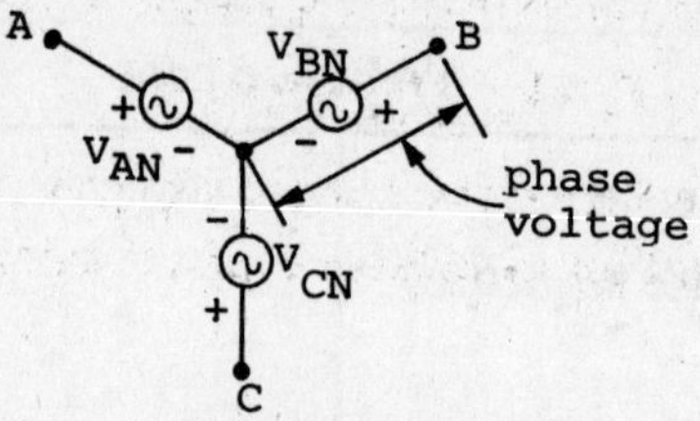

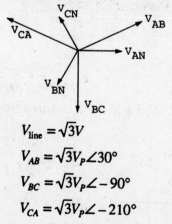

Figure 53. Wye-connected ideal voltage sources

Characteristics of balanced three-phase sources are as follows:

1) $|V_{AN}| = |V_{BN}| = |V_{CN}|$ and $V_{AN} + V_{BN} + V_{CN} = 0$

2) If $V_{AN} = V_P \angle 0°$ is the reference where $V_p = rms$ is the magnitude of any of the phase voltages, then $V_{BN} = V_P \angle -120°$ and $V_{CN} = V_P \angle -240°$ (positive phase or sequence ABC) or $V_{BN} = V_P \angle 120°$ and $V_{CN} = V_P \angle 240°$ (negative phase or sequence CBA).

Phasor diagrams of positive and negative sequences are shown in Figure 54 and line-phasor voltage relationships in Figure 55 for balanced three phase circuits.

$V_{CN} = V_P \angle -240°$

$(V_p = \text{phase voltage})$ N $V_{AN} = V_P \angle 0°$

$V_{BN} = V_P \angle -120°$

$V_{BN} = V_P \angle 120°$

N $V_{AN} = V_P \angle 0°$

$V_{CN} = V_P \angle 240°$

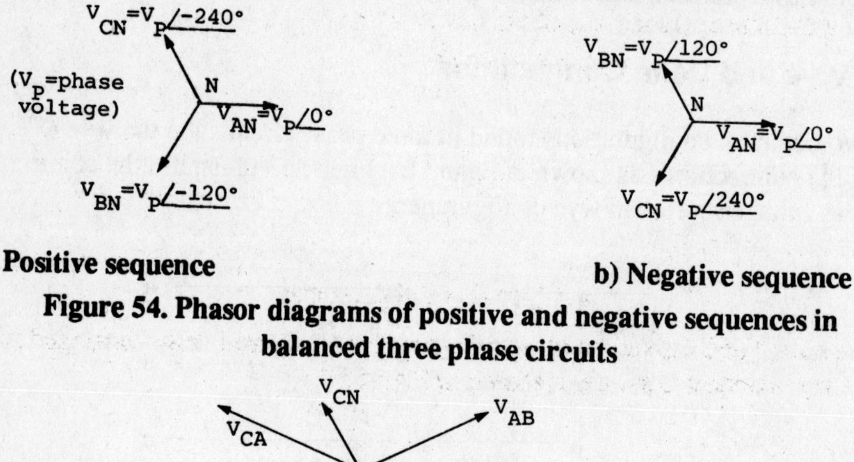

a) Positive sequence   **b) Negative sequence**

Figure 54. Phasor diagrams of positive and negative sequences in balanced three phase circuits

V_{CN} V_{AB}

V_{CA}

V_{AN}

V_{BN}

V_{BC}

$$V_{line} = \sqrt{3}V$$

$$V_{AB} = \sqrt{3}V_P \angle 30°$$

$$V_{BC} = \sqrt{3}V_P \angle -90°$$

$$V_{CA} = \sqrt{3}V_P \angle -210°$$

Figure 55. Phasor diagrams of line and phase voltage for balanced three phase circuits

A wye-connected load is shown in Figure 56. The phase power, P_p, is:

$$P_p - V_{\text{phase}} I_{\text{line}} \cos\theta,$$

where θ is the angle by which phase current lags phase voltage (for inductive loads, θ would be positive, and for capacitive loads, θ would be negative), and the total power P_t is:

$$V_P I_P = V_P I_L = \frac{V_L I_L}{\sqrt{3}},$$

$$P_t = 3P_P, \text{ or } P_t = \sqrt{3} \cdot V_L I_L \cos\theta \text{ where } V_L = \sqrt{3} V_P.$$

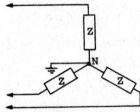

Figure 56. Wye-connected load

For a Δ-connected load, the phase power, P_p is:

$$= P_P = V_L I_P \cos\theta,$$

where

$$V_L I_P = V_P I_P = V_L \frac{I_L}{\sqrt{3}}.$$

The total power P_t is:

$$P_t = 3P_P \text{ or } P_t = \sqrt{3} V_L I_L \cos\theta.$$

The circuit and associated phasor diagram for a balanced delta-connected load with a wye-connected source is shown in Figure 57.

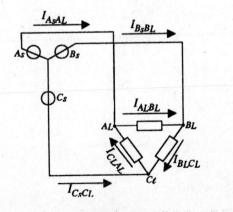

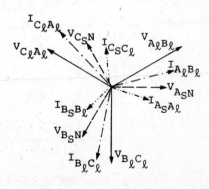

a) Circuit   b) Phasor diagram

Figure 57. Balanced delta-connected load with wye-connected source

Referring to Figure 57, the voltages are:

$$V_{\text{phase}} = |V_{A,N}| = |V_{B,N}| = |V_{C,N}|,$$

assuming that the line voltage:

$$V_{\text{line}} = |V_{A,B_s}| = |V_{A,C_s}| = |V_{C,A_s}|$$

where

$$V_L = \sqrt{3}V_A \text{ and } V_{A,B_s} = \sqrt{3}V_{A,N}\angle 30°.$$

Then the phase currents are:

$$I_{A_L B_L} = \frac{V_{A,B_s}}{Z_p}, \ I_{B_L C_L} = \frac{V_{B,C_s}}{Z_p} \text{ and } I_{C_L A_L} = \frac{V_{C,A_s}}{Z_p},$$

and the line currents are:

$$I_{A_L B_L} - I_{C_L A_L} = I_{A,A_L}, \text{ etc.}$$

The three-phase currents are equal in magnitude:

$$I_P = |I_{A,B_s}| = |I_{B,C_s}| = |I_{C,A_s}|,$$

$$I_L = |I_{A,A_L}| = |I_{B,B_L}| = |I_{C,C_L}|,$$

and

$$I_L = \sqrt{3}I_P.$$

REVIEW PROBLEMS
PROBLEM 1

Calculate the magnitude of a line current in the circuit shown in Figure 58.

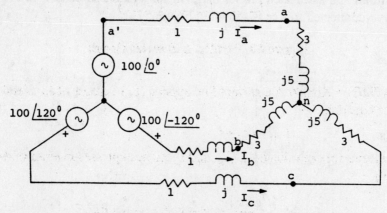

Figure 58. Line current

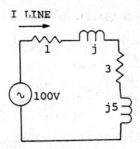

Figure 59. Simplified circuit

SOLUTION: The circuit is balanced because the load for each phase is the same, the magnitude of the source for each phase is the same, and the angle for each phase is displaced by 120°. Since the circuit is balanced, the magnitude of the line current in each phase is the same. To find the line current, Figure 58 is redrawn in Figure 59 to include only one phase.

The total impedance of one phase is $1 + j + 3 + j5 = 4 + j6\Omega$. The magnitude of the line current is:

$$|I_{\text{line}}| = \left|\frac{100}{4 + j6}\right| = \frac{100}{\sqrt{4^2 + 6^2}} = \frac{100}{7.21} = 13.85A.$$

PROBLEM 2

Use Kirchhoff's current law to write an integrodifferential equation for v(t) for the circuit shown.

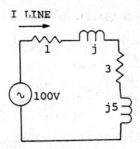

Figure 60. Parallel and series circuit

SOLUTION: Kirchhoff's current law applied to the upper node of the circuit yields the equation:

$$i = i_C + i_R + i_L.$$

The currents for each element i_C, i_R, and i_L can be expressed in terms of the same voltage v:

$$i_C = C\frac{dv}{dt}, \; i_R = \frac{1}{R}\text{v}, \; i_L = \frac{1}{L}\int_{-\infty}^{t}\text{v}\; d\tau.$$

Substituting these terms into the KVL equation, yields the required integrodifferential equation:

$$i = C\frac{dv}{dt} + \frac{1}{R}v + \frac{1}{L}\int_{-\infty}^{t} v\ d\tau.$$

PROBLEM 3

Given an infinitely long straight filament carrying a current I, find the field H at point P:

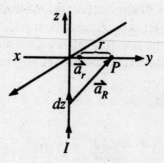

Figure 61. Field produced by filament with current I

SOLUTION:

$$dH = \frac{Idza_z \times (ra_r - za_z)}{4\pi(r^2 + z^2)^{3/2}}$$

$$= \frac{Idzra_\phi}{4\pi(r^2 + z^2)^{3/2}}$$

$$H = \frac{I}{4\pi}\int_{-\infty}^{\infty}\frac{rdza_\phi}{(r^2 + z^2)^{3/2}}$$

$$= \frac{I}{2\pi r}a_\phi.$$

PROBLEM 4

Find the voltage across an inductor, shown in Figure 62, whose inductance is given by:

$$L(t) = te^{-t} + 1$$

and the current through it is given by:

$$i(t) = \sin\omega t.$$

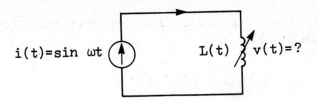

Figure 62. Simple inductor

SOLUTION: The voltage across an inductor is defined as $v(t) = \dfrac{d\phi}{dt}$ and $\phi = L\, i(t)$. In this problem L is a time-varying inductance $L(t)$:

$$\phi(t) = L(t)\, i(t) = (te^{-t} + 1)\,(\sin \omega t)$$

the voltage becomes:

$$v(t) = \frac{d}{dt}\,[(te^{-t} + 1)\,(\sin \omega t)]$$

$$v(t) = (1 + te^{-t})\,(\omega \cos \omega t) + (\sin \omega t)(e^{-t} - te^{-t})$$

$$v(t) = (\omega \cos \omega t)\,(1 + te^{-t}) + (1-t)e^{-t}\,\sin \omega t.$$

PROBLEM 5

Consider the capacitor shown in Figure 63. The capacitance $C(t)$ is given by:

$$C(t) = C_0(1 + 0.5\sin t).$$

The voltage across this capacitor is given by:

$$v(t) = 2\sin \omega t.$$

Find the current through the capacitor.

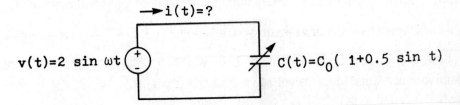

Figure 63. Simple capacitor

SOLUTION: We can find the charge on the capacitor $q(t)$ by using the definition $q(t) = C\,v(t)$. In this problem C is a time varying function $C(t)$:

$$q(t) = C(t)v(t)$$

$$q(t) = C_0(1 + 0.5\sin t)(2\sin \omega t).$$

Since $i(t) = \dfrac{dq}{dt}$, we have:

$$i(t) = \frac{d}{dt}\left[C_0(1+0.5\sin t)(2\sin \omega t)\right]$$
$$= (2\sin \omega t)(0.5C_0 \cos t) + C_0(1+0.5\sin t)(2\omega \cos \omega t)$$
$$i(t) = C_0 \sin \omega t \cos t + 2\omega C_0 \cos \omega t(1+0.5\sin t).$$

PROBLEM 6

For the circuit shown in Figure 64, find i and v as functions of time for $t > 0$.

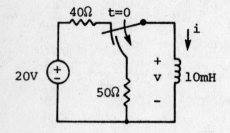

Figure 64. Circuit for Problem 6

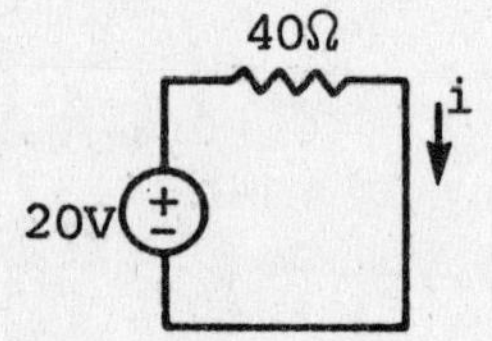

Figure 65. Circuit when $t < 0$

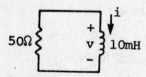

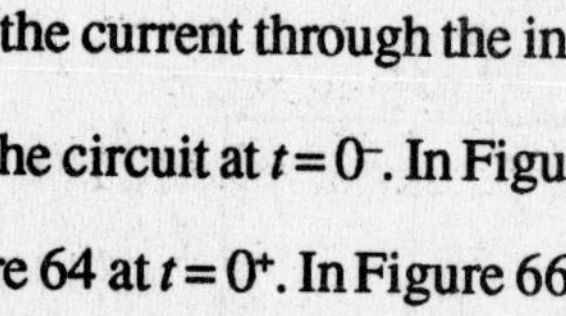

Figure 66. Circuit when $t \geq 0$

SOLUTION: First, find the current through the inductor i, just before the switch is thrown. Figure 65 shows the circuit at $t = 0^-$. In Figure 65, $i = \dfrac{20v}{40\Omega} = 0.5A$. Figure 66 shows the circuit in Figure 64 at $t = 0^+$. In Figure 66, in order for the voltage drops to sum to zero around loop, the voltage v must be $-(50\Omega)(0.5A) = -25V$. The time constant is found to be:

$$\frac{L}{R} = \frac{10mH}{50\Omega} = \frac{1}{5,000}$$

Write the response:

$$i(t) = 0.5e^{-5,000t} A; \ t > 0$$
$$v(t) = -25e^{-5,000t} V; \ t > 0.$$

PROBLEM 7

Find R, L, and C for the networks shown.

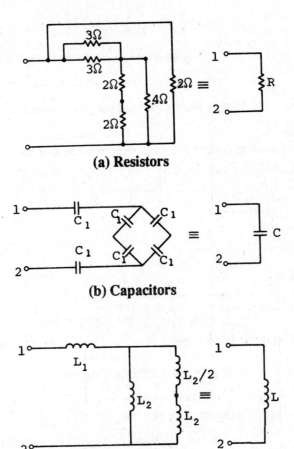

(a) Resistors

(b) Capacitors

c) Inductors

Figure 67. Component networks and equivalents

SOLUTION: (a) First, combine the parallel resistances, the two 3-Ω resistors and the two 4-Ω resistors.

Finally:

$$(1.5+2)\Omega \| 2\Omega$$

$$R = \frac{3.5(2)}{5.5} = 1.272\Omega$$

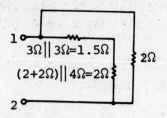

Figure 68. Combined parallel resistors

(b) Since capacitors combine the conductances, and finally,

$$C = \cfrac{1}{\cfrac{1}{C_1} + \cfrac{1}{C_1} + \cfrac{1}{C_1}} = \cfrac{1}{\cfrac{3}{C_1}}$$

$$C = \frac{C_1}{3} F.$$

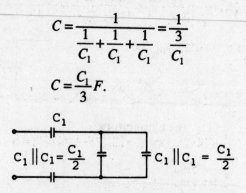

Figure 69. Combined capacitors

(c) Since inductors combine like resistances

$$L = (L_1 + 0.6L_2)H.$$

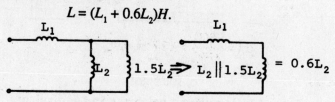

Figure 70. Combined inductors

PROBLEM 8

A balanced three-phase three-wire system supplies two balanced Y-connected loads. The first draws 6 kW at 0.8 PF lagging while the other requires 12 kW at 0.833 PF leading. If the current in each line is 8 A rms, find the current in the: (a) first load; (b) second load; (c) source phase.

SOLUTION: It does not matter how the load is connected. For any balanced 3-phase system,

$$P = |V| \, |I| \cos \theta$$

where P is the average power, $|V|$ is the rms voltage magnitude, $|I|$ is the rms current magnitude, and $\cos(\theta)$ is the power factor. θ is the angle by which the voltage leads the current in each line.

(a) For $P_1 = 6\text{ kW}$, $\cos(\theta_1) = 0.8$ lagging (that is, θ is negative),

$$\theta_1 = -\cos^{-1}(0.8) = 36.87°$$

$$|I_1| = \frac{P_1}{|V|\cos(\theta_1)} = \frac{(6000)}{|V|(0.8)} = \frac{7500}{|V|} A \text{ rms}.$$

For the second load, $P_2 = 12\text{ kW}$, $\cos(\theta_2) = 0.833$ leading

$$\theta_2 = +\cos^{-1}(0.833 = 33.59°$$

$$|I_2| = \frac{P_2}{|V|\cos(\theta_2)} = \frac{12,000}{|V|(0.833)} = \frac{14,406}{|V|} A \text{ rms}.$$

We are also given that $|I_1 + I_2| = 8A$ rms, so

$$\frac{1}{|V|}|7500\angle -36.87° + 14,406\angle +33.59°| = 8$$

$$\frac{1}{|V|}|6000 - j4500 + 12,000 + j7979| = 8$$

$$\frac{1}{|V|}|18,000 + j3470| = 8$$

$$|V| = \frac{18,331}{8} = 2291V \text{ rms}.$$

Finally,

$$|I_1| = \frac{7500}{2291} = 3.273A \text{ rms}.$$

(b) From (a)

$$|I_2| = \frac{14,406}{|V|} = \frac{14,406}{2291} = 6.288A \text{ rms}.$$

(c) The source phase current is the same as the line current, so it is 8 A rms.

PROBLEM 9

Find *i* in the circuits of Figure 71 a, b, and c.

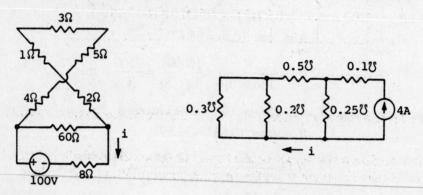

a) Loop resistors in series

b) Conductance

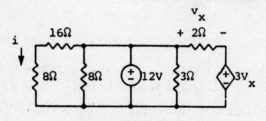

c) Series combination

Figure 71. Circuits

SOLUTION: (a) To find the current *i*, combine resistors to form a single loop circuit. Note that, despite its complicated appearance, the five resistors in the upper part of the circuit are all in series.

$$R_{eq} = \frac{60(15)}{75} = 12\Omega$$

Figure 72. Equivalent resistance

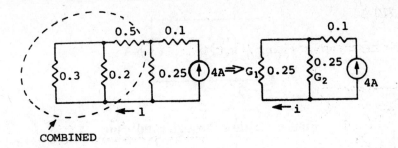

Figure 73. Combined resistors   **Figure 74. Equivalent circuit**

The sum of the resistors in the upper half of the network is in parallel with the 60Ω resistor. Figure 72 demonstrates the resulting circuit. KVL around the loop

$$100 = i\,(12 + 8)$$

gives

$$i = \frac{100}{20}\,5A.$$

(b) Figure 73 shows the conductances that are combined to form the circuit in Figure 74.

To find i, which is the current throught the 0.25 conductance G_1 on the left, it is necessary to know the voltage across G_1. Since the voltage across G_1 and G_2 are the same, one can combine them. Remembering that $4A$ flows through the combined conductance of 0.5, calculate the voltage.

$$v = \frac{1}{G} = \frac{4}{0.5} = 8V.$$

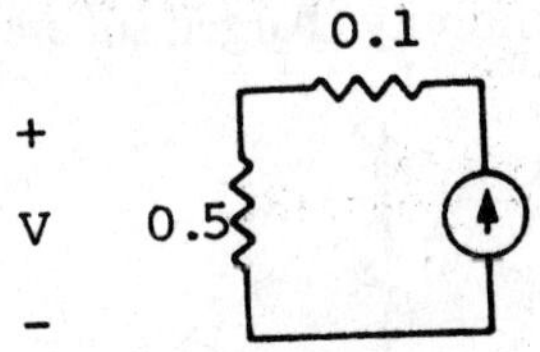

Figure 75. Simple conductance

The current sense of i is opposite the convention adopted, thus

$$i = -8(0.25)$$

$$i = -2A.$$

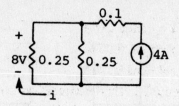

Figure 76. Flow through conductor

(c) Observe in Figure 71c that the 12-V source is connected across the 8 and 16Ω series combination on the left.

Hence,

$$i = \frac{V}{R} = \frac{12}{8+16} = \frac{12}{24} = 0.5A.$$

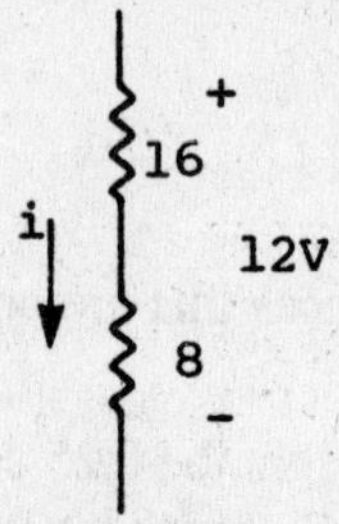

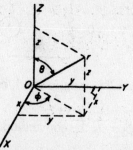

Figure 77. Series combination

PROBLEM 10

Find the E field due to a uniformly charged sphere by direct integration. FIG

Figure 78. Relation of spherical and rectangular coordinates

SOLUTION: For a sphere of radius R, the charge per unit area is $q_s = \dfrac{q}{4\pi R^2}$. For convenience in integration, use spherical coordinates r, θ, ϕ. These are defined in terms of the rectangular coordinates x, y, z by

$$x = r \sin \theta \cos \phi$$
$$y = r \sin \theta \sin \phi$$
$$z = r \cos \theta$$

and are illustrated in Figure 78. Then construct a thin ring on the surface of the charged sphere of Figure 79 such that the ring is symmetrical about the z-axis (note that the coordinate system has been rotated) and subtends a half-angle θ at the center. An element of area dS is chosen with sides given by $R\,d\theta$ and $R \sin \theta\,d\phi$. The charge on this area is $q_s R^2 \sin \theta d\,\theta d\phi$ and the field at P due to this charge has a magnitude

$$dE_P = \frac{q_s R^2 \sin \theta d\theta d\phi}{4\pi\varepsilon_0 a^2}$$

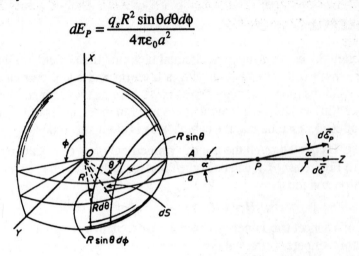

Figure 79. Surface element used to calculate field of charged sphere

Because of symmetry, the resultant field dE due to the ring is along OZ, so that

$$dE = dE_p \cos \alpha$$

Now integrate with respect to ϕ, using limits of 0 and 2π, to obtain

$$dE = \frac{q_s R^2 \sin \theta d\theta \cos \alpha}{2\varepsilon_0 a^2} \qquad (1)$$

Now

$$\cos \alpha = \frac{A - R \cos \theta}{a}$$

and

$$a^2 = A^2 + R^2 - 2AR \cos \theta$$

from which

$$a\,da = AR \sin \theta\,d\theta \qquad (2)$$

and

$$\cos\alpha = \frac{A^2 - R^2 + a^2}{2Aa} \tag{3}$$

Subsituting (2) and (3) into (1) gives

$$E = \frac{q_s R}{4\varepsilon_0 A^2} \int\limits_{A-R}^{A+R} \frac{A^2 - R^2 + a^2}{a^2}\, da = \frac{q_s R^2}{\varepsilon_0 A^2} = \frac{q}{4\pi\varepsilon_0 A^2}$$

PROBLEM 11

Through the use of Ampere's circuital law, find the H field in all regions of an infinite length coaxial cable carrying a uniform and equal current I in opposite directions in the inner and outer conductors. Assume the inner conductor to have a radius of a(m) and the outer conductor to have an inner radius of b(m) and an outer radius of c(m). Assume that the cable's axis is along the z-axis.

SOLUTION: Through the use of symmetrical pairs of filamentary currents as shown in Figure 80(b), it can be argued that the H field in all regions will be in the ϕ direction and thus

$$H = \phi H_\phi.$$

(a) For a concentric amperian closed loop drawn in the region $(r_c < a)$ of the inner conductor, Ampere's circuital law becomes

$$\oint_\ell H \bullet dl = \oint_\ell \left(\phi H_\phi\right) \bullet \left(\phi\, r_c d\phi\right)$$

$$= H_\phi r_c \int\limits_0^{2\pi} d\phi = H_\phi 2\pi r_c = I_{en}. \tag{1}$$

a) Graphical display for finding the *H* field inside and outside a conductor of finite cross section

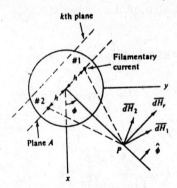

b) Symmetrical pairs of filamentary currents produce a resultant field in the ϕ direction

Figure 80. Magnetic field in a coaxial cable

Now,

$$I_{en} = \frac{I r_c^2}{a^2}$$

for the amperian closed loop inside the inner conductor. Thus, the last two terms in (1) become

$$H_\phi 2\pi r_c = I_{en} = \frac{I r_c^2}{a^2}$$

yielding

$$H_\phi = \frac{I r_c}{2\pi a^2}$$

and

$$H = \phi \frac{I r_c}{2\pi a^2} \left(Am^{-1}\right)(r_c < A). \qquad (2)$$

(b) For a concentric amperian closed loop drawn in the region $(a < r_c < b)$, Ampere's circuital law becomes

$$\oint_\ell H \bullet dl = \oint_\ell \left(\phi H_\phi\right) \bullet \left(\phi r_c d\phi\right)$$

$$= H_\phi r_c \int_0^{2\pi} d\phi = H_\phi 2\pi r_c = I \qquad (3)$$

where $I_{en} = I(A)$. Solving for H_ϕ from the last two terms in (3),

$$H = \frac{\phi I}{2\pi r_c} \left(Am^{-1}\right) \quad (a < r_c < b) \qquad (4)$$

(c) For a concentric amperian closed loop in the region ($b < r_c < c$), Ampere's circuital law becomes

$$\oint_\ell H \bullet dl = \oint_\ell \left(\phi H_\phi\right) \bullet \left(\phi r_c d\phi\right)$$

$$= H_\phi r_c \int_0^{2\pi} d\phi = H_\phi 2\pi r_c = I_{en}. \qquad (5)$$

Now,

$$I_{en} = I - I\left[\frac{\left(r_c^2 - b^2\right)}{\left(c^2 - b^2\right)}\right]$$

for the amperian closed loop inside the outer conductor. It should be noted that some of the enclosed current flows in the reverse direction. Substituting for I_{en} into (5) and solving for H_ϕ from the last two terms,

$$H = \phi \frac{I}{2\pi r_c}\left(\frac{c^2 - r_c^2}{c^2 - b^2}\right)\left(Am^{-1}\right) \quad (b < r_c < c)(-b).$$

(d) For a concentric amperian closed loop drawn in the region ($c < r_c$), the enclosed current is found to be zero, and thus H is zero.

PROBLEM 12

The force field

$$F = ya_x - xa_y$$

is nonconservative, and the work done in opposing the field,

$$-\int_B^A F \bullet dl$$

depends on the path followed from B to A. Let B be $(0,1,0)$ and let A be $(0,-1,0)$. Determone the work done in following these paths consisting of straight line segments:

(a) $(0, 1, 0)$ to $(1, 1, 0)$ to $(1, -1, 0)$ to $(0, -1, 0)$;

(b) $(0, 1, 0)$ to $(0, -1, 0)$;

(c) $(0, 1, 0)$ to $(-1, 1, 0)$ to $(-1, -1, 0)$ to $(0, -1, 0)$.

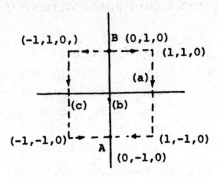

Figure 81. Work paths

SOLUTION: The paths are indicated as shown. Since
$$F = ya_x - xa_y,$$

and
$$dl = dx\, a_x + dy\, a_y,$$

and
$$W = -\int_B^A F \cdot dl,$$

(a) from (0, 1, 0) to (1, 1, 0)

$$W_1 = \int_{0,1,0}^{1,1,0} \left(y\, a_x - x\, a_y\right) \cdot \left(dx\, a_x + dy\, a_y\right)$$

$$= \int_{0,1,0}^{1,1,0} ydx - xdy = -\int_0^1 dx = -1$$

from (1, 1, 0) to (1, –1, 0)

$$W_2 = \int_{1,1,0}^{1,-1,0} ydx - xdy = -\int_1^{-1} -dy = -2$$

from (1, –1, 0) to (0, –1, 0).

$$W_3 = \int_{1,-1,0}^{0,-1,0} ydx - xdy = -\int_1^0 (-1)dx = -1$$

The total work done in path *a* is
$$W_1 + W_2 + W_3 = W_a$$
$$= -4 \text{ Joules.}$$

(b) From (0, 1, 0) to (0, −1, 0), $dx = 0$, $x = 0$

$$\therefore W_b = \int_{1}^{-1} -0 \, dy = 0$$

(c) From (0, 1, 0) to (−1, 1, 0) to (−1, −1, 0) to (0, −1, 0),

$$W_c = -\left[\int_{0}^{-1} dx - \int_{1}^{-1} -dy + \int_{-1}^{0} dx \right] = -[-1 - 2 - 1] = 4 \text{ Joules}.$$

CHAPTER FIVE

Statics

Chapter 5

STATICS

This chapter will review that portion of the study of mechanics that deals with bodies in equilibrium under the influence or action of forces, namely statics.

VECTOR FORCES

The action of one body on another produces force. A force has magnitude, direction, and sense and hence may be represented by a vector. There are three classifications of vectors:

1) When the action of a vector is not confined to or associated with a unique line in space, then it is a **free vector**.
2) When the quantity of a vector must maintain a certain unique line of action, then it is a **sliding vector**.
3) When a unique point of application is defined for a vector, and therefore it occupies a specific position in space, then that is called a **fixed vector**.

In Figure 1, we have vector **V** and its negative **–V**. The sense of direction is measured by the angle θ.

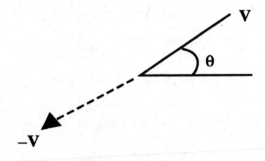

Figure 1. Vector V

Remember that vectors must obey the parallelogram law of equilibrium which states that two vectors **V₁** and **V₂** can be replaced by the resultant vector **V**: $\mathbf{V} = \mathbf{V_1} + \mathbf{V_2}$ (vector addition).

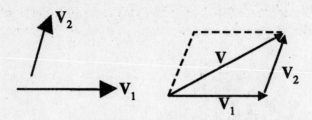

Figure 2. Vector addition

Figure 3 illustrates the use of the parallelogram method in vector subtraction.

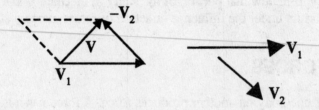

Figure 3. Vector subtraction

In three-dimensional problems, the rectangular components of a vector can be expressed in terms of i, j, and k unit vectors in the x, y, and z directions, respectively.

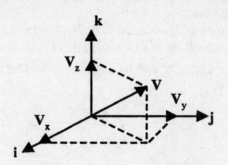

Figure 4. $V = i\,V_x + j\,V_y + k\,V_z$

The figure below shows $\mathbf{F_x}$ and $\mathbf{F_y}$, the rectangular components of a vector $\mathbf{F}$.

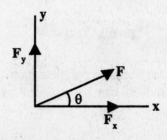

Figure 5. Rectangular components F_x and F_y of vector F

The magnitudes of the vectors, **F**, **F**$_x$ and **F**$_y$ are related as shown below, where F is the magnitude of **F**, F_x is the magnitude of **F**$_x$, and F_y is the magnitude of **F**$_y$:

$$F_x = F\cos\theta$$

$$F_y = F\sin\theta$$

$$F = \sqrt{F_x^2 + F_y^2}$$

$$\theta = \tan^{-1}\frac{F_y}{F_x}$$

In cases where the assignment of reference axes is not readily identifiable, use the following guide, as shown in Figure 6.

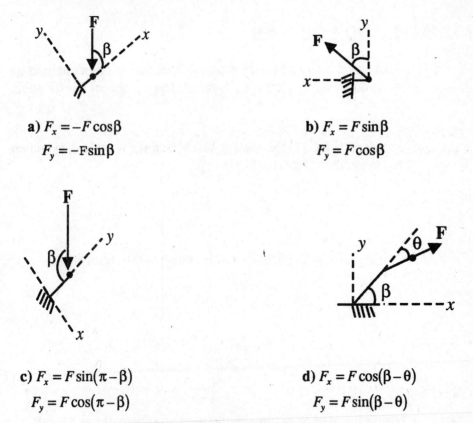

a) $F_x = -F\cos\beta$
   $F_y = -F\sin\beta$

b) $F_x = F\sin\beta$
   $F_y = F\cos\beta$

c) $F_x = F\sin(\pi-\beta)$
   $F_y = F\cos(\pi-\beta)$

d) $F_x = F\cos(\beta-\theta)$
   $F_y = F\sin(\beta-\theta)$

Figure 6. Assignment of reference axes

In the three-dimensional rectangular force system, we can resolve a force into its three components (as illustrated graphically in Figure 7):

$$F_x = F\cos\theta_x \qquad F = \sqrt{F_x^2 + F_y^2 + F_z^2}$$

$$F_y = F\cos\theta_y \qquad F = iF_x + jF_y + kF_z$$

$$F_z = F\cos\theta_z \qquad F = F(i\cos\theta_x + j\cos\theta_y + k\cos\theta_z)$$

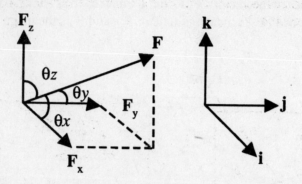

Figure 7. Rectangular components F_x, F_y, F_z of vector F

MOMENTS AND COUPLES

The tendency of a force to rotate a body about any axis that does not intersect its line of action and is not parallel to it, is a **moment M**. The moment of a force about a point A is equal to

$$M_A = r \times F$$

where **r** is a vector from point A to the point at which the force is applied, as shown in Figure 8. The magnitude of moment M_A is

$$M_A = rF\sin\theta$$

or

$$M_A = dF$$

where θ is the angle shown in the figure and d is known as the moment arm.

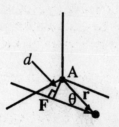

Figure 8. Geometry used to define the moment of a force about a point

The sense of a moment will be either clockwise or counterclockwise, as determined by the right-hand rule.

In a system where several forces are acting on a rigid body, we can move all the forces to an arbitrary point, A, provided a **couple** is introduced as well for each force that is moved. A couple is a pair of forces of equal magnitude, opposite direction, and parallel lines of action, which has a pure turning or moment effect. A system of non-concurrent forces such as that shown below can be replaced by a resultant force, *R*, and resultant couple, *M*.

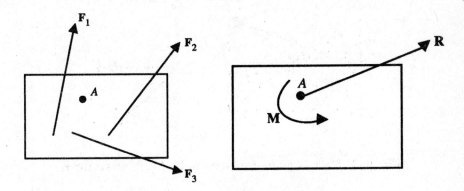

Figure 9. Resultant force and moment

The resultant force is

$$R = F_1 + F_2 + F_3 = \sum F$$

and the resultant couple is $M = M_1 + M_2 + M_3 = \sum d_x F$, where *d* is the moment arm of each force to point *A*.

TWO-DIMENSIONAL EQUILIBRIUM

The free-body diagram is the most useful tool in the solving of problems in mechanics. To review the construction of free-body diagrams:
  (1) Decide which combination of bodies (or body) is to be isolated and make sure that one or more of the desired unknowns is involved in the choice.
  (2) Isolate the bodies (or body) chosen by a diagram that show its complete external boundary.
  (3) Show all forces that act on the isolated bodies (or body).
  (4) Choose coordinate axes.

In two-dimensional (and three-dimensional) equilibrium, the resultant moment and force vectors must be zero:

$$\sum F_x = 0$$
$$\sum F_y = 0$$
$$\sum M_A = 0$$

The location of the axis, *A*, about which moments are summed, is arbitrary. The magnitude of the resultant force and moment vectors must be zero.

The different types of equilibrium in two dimensions are shown in Figure 10.

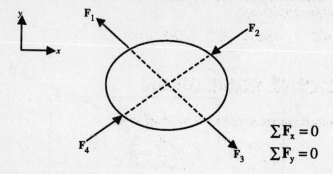

$$\Sigma F_x = 0$$
$$\Sigma F_y = 0$$

a) Concurrent at a specific point

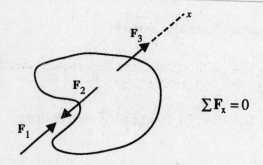

$$\Sigma F_x = 0$$

b) Collinear

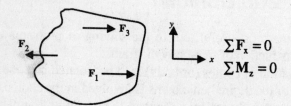

$$\Sigma F_x = 0$$
$$\Sigma M_z = 0$$

c) Parallel

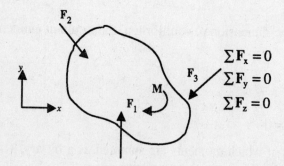

$$\Sigma F_x = 0$$
$$\Sigma F_y = 0$$
$$\Sigma F_z = 0$$

d) General

Figure 10. Types of force systems

In addition, where A, B, and C are any three points not lying in the same straight line on a body acted on by a force, then

$$\sum M_A = 0 \qquad \sum M_B = 0 \qquad \sum M_C = 0.$$

THREE-DIMENSIONAL EQUILIBRIUM

In the case of three-dimensional equilibrium we may write:

$$\sum F_x = 0$$

$$\sum F = 0 \qquad \text{or} \qquad \sum F_y = 0$$

$$\sum F_z = 0$$

$$\sum M = 0 \qquad \text{or} \qquad \sum M_x = 0$$

$$\sum M_y = 0$$

$$\sum M_z = 0$$

Complete equilibrium necessitates all six of these equations. Refer to the guide in Figure 11 to determine the actions of three-dimensional forces.

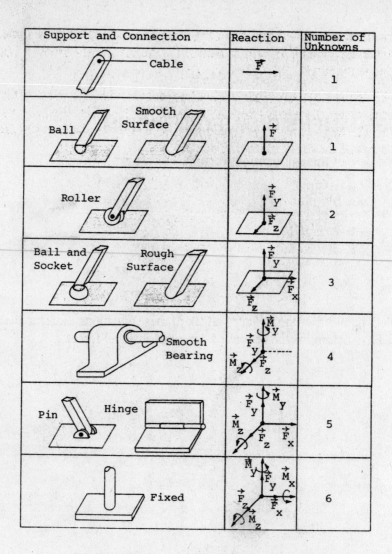

Figure 11. Reactions at supports and connections in three-dimensional structures

Two- and three-dimensional equilibrium equations, free-body diagrams, and the information in Figures 10 and 11 provide a foundation for problems involving determinate reactions.

EXAMPLE: A 1,000 lb weight is hung from the end of a pipe which is fastened to a ball and socket at the lower end and supported at the top by two cables as shown in Figure 12. Neglecting the weight of the pipe, determine the forces in each of the two cables and the reaction at point *A*.

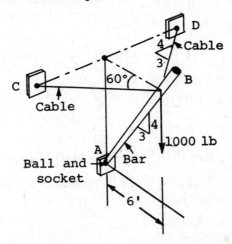

Figure 12. Weight hung from a pipe

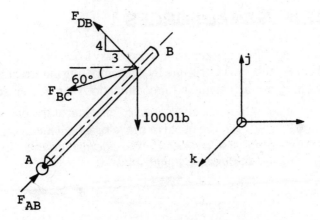

Figure 13. Reaction

SOLUTION: We isolate the pipe *AB* and obtain the free-body diagram shown in Figure 13. Recognizing that *AB* is two-force member, we obtain the following equations:

$$\mathbf{F}_{DB} = F_{DB}\left(-\frac{3}{5}\mathbf{i} - \frac{4}{5}\mathbf{k}\right)$$

$$\mathbf{F}_{CB} = F_{CB}\left(-\cos 60°\mathbf{i} + \sin 60°\mathbf{k}\right)$$

$$\mathbf{F}_{AB} = F_{AB}\left(\frac{3}{5}\mathbf{i} + \frac{4}{5}\mathbf{k}\right)$$

or

$$\mathbf{F}_{DB} = F_{DB}(-0.6\mathbf{i} - 0.8\mathbf{k})$$

$$\mathbf{F}_{CB} = F_{CB}(-0.5\mathbf{i} + 0.866\mathbf{k})$$

$$\mathbf{F}_{AB} = F_{AB}(0.6\mathbf{i} + 0.8\mathbf{j})$$

$$\mathbf{W} = -1,000\mathbf{j}\ \text{lb}$$

Applying the condition $\mathbf{F} = 0$, we have

$$\left(-0.6F_{BD} - 0.5F_{CB} + 0.6F_{AB}\right)\hat{i} + \left(0.8F_{AB} - 1,000\right)\hat{j} + \left(-0.8F_{BD} + 0.866F_{BC}\right)\hat{k} = 0$$

The components of this equation give us the three equations needed to find the three unknowns. Solving them yields,

$$F_{AB} = 1,250\ \text{lb}$$
$$F_{BD} = 705\ \text{lb}$$
$$F_{BC} = 652\ \text{lb}$$

ANALYSIS OF INTERNAL FORCES

To analyze the forces acting on a structure, it is necessary to analyze separate free-body diagrams of the members of the structure by dismembering the structure itself. The internal forces acting on each member (or combination) can be looked at.

A truss is made up of straight members connected at joints at the ends of each member. Hence, in trusses, the basic element is a triangle, and all members are two force members in either compression or tension. The external forces on a member in compression or tension are as shown in Figure 14.

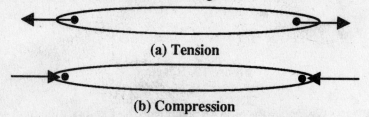

(a) Tension

(b) Compression

Figure 14. External forces on a member

Assume in simple trusses that all external forces are applied at the pin connections.

Examine the truss structure at each pin location to determine the direction and application of the forces acting at that location.

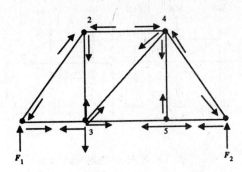

Figure 15. Truss

Apply equilibrium equations at each joint to determine the reactions. Examining the internal forces of a structure at each joint in this manner is called the method of joints.

We can also use the method of sections to look at the forces acting on an entire section. This method employs equilibrium equations on a section of a truss, as shown in Figure 16.

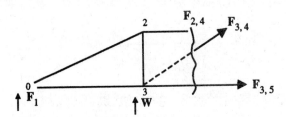

Figure 16. Section of truss

EXAMPLE: In the bridge truss of Figure 17, calculate the forces in members *UV* and *DE*.

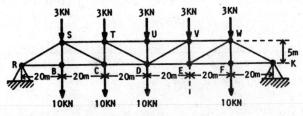

Figure 17. Bridge truss

SOLUTION: The free-body diagram is shown in Figure 18. The method of section will again be used. The section *aa* cuts through the truss intersecting members *UV* and *DE*. The reaction at *K* is determined using the equation

$$+\, \circlearrowleft \, \Sigma M_R = 0.$$

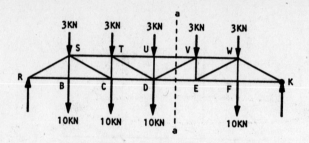

Figure 18. Free-body diagram

$$-(13KN)(20m) - (13KN)(40m) - (13KN)(60m) - (3KN)(80m) - (13KN)(100m)$$
$$+ K(120m) = 0$$

or

$$K = 25.8KN$$

The right portion of the truss *VWKE* cut off by the line *aa* in Figure 19 will be used as a free body to calculate the force in the three members. For force in member *DE*

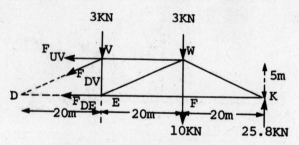

Figure 19. Right portion of truss

$$+ \text{)} \sum M_V = 0$$
$$(25.8KN)(40m) - (13KN)(20m) - F_{DE}(5m) = 0$$
$$\therefore F_{DE} = +154KN = 154KN \text{ tension}$$

The force in member *UV* is determined by taking moments about joint *D*. From Figure 19,

$$+ \text{)} \sum M_D = 0$$
$$(25.8KN)(60m) - (13KN)(40m) - (3KN)(20m) + \left(F_{UV}\right)(5m) = 0$$
$$\therefore F_{UV} = -194KN$$

The negative sign indicates F_{UV} to be in compression.

In many structures, multiforce members are present. These members have three or more forces acting on them, as in Figure 20.

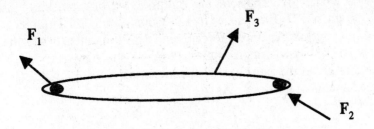

Figure 20. Multiforce members

DISTRIBUTED FORCES, CENTROIDS, AND CENTERS OF GRAVITY

Line distribution, in Newtons per meter $\left(\dfrac{N}{m}\right)$, occurs when a force is distributed along a line such as a suspended wire.

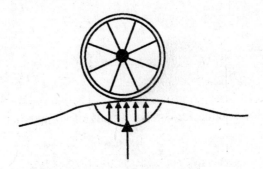

Figure 21. Line distribution

Area distribution, in $\dfrac{N}{m^2}$, occurs when force is distributed over an area. **Volume distribution** occurs when force is distributed over the volume of a body, such as the earth's gravitational pull.

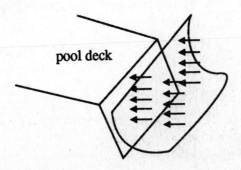

Figure 22. Area distribution - water against the wall of a pool

To find the center of gravity, apply the principle of moments—Varignon's Theorem. Varignon's Theorem states that the sum of individual moments about a point caused by multiple concurrent forces is equal to the moment of a resultant force about that point. The resultant gravitational force W about any axis equals the sum of the moments about the same axis of the incremental force dw acting on all infinitesimal elements of the body. As

$$W = mg$$

$$dw = gdm$$

the coordinates of the center of gravity are:

$$\bar{x} = \frac{\int xdm}{m} \qquad \bar{y} = \frac{\int ydm}{m} \qquad \bar{z} = \frac{\int zdm}{m},$$

where $\int xdm$, $\int ydm$ and $\int zdm$ are the sum of the moments.

The **centroid** of common shapes are listed in most engineering or mathematics handbooks.

The centroid of a line such as a wire is

$$\bar{x} = \frac{\int xdL}{L} \qquad \bar{y} = \frac{\int ydL}{L} \qquad \bar{z} = \frac{\int zdL}{L},$$

the centroid of areas is

$$\bar{x} = \frac{\int xdA}{A} \qquad \bar{y} = \frac{\int ydA}{A} \qquad \bar{z} = \frac{\int zdA}{A},$$

and the centroid of volume for a general body of volume is

$$\bar{x} = \frac{\int xdV}{V} \qquad \bar{y} = \frac{\int ydV}{V} \qquad \bar{z} = \frac{\int zdV}{V}.$$

For ease of calculation, use the elements that can be integrated in one continuous operation in the equation. Disregard higher order terms, choose a coordinate that best matches the region's boundaries, use first order system elements, and use a coordinate to the centroid of the element.

EXAMPLE: (a) Locate the centroid of the T-section shown in Figure 23.

(b) Also locate the centroid of the volume of the cone and hemisphere shown in Figure 24, the values of r and h being 6 in. and 18 in., respectively.

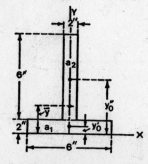

Figure 23. T section

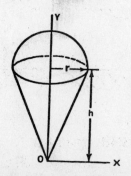

Figure 24. Cone and hemisphere

SOLUTION: (a) If axes be selected as indicated it is evident from symmetry that $x = 0$. By dividing the given area into areas a_1 and a_2 and by taking moments about the bottom edge of the area, $\bar{y}$ may be found as follows:

$$A\bar{y} = \Sigma(ay_0),$$

$$\bar{y} = \frac{12 \times 1 + 12 \times 5}{6 \times 2 + 6 \times 2} = 3 \text{ in.}$$

(b) The axis of symmetry will be taken as the y-axis. From symmetry then $\bar{x} = 0$. By taking the x-axis through the apex of the come as shown, the equation $V\bar{y} = \Sigma(vy_0)$ becomes

$$\frac{1}{3}\left(\pi r^2 h + \frac{2}{3}\pi r^3\right)\bar{y} = \frac{1}{3}\pi r^2 h \times \frac{3}{4}h + \frac{2}{3}\pi r^3\left(h + \frac{3}{8}r\right).$$

That is,

$$\frac{1}{3}\pi r^2(h + 2r)\bar{y} = \frac{1}{3}\pi r^2\left(\frac{3}{4}h^2 + 2rh + \frac{3}{4}r^2\right).$$

Therefore,

$$\bar{y} = \frac{\dfrac{3}{4}h^2 + 2rh + \dfrac{3}{4}r^2}{h + 2r}$$

$$= \frac{\dfrac{3}{4} \times (18)^2 + 2 \times 6 \times 18 + \dfrac{3}{4} \times (6)^2}{18 + 2 \times 6} = 16.2 \text{ in.}$$

FRICTION

Let's examine the effects of dry friction acting on rigid bodies and their exterior surfaces. Consider Figure 25, and the associated free-body diagram, where N is the normal force and F is the friction force.

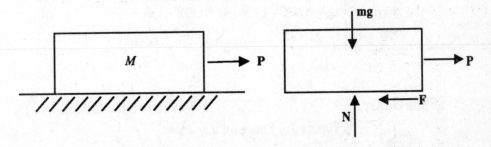

a) Block schematic	b) Free-body diagram

Figure 25. Sliding block

Obviously greater force is required to overcome friction and get the block moving than to keep it in motion once it is sliding across the plane surface. The frictional resistance up to the point of slippage is called static friction and is defined by

$$f_{s\,max} = \mu_s N,$$

where μ_s is the coefficient of static friction and N is the magnitude of the normal force on the block. Once the block is moving, kinetic friction takes over:

$$f_k = \mu_k N,$$

where μ_k is the coefficient of kinetic friction.

For a body on an incline, the body will not slide until the parallel component of the weight is greater than the friction force. The body will slide when the angle of incline equals a critical angle which can be related to the coefficient of friction:

$$\tan\theta = \frac{f}{N}$$

or

$$\tan\theta = \mu$$

Depending on whether we use μ_k or μ_s, θ is called the angle of static friction or kinetic friction.

EXAMPLE: A ladder of length L = 10m and mass M = 10kg leans against a frictionless vertical wall at an angle of 60° from the horizontal. The coefficient of static friction between the horizontal floor and the foot of the ladder is μ_s = 0.25. A man of mass M = 70kg starts up a ladder. How far along the ladder does he get before the ladder begins to slide down the wall?

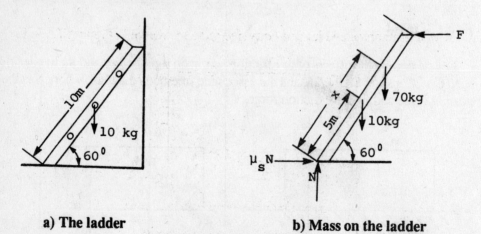

a) The ladder   b) Mass on the ladder

Figure 26. Force on a ladder

SOLUTION: The forces on the ladder are shown. The horizontal force on the foot of the ladder is equal to $\mu_s N$ only at the instant before the ladder begins to slide. We wish to find the position of the man at this instant. Because the vertical wall is frictionless, it can exert a force on the ladder only perpendicular to itself as shown.

Since the net force on the ladder is zero, we find, taking vertical components,

$$N - 10g - 70g = 0$$

and, from horizontal components,

$$\mu_s N - F = 0$$

Let x be the distance of the man along the ladder from the foot at the instant the ladder begins to slide. Equate the torque about the foot of the ladder to zero.

$$\frac{L}{2} mg \cos\theta + xMg \cos\theta - FL\sin\theta = 0$$

From the equation above,

$$N = 80g \text{ newtons.}$$

From this and the second equation,

$$F = 0.25 \times 80g = 20g \text{ newtons.}$$

From this and the third equation,

$$\frac{1}{2} mLg \cos\theta + xMg \cos\theta - 20gL\sin\theta = 0$$

$$x = \frac{20gL\sin\theta - \frac{1}{2} mLg \cos\theta}{Mg \cos\theta}$$

$$= \frac{200 \times 0.866 - \frac{1}{2} \times 10 \times 10 \times \frac{1}{2}}{70 \times \frac{1}{2}} = 4.2 \text{ meters.}$$

A wedge machine and its free-body diagrams are shown in Figure 27.

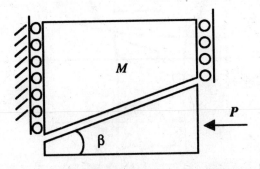

a) Wedge machine schematic

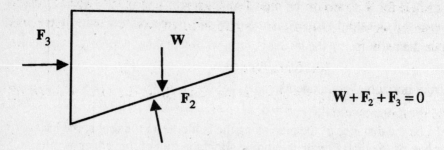

$$W + F_2 + F_3 = 0$$

b) Free-body diagram on top component

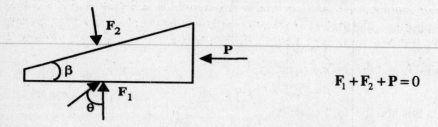

$$F_1 + F_2 + P = 0$$

c) Free-body diagram on bottom component
Figure 27. Wedge machine

A screw is a common wedge machine, as shown in Figure 28.

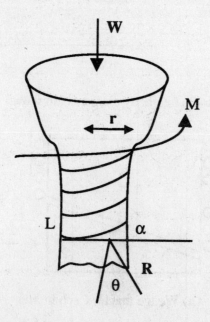

Figure 28. Common wedge machine

θ is the angle for **R** normal to the thread so: $\tan \theta = M$ will be the angle of friction. The moment of equilibrium necessary to keep the screw from unwinding downward under the force **W** is

$$M = Wr\tan(\alpha + \theta).$$

The screw will remain in place provided that $\alpha < \theta$. If $\alpha > \theta$, then the screw will unwind itself downward.

MOMENT OF INERTIA

The area moment of inertia is used to determine the deflection and stresses of shafts and beams, and the buckling loads of columns. Moments of inertia are always positive, and have units of length to the fourth power. An elementary area A composed of i area elements is shown in Figure 29.

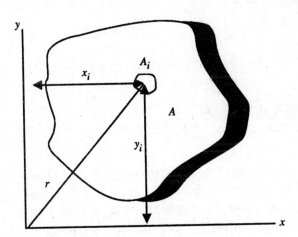

Figure 29. Area A

For area A, the moment of inertia with respect to the x-axis is

$$I_x = \sum_i y_i^2 A_i,$$

and the moment of inertia with respect to the y-axis is

$$I_y = \sum_i x_i^2 A_i.$$

The formal definitions of moment of inertia of a plane area are:

$$I_x = \int_A y^2 dA$$

which is the moment of inertia about the x-axis, and:

$$I_y = \int_A x^2 dA$$

which is the moment of inertia about the y-axis.

The polar moment of inertia may be informally defined as the resistance of an area to torsion. It is denoted by J, and like the area moment of inertia, has units of length to the fourth power. The Perpendicular Axis Theorem relates polar and area moments of inertia as follows:

$$J = I_x + I_y.$$

The polar moment of inertia in integral form is

$$J = \int_A (x^2 + y^2) dA.$$

The imaginary distance from a centroidal axis at which the area would not affect the moment of inertia is the radius of gyration, denoted by K, and defined as

$$K_x = \sqrt{\frac{I_x}{A}}$$

$$K_y = \sqrt{\frac{I_y}{A}}.$$

REVIEW PROBLEMS

PROBLEM 1

A 200 lb man is standing 3 feet from the top of an 8 foot ladder that leans against a smooth wall as shown in the figure. If the ladder is in equilibrium, determine the forces at A and D in Figure 30. Neglect the weight of the ladder.

SOLUTION: This problem can be solved by determining the line of action of F_D and then setting the x and y components of the force to zero. The lines of action of F_A and the 200 lb gravity force are known and intersect at point C. (See Figure 30.)

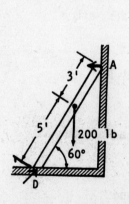

a) Force equilibrium diagram

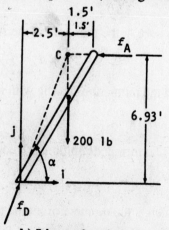

b) Lines of action

Figure 30. Schematic of ladder in equilibrium

Since the ladder is a three-force member, the line of action of F_D must pass through point C. First we determine the angle α:

$$\tan\alpha = \frac{6.93}{2.5} = 2.772$$

$$\alpha = 70.16° \qquad \cos\alpha = 0.339 \qquad \sin\alpha = 0.94$$

$$(-F_A + 0.339F_D)i - (0.94F_D - 200)j = 0,$$

which yields:

$$-F_A = -0.339\,F_D$$

$$0.94\,F_D = 200 \qquad F_D = \frac{200}{0.94}$$

$$F_A = 0.339\left(\frac{200}{0.94}\right)$$

$$F_D = 213 \text{ lbs.}$$
$$F_A = 72.2 \text{ lbs.}$$

PROBLEM 2

In Figure 31, a 50-N tension is required to maintain the box B in equilibrium with force **F**. Calculate the magnitude of F given that $d = 10$ cm and $r = 5$ cm.

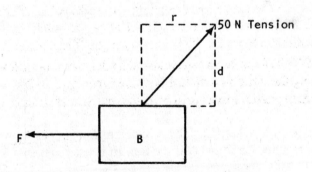

Figure 31. Forces on box

SOLUTION: The free-body diagram of the box is shown in Figure 32. It accounts for all forces acting on the box. Since only **F** is required, it is sufficient to consider only the x-direction. It is given that the box is in equilibrium, thus the summation of all the forces in the x-direction must be zero.

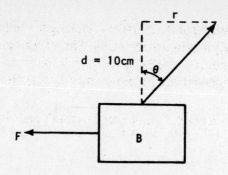

Figure 32. Free-body diagram of box

$$-F + (50N)(\sin\theta) = 0 \tag{1}$$

From trigonometry,

$$\sin\theta = \frac{r}{\sqrt{10^2 + r^2}}.$$

Substituting for $\sin\theta$ in equation (1) gives

$$F = (50N)\frac{r}{\sqrt{10^2 + r^2}}$$

But

$$r = 5 \text{ cm}$$

$$F = (50N)\left(\frac{5 \text{ cm}}{\sqrt{10^2 + 5^2}}\right) = (50N)\left(\frac{5}{\sqrt{100 + 25}}\right)$$

$$F = 22.4N$$

PROBLEM 3

A 2,000 kg member is held in place at locations P and Q as shown in Figure 33. Determine the reactions at P and Q if the forklift is used to haul a 3,000 kg load.

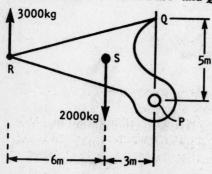

Figure 33. Position and forces on member

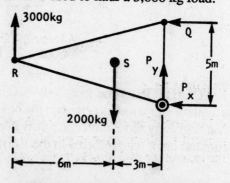

Figure 34. Free-body diagram

SOLUTION: The weight of the member acts along the center of gravity located at point S on the free-body diagram of Figure 34. The reaction at Q has been resolved into its x and y components. Assume the forces to act in the direction shown. The final sense of each force component will be known after the magnitudes of the forces are determined.

Taking moments about point P eliminates P_x and P_y and allows for the value of Q to be determined.

$$\Sigma M_P = 0$$
$$(3,000 \text{ kg}) (9 \text{ m}) - (2,000 \text{ kg}) (3 \text{ m}) - Q (5 \text{ m}) = 0$$
$$Q = 4,200 \text{ kg}$$

P_x is determined by summing the horizontal components of all external forces and setting it equal to zero.

$$\therefore \Sigma F_x = 0: -P_x - Q \text{ or } -P_x - 4,200 \text{ kg} = 0$$
$$P_x = -4,200 \text{ kg}$$

Since P_x is negative, it means that its sense as shown in Figure 34 should be in the opposite direction. To determine P_y, all vertical forces are summed equal to zero.

$$\therefore \Sigma F_y = 0: 3,000 \text{ kg} - 2,000 \text{ kg} + P_y = 0$$
$$P_y = -1,000 \text{ kg}$$

P_y is negative, thus it has an opposite sense to that shown in Figure 34. The forces acting on the member and their directions as determined above are shown in Figure 35.

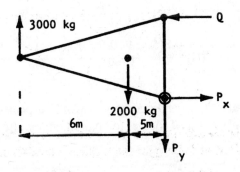

Figure 35. Resultant forces

PROBLEM 4

A crane lifts a load of 10,000 kg mass. The boom of the crane is uniform, has a mass of 1,000 kg, and a length of 10 m. Calculate the tension in the upper cable and the magnitude and direction of the force exerted on the boom by the lower pivot.

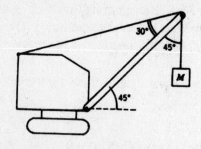
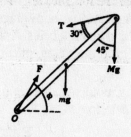

a) Schematic of crane   **b) Free-body diagram**

Figure 36. Crane lifting load

SOLUTION: Isolate the boom analytically and indicate all forces on it as in the right-hand portion of the figure, where **T** is the tension in the upper cable, **F** is the force exerted on the boom by the lower pivot, m is the mass of the boom, and M is the mass of the load being lifted by the crane. The magnitude of **T** is unknown and both the magnitude and direction of **F** are unknown. Set the net torque about point O equal to zero. If the length of the boom is S, this net torque is given by the equation:

$$\frac{S}{2}mg\sin 45° + SMg\sin 45° - ST\sin 30° = 0$$

$$\frac{g(m/2 + M)\sin 45°}{\sin 30°} = T$$

Substitute the values given above.

$$T = \frac{9.8(500 + 10{,}000)(1/\sqrt{2})}{\dfrac{1}{2}} = 1.46 \times 10^5 \, \text{N}$$

We can find F_x and F_y, the x and y components of **F** respectively, by requiring that both the x and y components of the net force on the boom be equal to zero.

$$\overset{+}{\rightarrow} \Sigma F_x = 0$$

$$F_x - T\cos 15° = 0$$

$$F_x = 1.46 \times 10^5 \, (\cos 15°)$$

$$F_x = 1.41 \times 10^5 \, \text{N}$$

$$+\uparrow \Sigma F_y = 0$$

$$F_y - T\sin 15° - mg - Mg = 0$$

$$F_y = 1.46 \times 10^5 \, (\sin 15°) + 9.8(1{,}000 + 10{,}000)$$

$$F_y = 1.46 \times 10^5 \, \text{N}$$

So that the magnitude of **F** is:

$$F = \sqrt{F_x^2 + F_y^2} = 2.03 \times 10^5 \, N.$$

The angle ϕ that **F** makes with the horizontal is given by:

$$\tan\phi = \frac{F_x}{F_y} = \frac{1.46}{1.41} = 1.035$$

$$\phi = 46°$$

PROBLEM 5

Determine the forces acting on each member of the frame in Figure 37.

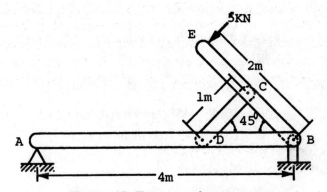

Figure 37. Frame and components

SOLUTION:

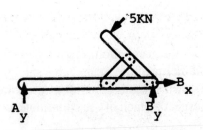

Figure 38. Forces on entire frame

Determine the reaction at A by calculating the moment at point B.

$$+\, \text{\Large\textcircled{}}\; \Sigma M_B = 0; \; -(Ay)(4m) + (5kN)(2M) = 0$$

$$A_y = 2.5 kN \uparrow$$

Determine the remaining reactions by taking each member as a free body. Figure 39 shows a free-body diagram for each member.

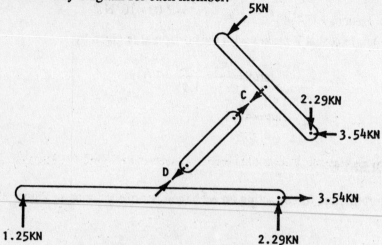

Figure 39. Forces on individual components

$$+\circlearrowleft \Sigma M_B = 0;\ (-2.5\text{kN})\,(4\text{m}) - (D\sin45°)\,(\sqrt{2}\text{m}) = 0$$

$$D = 10\text{kN}\quad 45°\ (\text{on member } AB)$$
$$C = 10\text{kN}\quad 45°\ (\text{on member } BE)$$

$$\overset{+}{\rightarrow}\Sigma F_x = 0;\ B_x - 10\cos45° = 0$$
$$B_x = 7.07\text{kN}$$

$$+\uparrow\Sigma F_y = 0;\ 2.5 + B_y - 10\cos45° = 0$$
$$B_y = 4.57\text{kN}$$

PROBLEM 6

(a) Find the centroid of the parabolic section shown in Figure 40. (b) Also, determine the moments of inertia of the shaded area shown with respect to each of the coordinate axes. (c) Using the results of part (b), determine the radius of gyration of the shaded area with respect to each of the coordinate axes.

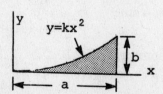

Figure 40. Parabolic section

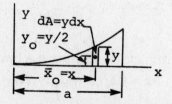

Figure 41. Vertical differential element

SOLUTION: Given $y = kx^2$, the value of k can be obtained from the point $x = a, y = b$. Hence, $k = \dfrac{b}{a^2}$.

$$y = \frac{b}{a^2}x^2 \tag{1}$$

or

$$x = \frac{a}{b^{\frac{1}{2}}}y^{\frac{1}{2}} \tag{2}$$

Choose a vertical differential element as shown in Figure 41. Then the total area A is

$$A = \int dA = \int y\,dx = \int_0^a \frac{b}{a^2}x^2\,dx = \left[\frac{b}{a^2}\frac{x^3}{3}\right]_0^a$$

$$= \frac{ab}{3} \tag{3}$$

The moment of the differential element with respect to the y-axis is $\overline{x}_o dA$. The moment of the total area with respect to this axis is, therefore,

$$\int \overline{x}_o dA = \int xy\,dx = \int_0^a x\left(\frac{b}{a^2}x^2\right)dx$$

$$= \left[\frac{b}{a^2}\frac{x^4}{4}\right]_0^a = \frac{a^2 b}{4} \tag{4}$$

The centroid $\overline{x}$ is that point which, when multiplied by the area, gives the same result for the moment as the integration of the differential moments $x_o dA$.

Thus, $\qquad\qquad \overline{x}A = \int \overline{x}_o dA$

$$\overline{x}\frac{ab}{3} = \frac{a^2 b}{4}$$

$$\overline{x} = \frac{3}{4}a \tag{5}$$

Similarly, the moment of the differential element with respect to the x-axis is $\overline{y}_o dA$, and the moment of the total area is

$$\int \overline{y}_o dA = \int \frac{y}{2}y\,dx = \int_0^a \frac{1}{2}\left(\frac{b}{a^2}x^2\right)^2 dx$$

$$= \frac{ab^2}{10}$$

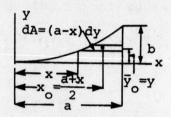

Figure 42. Horizontal element

Thus,
$$\bar{y}_o = \int \bar{y}_{e\ell} dA$$

$$\bar{y}\frac{ab}{3} = \frac{ab^2}{10}$$

$$\bar{y} = \frac{3}{10}b \tag{6}$$

The same result will be obtained by using a horizontal element as shown in Figure 42. For example, the moments with respect to the coordinate axes are:

$$\int \bar{x}_o dA = \int \frac{a+x}{2}(a-x)dy = \int_0^b \frac{a^2 - x^2}{2}dy$$

$$= \frac{1}{2}\int_0^b \left(a^2 - \frac{a^2}{b}y\right)dy = \frac{a^2 b}{4},$$

$$\int \bar{y}_o dA = \int y(a-x)dy = \int y\left(a - \frac{a}{b^{\frac{1}{2}}}y^{\frac{1}{2}}\right)dy$$

$$= \int_0^b \left(ay - \frac{a}{b^{\frac{1}{2}}}y^{\frac{3}{2}}\right)dy = \frac{ab^2}{10}$$

Hence equations (5) and (6) follow.

(b) Using the vertical differential element, compute the moment of inertia. Since all portions of this element are not at the same distance from the x-axis, treat the elements as a thin rectangle.

The moment of inertia of the element with respect to the x-axis is

$$dI_x = \frac{1}{3}y^3 dx = \frac{1}{3}\left(\frac{b}{a^2}x^2\right)^3 = \frac{1}{3}\frac{b^3}{a^6}x^6 dx$$

$$I_x = \int dI_x = \int_0^a \frac{1}{3}\frac{b^3}{a^6}x^6 dx = \frac{ab^3}{21} \tag{7}$$

Similarly,

$$dI_y = x^2 dA = x^2(ydx) = x^2\left(\frac{b}{a^2}x^2\right)dx$$

$$= \frac{b}{a^2}x^4 dx$$

$$I_y = \int dI_y = \int_0^a \frac{b}{a^2}x^4 dx = \frac{a^3 b}{5}. \tag{8}$$

The radii of gyration are

$$k_{x^2} = \frac{I_x}{A} = \frac{\dfrac{ab^3}{21}}{\dfrac{ab}{3}} = \frac{b^2}{7}$$

$$k_x = \sqrt{\frac{1}{7}}b,$$

$$k_{y^2} = \frac{I_y}{A} = \frac{\dfrac{a^3 b}{5}}{\dfrac{ab}{3}} = \frac{3}{5}a^2$$

$$k_y = \sqrt{\frac{3}{5}}a \tag{9}$$

PROBLEM 7

The wheels of a small wagon are separated by a distance d, and the center of mass is a distance h above the ground. The wagon is at rest on a hill of slope angle θ, and between the wheels and the surface of the hill the coefficient of static friction is μ. How steep a hill can the wagon rest on without tipping over or sliding?

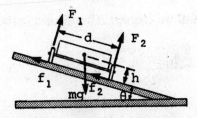

Figure 43. Forces on wagon

SOLUTION: The wagon is acted on by three forces: gravity and two contact forces at the wheels. Both the direction and the magnitudes of the contact forces are unknown. Consequently, we have four unknown force components, which are denoted by F_1, F_2, f_1, and f_2 in Figure 43. Since we have only three equilibrium equations, this problem normally is statically indeterminate in the sense that f_1 and f_2 cannot be determined separately. However, we are now interested only in the condition when the wagon is on the verge of sliding or on the verge of tipping. In these cases, we have additional relations imposed between the variables. When the wagon is on the verge of sliding, the friction forces are fully developed at both wheels, and, as we recall, are then simply related to the force components normal to the plane. If we let the symbols F_1, F_2, f_1, and f_2 denote the forces that are obtained when the wagon is on the verge of sliding, we have

$$f_1 = \mu F_1 \quad \text{and} \quad f_2 = \mu F_2.$$

If we combine these relations with the general conditions for equilibrium,

$$+$$
$$\underleftarrow{} \sum F_x = f_1 - f_2 + mg$$
$$\overline{\sin\theta_1 = 0}$$
$$\underuparrow{} + \sum F_y = F_1 - F_2 - mg$$
$$\overline{\cos\theta_1 = 0},$$

we obtain $\mu\, mg \cos\theta_1 = mg \sin\theta_1$ or

$$\tan\theta_1 = \mu.$$

In other words, in order to prevent sliding, we must have $\theta < \theta_1$.

When the wagon is on the verge of tipping, the forces F_1 and f_1 will be zero since then the contact between the upper wheel and the plane will be broken. Then, if we consider the torque with respect to the contact point of the lower wheel, we see that for the total torque to be zero, the lever arm of the gravitational force with respect to this point must be zero. This will occur when the angle of inclination of the plane has the value given by

$$\tan\theta_2 = \frac{d}{2h}$$

and tipping will be prevented if $\theta < \theta_2$. If $\theta_1 = \theta_2$, it follows that tipping and sliding will occur simultaneously when $\mu = \dfrac{d}{2h}$. If μ is less than $\dfrac{2}{dh}$, sliding will occur before tipping as the angle θ is increased. The opposite, of course, occurs if μ is larger than $\dfrac{d}{2h}$.

PROBLEM 8

A small pump is mounted on a concrete slab which, in turn, rests on four concrete posts. Two of the posts have settled slightly so that the slab must be leveled. The foreman decides to raise the low side with two large wedges located as drawn on Figure 43. Fortunately, there is a concrete floor on which the bottom wedge can be fastened. If μ_s between the two wedges is 0.4 and between the wedge and the concrete 0.6, and if the motor slab transmits a 1,000 lb. vertical force to the wedge, determine the force **P** required for lifting the slab.

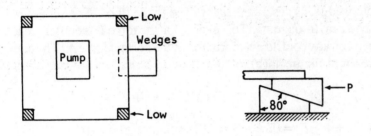

a) Top view   b) Side view

Figure 44. Pump mounted on concrete slab

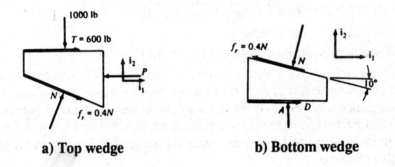

a) Top wedge   b) Bottom wedge

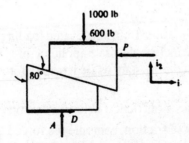

c) Both wedges

Figure 45. Free-body diagrams

SOLUTION: The free-body diagrams are shown in Figure 45. Notice that Figure 45c, showing the two wedge system, has only the external forces shown. The forces between the wedges are equal and opposite and canceled out by Newton's Third Law. If the sum of all external forces are set equal to zero, the system is in equilibrium, that is, nothing moves. Any greater force will start lifting the slab.

$$\xrightarrow{+}\sum F_{i_1} = 0;\ 600 + D - P = 0 \tag{1}$$

$$+\uparrow\sum F_{i_2} = 0;\ -1,000 + A = 0$$

so
$$A = 1,000\ \text{lb}.$$

The bottom wedge, as described, is fastened to the floor. The sum of the forces must be equal to zero throughout. This means that the force **D** is entirely reactive and does not depend on any coefficient of friction or the normal force **A**. The equilibrium equations are, from Figure 45(b),

$$\xrightarrow{+}\sum F_{i_1} = 0;\ D - 0.4N\cos10° - N\sin10° = 0 \tag{2}$$

$$+\uparrow\sum F_{i_2} = 0;\ A - N\cos10° + 0.4N\sin10° = 0$$

or,
$$1,000 - N\cos10° + 0.4N\sin10° = 0. \tag{3}$$

Solving equation (2) and (3) simultaneously for *D* yields

$$D = 1,000\frac{0.4\cos10° + \sin10°}{\cos10° - 0.4\sin10°}$$

$$= 621\ \text{lb}.$$

Putting this value into equation (1) yields
$$P = 1,221\ \text{lb}.$$

The force **P** needed by this method is greater than the 1,000 lb needed for a direct lift.

PROBLEM 9

A capstan is used to lower a crate from an elevated loading platform to the ground. When held stationary on the incline, the crate applies a force F_2 of 5,000N to the rope wound with two turns about the capstan, while it takes a force F_1 of 100N to hold the crate in place.

a) Calculate the coefficient of friction between the rope and the capstan.

b) Using this coefficient of friction, compute the force F_2 that will be held when the rope is wound with three turns.

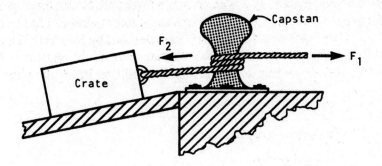

Figure 46. Capstan lowering a crate

SOLUTION: The equation relating the tensions at the beginning and end of a rope wrapped around a fixed circular cylinder is

$$ln\frac{T_2}{T_1} = \mu(\theta_2 - \theta_1). \qquad (1)$$

Since the rope is wrapped two full turns around the capstan, we have:

$$(\theta_2 - \theta_1) = 2(2\pi) = 4\pi \text{ rad.}$$

Given are:

$$F_1 = 100\text{N}, F_2 = 5,000\text{N}$$

so

$$4\pi\mu = ln\frac{5,000}{100} = ln50 = 3.91.$$

Solving for μ, we get

$$\mu = \frac{3.91}{4\pi} = 0.31.$$

To analyze the problem when the rope is wrapped three times about the capstan, we first modify equation (1) by taking antilogs, yielding

$$\frac{T_2}{T_1} = e^{\mu(\theta_2 - \theta_1)}.$$

So, using $\theta_2 - \theta_1 = 3(2\pi) = 6\pi$ rad., we get

$$T_2 = (100\text{N})e^{(0.31)(6\pi)}$$
$$= 34,493\text{N}$$

CHAPTER SIX

Dynamics

Chapter 6

DYNAMICS

Dynamics is divided into areas: **kinematics** (geometry of motion) and **kinetics** (force-acceleration relationships). Kinematics will be limited to particle and planar (two dimensional) motion of vehicles, projections, and rotating bodies. Kinetics will include force, acceleration, work, energy, impulse, momentum, and vibration.

KINEMATICS

Linear Particle Motion

Velocity (v) is the rate of change of the position (s) of a particle with time (t).

$$v = \frac{ds}{dt}.$$

Acceleration is the rate of change of velocity

$$a = \frac{dv}{dt} = \frac{d^2s}{dt^2}, \text{ and also } a = \frac{vdv}{ds}.$$

As an example, let

$$s = 18 + 8t - 2t^2$$

where s = position in feet and t = time in seconds.

Position at t = 10 sec is

$$s = 18 + 8 \times 10 - 2 \times 10^2 = -102 \text{ ft}.$$

Velocity at t = 10 sec is

$$v = \frac{ds}{dt} = 8 - 4t = -32 \text{ ft/sec}.$$

Acceleration at t = 10 sec is

$$a = \frac{dv}{dt} = -4 \text{ ft/sec}^2.$$

Distance traveled from t = 0 to t = 10 seconds is a little more involved since the particle reverses directions when the velocity ($v = 8 - 4t$) becomes 0 at $t = \frac{8}{4} = 2$ sec.

The velocity is zero at $t = \dfrac{8}{4} = 2$ sec, thus the particle changes direction. The distance traveled from $t = 0$ to $t = 10$ sec is the sum of the distances in each direction.

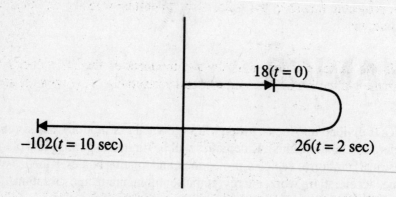

18($t = 0$)

−102($t = 10$ sec)

26($t = 2$ sec)

Figure 1. Movement of a particle

Total distance = 8 + 26 + 102 = 136 feet

Constant Velocity and Constant Acceleration

Constant velocity and constant acceleration are frequent problems in dynamics. At constant velocity (v), integrating

$$ds = vdt$$

$$s = s_o + vt$$

where s_o = position at $t = 0$.

The position of particle (s) at time (t) is equal to its initial position (s_o) plus its velocity (v) multiplied by time.

Constant acceleration (a), integrating

$$dv = adt$$

$$v = v_o + at$$

where v_o = initial velocity at $t = 0$. Substituting s for v and rearrange terms,

$$s = s_o + v_o t + \left(\frac{1}{2}\right)at^2$$

$$v^2 = v_o^2 + 2a(s - s_o)$$

$$(v + v_o)t = 2\,(s - s_o)$$

where v_o = velocity at $t = 0$.

Projectile Motion

Projectiles have motion in two-dimensions simultaneously, the horizontal x-direction and the vertical y-direction. These components are independent of one another. The projectile travels in the x-direction at constant velocity, and thus its position at time, t is

$$x = x_o + v_x t$$

where (x_o) is the initial position, v_x is the x-component of the the velocity v. Movement in the y-direction occurs with constant acceleration, or gravity (g) and

$$v_y = v_{yo} - gt$$

$$y = y_o + v_{yo}t - \left(\frac{1}{2}\right)gt^2$$

$$v_y^2 = v_{yo}^2 - 2g(y - y_o)$$

Projectile problems may be presented with various surface geometries. Some examples are: (a) horizontal surface requiring horizontal and vertical positions, maximum horizontal distance, or maximum vertical height; (b) surface at an angle requiring the horizontal and vertical distances; (c) point of impact fixed, more than one solution may be possible; and (d) object in projectile path, many solutions are possible.

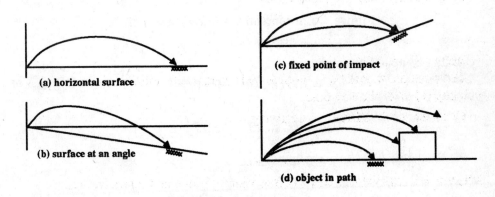

(a) horizontal surface

(b) surface at an angle

(c) fixed point of impact

(d) object in path

Figure 2. Projectile motion

As an example of projectile motion, consider the problem given below. Initial velocity is 100 ft/sec at an upward slope of 30°. The ground slopes downward at an angle of 12° to a horizontal distance of 500 feet. Determine the horizontal distance to projectile impact.

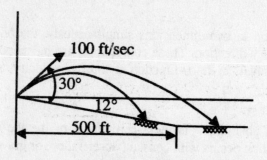

Figure 3. Projectile impact

There are two possible flight patterns: the projectile could hit the horizontal surface, or the projectile could hit the sloping surface. First, assume the projectile hits the horizontal surface.

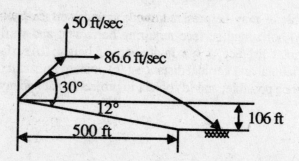

Figure 4. Landing on the horizontal surface

Initial computations show

$$v_x = 100 \cos 30° = 86.6 \text{ ft/sec}$$
$$x = 86.6t$$
$$v_{yo} = 100 \sin 30° = 50.0 \text{ ft/sec}$$

$$y = y_o + v_{yo}t - \left(\frac{1}{2}\right)gt^2$$

At x 500 ft $y = -500 \tan 12° = -106$ ft. Substituting for y and solving for t,

$$-106 = 0 + 50t - \left(\frac{32.2}{2}\right)t^2.$$

$$t = -1.45 \text{ sec}, 4.555 \text{ sec}.$$

Reject $t < 0$, thus at $t = 4.555$ sec results in $x = 394$ ft. Since 394 ft < 500 ft, the assumption used is incorrect, and the projectile hits on the slope.

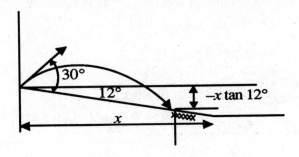

Figure 5. Hitting a slope

Next, assume that the projectile hits the slope. The equations of motion are

$$x = 86.6t$$

$$y = -x \tan 12° \text{ at impact found from geometry, and}$$

$$y = y_o + v_{yo}t - \left(\frac{1}{2}\right)gt^2$$

Substituting, we get

$$y = -x \tan 12° = (-86.6t) \tan 12° = 0 + 50t - \left(\frac{32.2}{2}\right)t^2.$$

Solving $\qquad$ $t = 0$ sec, 4.25 sec

Substituting $\qquad$ $x = 86.6(4.25) = 368$ ft

The correct solution, therefore, is $x = 368$ ft, and the projectile hits the sloping surface.

Rotational Motion

Rotation has similarities to linear motion in that essentially the same equations may be used with the variables having rotational meanings.

Rotational motion is movement around a circle, the change in position (α) is measured in angles and defined as angular velocity (ω).

$$\omega = \frac{d\theta}{dt}$$

Angular acceleration (α) is the rate of change of angular velocity.

$$\alpha = \frac{d\omega}{dt} = \frac{d^2\theta}{dt^2}$$

When the angular velocity is constant (acceleration is 0), then

$$\theta = \theta_o + \omega t$$

where θ_o = position at $t = 0$.

When the angular acceleration is constant, then the equations of motion are

$$\omega = \omega_o + \alpha_o t$$

$$\theta = \theta_o + \omega_o t + \left(\frac{1}{2}\right)\alpha t^2$$

$$\omega^2 = \omega_o^2 + 2\alpha\,(\theta - \theta_o)$$

where θ_o = position at $t = 0$, ω_o = angular velocity at $t = 0$, α = angular acceleration and t = time.

Tangential Velocity and Acceleration

The velocity of a particle rotating a distance (r = constant) about a fixed point is the tangential velocity

$$v_t = r\omega$$

where r = radius and ω = rotational velocity.

Acceleration has two components: tangential and radial. The tangential component of acceleration (a_t) is

$$a_t = r\alpha.$$

The radial component, (a_r), is directed back toward the center of rotation and has a magnitude of

$$a_r = r\omega^2 = \frac{v_t^2}{r}$$

where α = angular acceleration, ω = angular velocity, and v_t = tangential velocity.

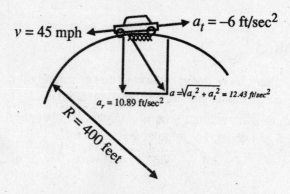

Figure 6. Motion on a curvature

As an example, suppose a car is traveling around a curve. The radius of curvature is 400 ft and the car is traveling at 45 miles/hr while decelerating at 6 ft/sec². Determine the tangential velocity.

$$v_t = 45 \text{ miles/hr } (5280 \text{ ft/mile})/(3{,}600 \text{ sec/hr}) = 66 \text{ ft/sec.}$$

Radial acceleration is

$$a_r = \frac{v_t^2}{r} = \frac{66^2}{400} = 10.89 \text{ ft/sec}^2.$$

Tangential acceleration is

$$a_t = -6.0 \text{ ft/sec}^2$$

Total acceleration is the resultant of both radial and tangential acceleration.

$$a = \sqrt{a_r^2 + a_t^2} = \sqrt{10.89^2 + (-6.0)^2} = 12.43 \text{ ft/sec}^2$$

Motion in Polar Coordinates (*r* = variable)

Equations of motion in polar coordinates with the radius of curvature variable are

$$a_r = \frac{d^2 r}{dt^2} - r\left(\frac{d\theta}{dt}\right)^2 = \frac{d^2 r}{dt^2} - r\omega^2$$

$$a_\theta = r\left(\frac{d^2\theta}{dt^2}\right) + 2\left(\frac{dr}{dt}\right)\left(\frac{d\theta}{dt}\right) = r\alpha + 2\left(\frac{dr}{dt}\right)\omega$$

$$v_r = \frac{dr}{dt}$$

$$v_\theta = r\left(\frac{d\theta}{dt}\right) = r\omega$$

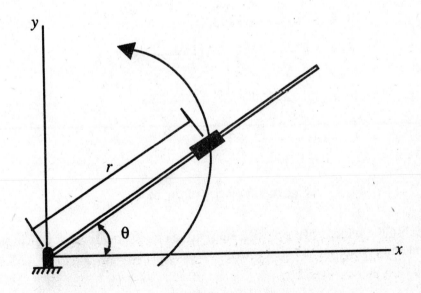

Figure 7. Motion in polar coordinates

Relative and Related Motion

Relative motion is the movement of one object A with respect to a second object B. The motion of observer A = motion of observer B + the difference between A and B.

acceleration $\qquad a_B = a_A + a_{B/A}$

velocity $\qquad v_B = v_A + v_{B/A}$

position $\qquad x_B = x_A + x_{B/A}$

These equations apply to linear, two-dimensional, and rotational motion.

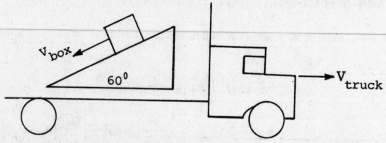

Figure 8. Relative motion inside and outside a truck

Suppose a box starts sliding at a constant velocity down the ramp of a truck inclined at 60° while the truck is still moving. If an observer outside the truck sees the box moving down vertically, and the speed of the box with respect to the moving ramp is 4 m/sec, then what is the speed of the truck?

FIG

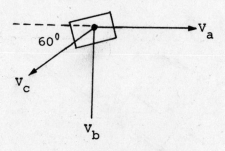

Figure 9. Velocity of the block

The absolute velocity of the block is equal to the velocity of the block with respect to the moving ramp, V_c, summed with the velocity of the truck, V_a.

$$V_b = V_a + V_c$$
$$\rightarrow \Sigma V_x = 0$$
$$V_a - V_c \cos 60° = 0$$
$$\therefore V_a = 4\left(\frac{1}{2}\right) = 2m/sec$$

The speed of the truck is 2 m/sec.

Linear motion of two objects may be shown as follows. The position of object *B* with respect to object *A* is a vector to the right while the position of *A* with respect to *B* is a vector to the left.

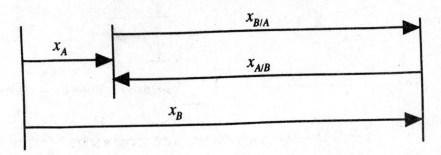

Figure 10. Vector diagram of linear motion

With **two-dimensional motion**, two objects are traveling in different directions. The relative position of object *B* with respect to object *A* is a vector from *A* to *B*. Likewise, the relative position of object *A* with respect to object *B* is a vector from *B* to *A*.

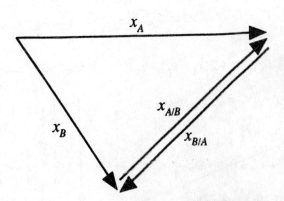

Figure 11. Two-dimensional motion

With **dependent motion**, two objects are connected by a computation. The When object *A* moves, then object *B* must move by a amount of movement is determined by either obse...

related motion of the first set of objects can usually be determined by observation. When object A moves downward then object B moves upward by the same amount. The related motion of the second and third set of objects is determined by finding the rope length from some fixed point. The derivative of the rope length with respect to time gives the velocity and acceleration. The constant is the sum of all irrelevant lengths of rope.

Figure 12. Rope length $= x_A + x_B + \text{constant}$

$$\frac{dL}{dt} = v_A + v_B = 0, \; v_A = -v_B$$

$$\frac{d^2L}{dt^2} = a_A + a_B = 0, \; a_A = -a_B$$

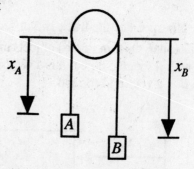

Figure 12. Rope length (a)

Figure 13. $L = x_A + 2x_B + c$

$$\frac{dL}{dt} = v_A + 2v_B = 0, \; v_A = -2v_B$$

$$\frac{d^2L}{dt^2} = a_A + 2a_B = 0, \; a_A = -2a_B$$

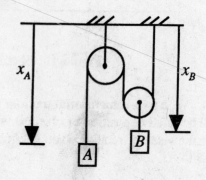

Figure 13. Rope length (b)

(c)   Figure 14. $L_1 = x_A + 2(x_A - x_C) + c_1$

$$\frac{dL_1}{dt} = 3v_A - 2v_C = 0, \; 3v_A = 2v_C$$

$$L_2 = x_B + x_C + c_2$$

$$\frac{dL_2}{dt} = v_B + v_C = 0, \; v_C = -v_B$$

bstituting we get $3v_A = -2v_B$.

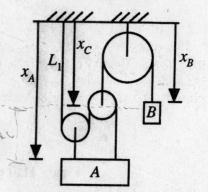

Figure 14. Rope length (c)

Rotational motion of gears and belts is also dependent. Again, the relation of position, velocity, or acceleration depends on the geometry. Two different examples are shown. In the first example the distance of movement along the rim surface (or number of gear teeth) is the same for all gears. In the second example the radius changes, and thus geometry change must be considered.

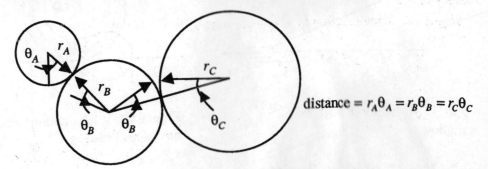

$$\text{distance} = r_A\theta_A = r_B\theta_B = r_C\theta_C$$

Figure 15. Rotational motion

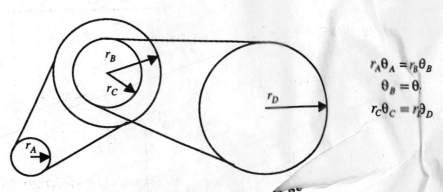

$$r_A\theta_A = r_B\theta_B$$
$$\theta_B = \theta.$$
$$r_C\theta_C = r_D\theta_D$$

Figure 16. Radi

...topic of discussion in the next set of ...e thorough study. The first problem shows

Plane motion of rigid bo~~dy~~ a link moving along definite paths. The second paragraphs. There, we wi~~ll~~ without sliding along a horizontal surface. two objects, *A* and *B* problem shows

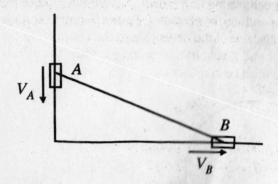

Figure 17. Two linked objects in motion

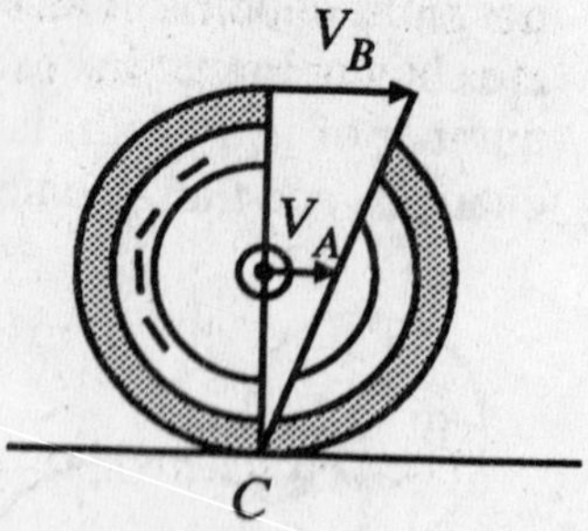

Figure 18. Wheel rolling w. horizontal sliding

Plane Motion of Rigid Bodies

Plane motion is rigid body motion in two-dimensions; rolling wheels are an example. Rigid body motion is the sum of a translational component and the rotation around a fixed axis.

Plane motion	=	Translation	+	Rotation
x_B	=	x_A	+	$x_{B/A}$
v_B	=	v_A	+	$v_{B/A}$
a_B	=	a_A	+	$a_{B/A}$

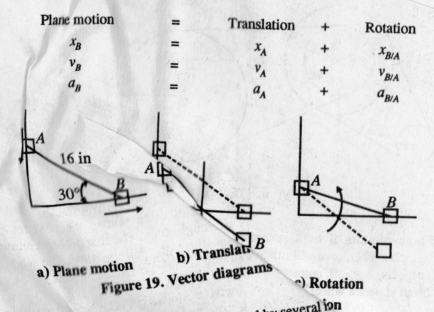

a) Plane motion   b) Translation   c) Rotation

Figure 19. Vector diagrams

Velocity for plane motion may be found by several ion working with the horizontal and vertical portion (x and y com equation. A second approach is to use the vector polygon. A th the instantaneous center of rotation. A fourth approach, not covehod is by vector mathematics (i, j, k unit vectors) to determine both velocity aocity

As an example, suppose that the connecting link shown above moves along the set paths at points A and B. Find the velocity of point B when the velocity of point A is a constant 24 in/sec downward. The velocity of point B is

$$v_B = v_A + v_{B/A}$$
$$v_B = v_A + r_{AB}\,\omega_{AB}$$
$$v_B = 24 + v_{B/A} = 24 + 16\omega_{AB}$$

The vertical and horizontal components of the equation give

$$\text{vertical} \qquad 0 = 24 + v_{B/A}\cos 30°$$
$$\text{horizontal} \quad v_B = v_{B/A}\sin 30°$$

The results are $v_{B/A} = 27.7$ in/sec 30°, $\omega_{AB} = 1.73$ radians/sec counterclockwise and $v_B = 13.86$ in/sec .

The vector polygon for the velocity equation is shown as follows and can be solved by trigonometry. The velocity vectors from either side of the equal sign begin from the same starting point and end at the same stopping point. This concept can be visualized more clearly when there are more vectors as in the acceleration polygon discussed later. This problem can also be solved with the instantaneous center method.

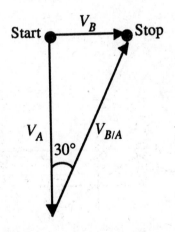

Figure 20. Vector ploygon method

The instantaneous center, point C is found by drawing radial lines from both points A and B. The body is temporarily hinged and rotated about point C. The rotational velocity vector is

$$\omega_{AB} = v_A/r_{CA} = 24/(16\cos 30°) = 1.73 \text{ rad/sec.}$$

The linear velocity of point B is

$$v_B = \omega_{AB}\,r_{CB} = 1.73(16\sin 30°) = 13.86 \text{ in/sec.}$$

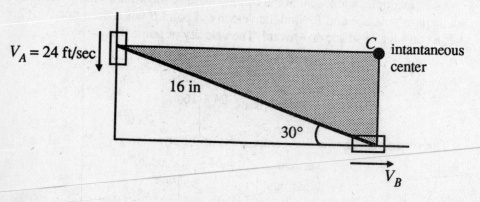

Figure 21. Instantaneous center method

As a second example, consider the large spool rolling to the right without slipping. The point of contact with the ground is the instantaneous center. If the rope is moving to the right with a velocity of 4 ft/sec, then the rotational velocity is

$$\omega = 4 \text{ ft/sec}/(2 \text{ ft}) = 2 \text{ rad/sec.}$$

The velocity of any point is equal to the rotational velocity times the distance from that point to the instantaneous center (point C).

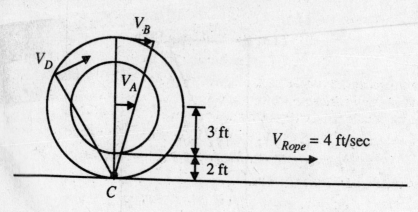

Figure 22. Large spool rolling without slipping

As a third example, consider the mechanism $ABDE$ shown. Some of the necessary calculations are superimposed on the same sketch. The instantaneous center of member BD at point C is also shown. Assume that the member AB is rotating at 6 rad/sec clockwise. Find the rotational velocity of BD and DE, and find the linear velocity of point D.

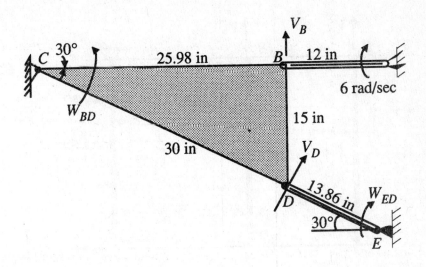

Figure 23. Diagram of mechanism *AB DE*

The instantaneous center for member *AB* is at point *A*, and the instantaneous center for member *ED* is at point *E*.

The linear velocity of each point is

$$v_B = r_{AB}\,\omega_{AB} = 12\ in(6\ rad/sec) = 72\ in/sec$$
$$v_D = r_{ED}\,\omega_{ED} = (13.86\ in)\omega_{ED}\ \sin 30°$$

For a mechanism such as this, we focus our efforts on the middle member *BD*, which has the instantaneous center at point *C*. Since the velocity of point *B* is 72 in/sec, the rotational velocity of *BD* and the linear velocity of *D* are:

$$\omega_{BD} = v_B/r_{CB} = (72\ in/sec)/25.98\ in = 2.77\ rad/sec$$
$$v_D = \omega_{BD}r_{CD} = (2.77\ rad/sec)30\ in = 83.14\ in/sec\ \sin\ 30°$$

The rotational velocity of *ED* is

$$\omega_{ED} = v_D/r_{ED} = (83.14\ in/sec)/(13.86\ in)$$
$$\omega_{ED} = 6.0\ rad/sec\ clockwise$$

The mechanism of example three can also be solved by either the plane motion equation or by the vector polygon. Both of the methods are abbreviated below. To solve the mechanism, we focus our attention on the middle member *BD*.

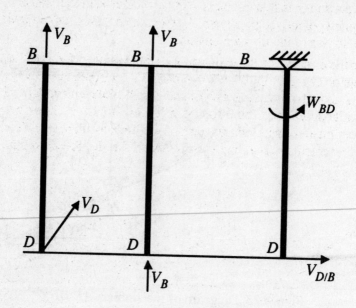

Figure 24. Plane motion

Solving this problem by the plane motion equation for *BD*

$$(13.86 \text{ in}) \; \omega_{ED} \; \sin \; 30° = 72 \text{ in/sec} + (15 \text{ in}) \; \omega_{BD}$$

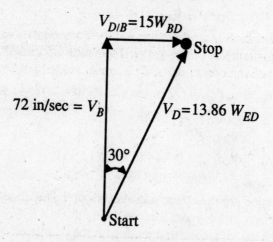

Figure 25. Vector polygon

Acceleration for plane motion can have six components whereas velocity has three components. Each term of the acceleration equation can have both radial and tangential components. To find acceleration, the velocity as discussed previously must first be determined. Acceleration may be found by working with the horizonal and vertical portion (x and y components) of the acceleration equation or by working with the geometry of the vector polygon. Vector mathematics, not covered here, may also be used to determine the acceleration.

Consider the example of the connecting link previously discussed for velocity. Values for velocity were: $v_A = 24$ in/sec (constant), $v_B = 13.86$ in/sec, and $\omega_{AB} = 1.73$ rad/sec. Even though point A has a constant velocity, the member AB will normally have an angular acceleration, and point B will normally have a linear acceleration. The line of action of the acceleration components are known, but the direction is not known. Further calculations are necessary to determine the direction of the acceleration components.

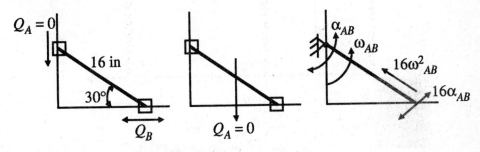

Figure 26. Plane acceleration

$$a_B = a_A + a_{B/A}$$
$$a_B = 0 + r_{AB}\,\omega_{AB}^2\,\sin 30° + r_{AB}\,\alpha_{AB}\,\sin 30°$$
$$a_B = 0 + 16\text{in}(1.73\text{ rad/sec})^2 + 16\text{ in}(\alpha_{AB})$$

Using either the x and y components or the vector polygon gives the solution as: $a_B = 55.43$ in/sec² to the left and $\alpha_{AB} = 1.73$ rad/sec² counterclockwise. The directions of the acceleration components may be found most easily by referring to the vector polygon.

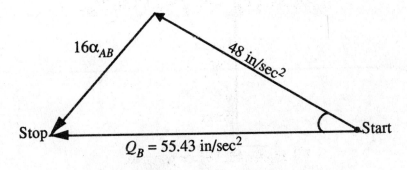

Figure 27. Vector polygon

Assume the angular acceleration of *AB* is 8 rad/sec² clockwise. Find the radial and tangential acceleration of points *B* and *D*.

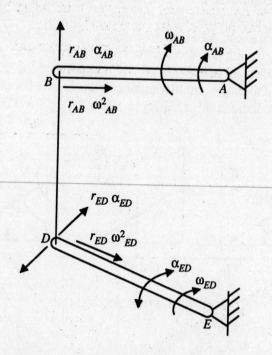

Figure 28. Acceleration of points *B* and *D*

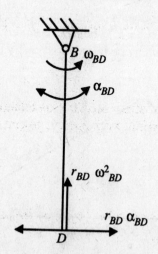

Figure 29. Relative acceleration of member *BD*

The plane motion equation for member *BD* is

$$a_D = a_B + a_{D/B}$$

$$r_{ED}\,\alpha_{ED} + r_{ED}\,\omega_{ED}^2 = r_{AB}\,\alpha_{AB} + r_{AB}\,\omega_{AB}^2 + r_{BD}\,\alpha_{BD} + r_{BD}\,\omega_{BD}^2$$

$$13.86\alpha_{ED} + 13.86(6^2) = 12(8) + 12(6^2) + 15\alpha_{BD} + 15(2.77^2)$$
$$13.86\,\alpha_{ED}\sin 30° + 498.8 = 96 + 432 + 15\alpha_{BD} + 115.2$$

The directions of the acceleration components are found by comparing the directions on the vector polygon in the previous Figure 29.

$$\alpha_{BD} = 17.73 \text{ rad/sec}^2 \text{ counterclockwise}$$
$$\alpha_{ED} = 38.40 \text{ rad/ sec}^2 \text{ clockwise}$$

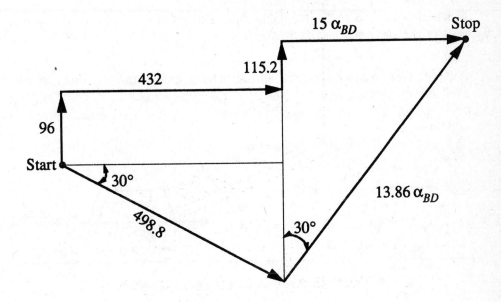

Figure 30. Resultant vector polygon

KINETICS

The study of kinetics covers force = acceleration relationships. Topics studied in kinetics include: force acceleration, work energy, power, impulse momentum, and impact.

Mass and Weight

Weight (w) relates the mass (m) of an object by the acceleration of gravity (g).

$$w = mg$$

On the Earth's surface the average acceleration of gravity is

$$g = 32.2 \text{ ft/sec}^2 = 9.81 \text{ m/sec}^2.$$

Force and Acceleration of Particles

Force-acceleration of particles are related by Newton's second law. When there is an unbalanced force on an object, the object will accelerate in the direction of the unbalanced vector force. The relationship is

$$F = ma$$

where F = unbalanced force in any direction, m = mass and a = acceleration in the direction of the unbalanced force (x, y, or z).

Friction (F_f) is a force that resists motion. Its magnitude depends on the normal force (N) and the coefficient of friction (μ).

$$F_f = \mu N$$

Suppose that an automobile has all four wheels locked and is sliding, friction causes the vehicle to decelerate. The frictional force is equal to the coefficient of friction (μ) times the weight (w) of the automobile. If $\mu = 0.4$, the deceleration is found by

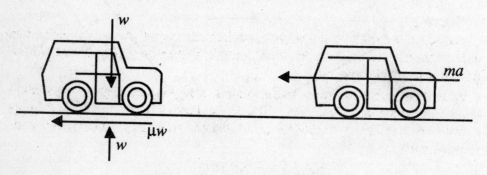

Figure 31. Sliding automobile

$$F = ma$$
$$\mu w = (w/g)a$$
$$a = \mu g$$

If the coefficient of friction is $\mu = 0.4$, then the deceleration is

$$a = 0.4(32.2 \text{ ft/sec}^2) = 12.88 \text{ ft/sec}^2$$

or

$$a = 0.4(9.81 \text{ m/sec}^2) = 3.92 \text{ m/sec}^2$$

The problem of the sliding automobile becomes much more complicated when ideal conditions are not present. The first difficulty is that the assumed coefficient of friction varies considerably depending on the surface condition. Also, the solution becomes more difficult when all wheels are not sliding or when part of the wheels are sliding part of the time. To solve the problem some questionable assumptions must be made.

Two equations are needed when two weights are connected by a rope as shown in the sketch below. The $F = ma$ equation is applied to the free body diagram of each weight.

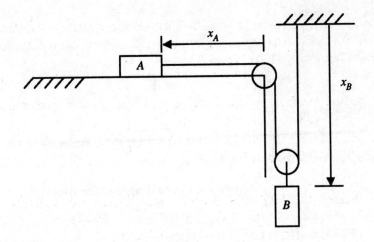

Figure 32. Weights connected by rope

Assume μ = coefficient between the surface and the weight w_A. For block A to move from rest or accelerate, the tensile force (T) in the rope must be larger than the frictional resistance, μw_A.

After it has been established that the weights are moving, the relative acceleration must be determined. The relation of the acceleration of block A to block B is determined by considering dependent motion studied previously. Since the length of the rope remains constant

$$L = x_A + 2x_B + \text{constant}$$

$$\frac{dL}{dt} = v_A + 2v_B = 0$$

$$\frac{d^2L}{dt^2} = a_A + 2a_B = 0, \; a_A = -2a_B$$

Assume the direction of the force on each block is positve and write the force equation for each.

Block A

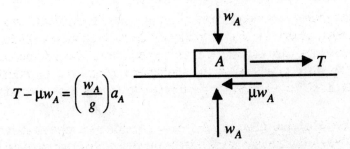

$$T - \mu w_A = \left(\frac{w_A}{g}\right)a_A$$

Figure 33. Force diagram for block A

Block *B*

$$w_B - 2T = \left(\frac{w_B}{g}\right) a_B$$

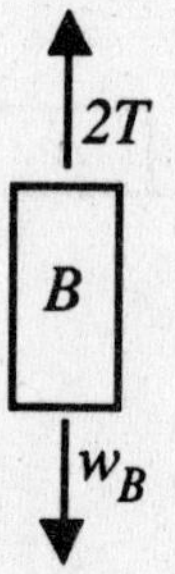

Figure 34. Force diagram for block B

Solving provides

$$a_A = \frac{2(w_B - 2\mu w_A)g}{(4w_A + w_B)}; a_B = \frac{2a_A}{2}; T = \mu w_A + \frac{2w_A(w_B - 2\mu w_A)}{(4w_A + w_B)}$$

Centripetal Force

Rotational motion can be expressed as the sum of two forces: tangential and centripetal (directed inward). **Centripetal force**, F_c, is the force associated with the normal of the path of the object.

$$F_c = ma_n = \frac{mv_t^2}{r}$$

where m = mass of the object, a_n = normal (radial) acceleration, v_t = tangential velocity, and r = radius of the rotation.

In the diagram is a simple pendulum consisting of a small mass, m, attached to the end of a wire length l; the other end of the wire is attached to a fine point A. When the mass is displaced slightly it oscillates with simple harmonic motion along the arc of a circle with center A. (Assume: arc $OB = x$, where x is measured from 0; $\angle OAB = \theta$ and θ is small.)

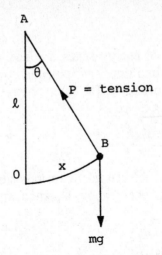

Figure 35. Simple pendulum

To find the acceleration along arc *OB*, no line the force *mg* sin θ, directed along the tangent at *B* is responsible for the acceleration along the arc *OB*. The tension *P* has no component in this direction.

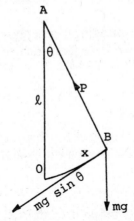

Figure 36. Acceleration component of simple pendulum

Therefore −*mg* sin θ = *ma* where *a* is the acceleration along the arc *OB*.

Note, the reason for the minus sign is that the force *mg* sin θ is directed toward *O* while the displacement *x* is measured along the arc from *O*.

Since θ is small sin θ ≈ θ so that −*mg* θ = *ma*

$$a = -g\theta$$

Since

$$\ell\theta = x$$

$$a = -g\left(\frac{x}{\ell}\right)$$

Work and Energy

The law of conservation of energy states that energy cannot be created nor destroyed. **Energy** is divided into kinetic and potential components. **Kinetic energy** is associated with movement of a body. The kinetic energy (KE) of an object of mass (m) moving at velocity (v) is defined as

$$KE = \left(\frac{1}{2}\right)mv^2$$

Potential energy (PE) of an object is relative to its position in a gravitational field. An object mass (m) raised a height (y) above a particular point has potential energy of

$$PE = mgy$$
$$PE = wy$$

where $w = mg$ is the weight of the object.

The potential energy of a linear spring, with a constant (k) is the work needed to compress the spring a distance (x).

$$PE = \left(\frac{1}{2}\right)kx^2$$

Work is defined as the process of changing the energy of an object. The work (W) done by an object is the integral of the force component (F) over all the increments of distance (ds).

Work of a force is

$$\text{Work} = \int F\,ds$$

where Work = total work added to all objects, F = force component in the direction of movement, and ds = the increment of movement. When the force is constant over a distance (d), then

$$\text{Work} = Fd$$

The work done in moving an object from point 1 to point 2 is denoted by W_{1-2}. During any process, the change in total energy is zero.

$$KE_1 + \text{Work}_{1-2} = KE_2$$

where KE_1 = kinetic energy of the object at position 1, KE_2 = kinetic energy at position 2, and Work$_{1-2}$ = total work involved from position 1 to position 2. To note the ease of the energy solution approach to certain problems, consider the two blocks shown in the sketch. Weight A is 100 lb, weight B is 200 lb, the coefficient of friction is 0.4, and the initial velocity of B is 4 ft/sec downward. Find the velocity of both blocks after block B has moved downward 10 feet.

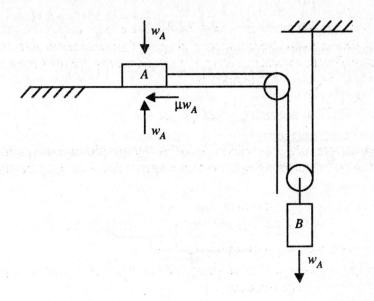

Figure 37. Schematic of two block system

Using dependent motion studied previously, we determine the related movement and velocity of both blocks. When block B moves downward 10 feet, then block A moves horizontal to the right 20 feet. Also, when block B is moving downward 4 ft/sec, then block A is moving to the right at 8 ft/sec. The work of block B moving down is positive, but the work of block A is the negative frictional resistance. Internal work is not considered. Tensile force in the rope is found later by working separately with the free body diagram of either weight. Substituting into the work-energy equation gives

$$KE_1 + \text{Work}_{1-2} = KE_2$$

$$KE_1 = \left\{ \left(\frac{1}{2}\right)\left(\frac{w_A}{g}\right)v_{A1}^2 + \left(\frac{1}{2}\right)\left(\frac{w_B}{g}\right)v_{B1}^2 \right\}$$

$$\text{Work}_{1-2} = \left\{ -\mu w_A d_A + w_B d_B \right\}$$

$$KE_2 = \left\{ \left(\frac{1}{2}\right)\left(\frac{w_A}{g}\right)v_{A2}^2 + \left(\frac{1}{2}\right)\left(\frac{w_B}{g}\right)v_{B2}^2 \right\}$$

Substituting

$$KE_1 = \left\{ \left(\frac{1}{2}\right)\left(\frac{100}{32.2}\right) \times 8^2 + \left(\frac{1}{2}\right)\left(\frac{200}{32.2}\right) \times 4^2 \right\} = \{99.4 + 49.7\} \text{ ft-lb}$$

$$\text{Work}_{1-2} = \{-0.4 \times 100 \times 20 + 200 \times 10\} = \{-800 + 2,000\} \text{ ft-lb}$$

$$KE_2 = \left\{ \left(\frac{1}{2}\right)\left(\frac{100}{32.2}\right)v_{A2}^2 + \left(\frac{1}{2}\right)\left(\frac{200}{32.2}\right)v_{B2}^2 \right\} = \left\{ 1.55v_{A2}^2 + 3.11v_{B2}^2 \right\}$$

Substituting into the energy equation gives

$$\{99.4 + 49.7\} + \{-800 + 2{,}000\} = \{1.55v_A^2 + 3.11v_B^2\}.$$

Using $v_A = 2v_B$ and solving

$$v_A = 24.07 \text{ ft/sec and } v_B = 12.03 \text{ ft/sec.}$$

The tensile force in the rope connecting the two weights may be found by dealing with either weight. When weight A is used, the tensile force is found as follows.

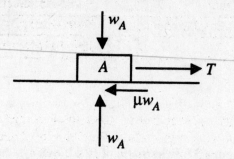

Figure 38. Force diagram for block A

$$KE_1 + \text{Work}_{1-2} = KE_2$$

$$\frac{1}{2}\left(\frac{w_A}{g}\right)v_{A1}^2 + (-\mu w_A + T)d = \frac{1}{2}\left(\frac{w_A}{g}\right)v_{A2}^2$$

Using the values found previously and substituting gives

$$\frac{1}{2}\left(\frac{100}{32.2}\right)8^2 + (-0.4\times100+T)20 = \frac{1}{2}\left(\frac{100}{32.2}\right)24.07^2$$

$$99.3 + (-40 + T)20 = 899.6 \text{ ft-lb}$$

Solving gives $T = 80.0$ lb.

The tensile force in the rope of 80 lb is reasonable when compared to the boundary possibilities. A force of 40 lb (μw_A) is needed to move weight A. If weight B is stationary, the tensile force is 100 lb $\left(\dfrac{w_B}{2}\right)$. When the weights are moving, the actual tensile force must be between these two bounds, and in this case the tensile force must be 80 lb.

$$40 \text{ lb} \leq T \leq 100 \text{ lb}$$

The previous problem involving the two weights could have been solved by using $F = ma$. The accelerations could have been found as discussed previously. After the accelerations were found, the velocity could have been found by using the equations discussed at the beginning of the chapter. Impulse-momentum, discussed later, could also have been used. When velocity, distance, and work are involved, the energy method is usually preferred.

As a generalization, when either force or acceleration is needed, then the force-acceleration method is usually the most convenient solution approach. When velocity and work are related, then work-energy is usually the preferred solution approach. Lastly, when velocity and time are related directly then impulse-momentum is usually the preferred solution approach.

Related	Preferred Solution Approach
Force, Acceleration	Force-Acceleration
Work, Velocity, Displacement	Work-Energy
Time, Velocity	Impulse-Momentum

Potential Energy

The potential energy concept can be used to solve certain problems more conveniently than the work-energy method. Problems involving elastic springs can often be solved more easily by using potential energy. For other engineering applications potential energy can be extended beyond the mechanical energy definition used here. Potential energy can be in the form of voltage, water pressure, water head, or chemical as in explosives.

Potential energy of a compressed linear spring is the work needed to compress the spring and is

$$PE = \left(\frac{1}{2}\right)kx^2$$

where PE = potential energy, k = linear spring constant, and x = amount of spring compression. Potential of a weight at a certain distance above a particular point is

$$PE = wy$$

where w = weight and y = vertical distance.

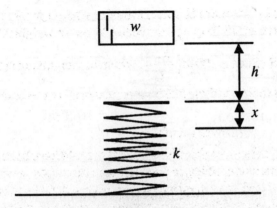

Figure 39. Spring compression

Suppose a weight of w is dropped a height of h on a spring having a spring constant of k. Determine the amount of spring compression (x). The potential energy equation can be written as

$$KE_1 + PE_1 = KE_2 + PE_2$$

where KE = kinetic energy and PE = potential energy. Continuing with the solution gives

$$0 + w(h + x) = 0 + \frac{1}{2}\left(kx^2\right)$$

where the KE = initial and final kinetic energies are 0, and x = compression of the spring. Solving gives

$$x = \left(\frac{w}{k}\right) \pm \sqrt{\left(\frac{w}{k}\right)^2 + \left(\frac{2wh}{k}\right)}$$

where $\dfrac{w}{k}$ = the static deflection.

If the weight is attached permanently to the spring so it will not "hop off," then the weight will vibrate in harmonic motion, as discussed later. Vibration will be around static equilibrium $\left(x = \dfrac{w}{k}\right)$ with an amplitude of $\sqrt{\left(\dfrac{w}{k}\right)^2 + \left(\dfrac{2wh}{k}\right)}$.

As a second example, determine the additional spring compression of the problem below. A weight is dropped on a linear spring as shown. Assume the necessary data as follows: 5 lb weight, dropped 3.0 ft, initial velocity of the weight is 10 ft/sec downward, spring has an initial compression of 3.0 in, and the spring constant is 10 lb/in.

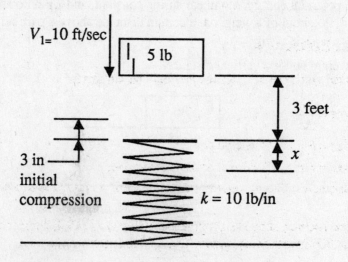

Figure 40. Additional spring compression

Converting all values into the same units

$$k = 10 \text{ lb/in } (12 \text{ in/ft}) = 120 \text{ lb/ft}$$

$$x_1 = \frac{3}{12} \text{ft}$$

$$x_2 = x_1 + \frac{3}{12} \text{ft}$$

$$v_1 = 10 \text{ft / sec}$$

$$v_2 = 0 \text{ft / sec}$$

The energy equation is written as

$$KE_1 + PE_1 = KE_2 + PE_2$$

$$\left(\frac{1}{2}\right)\left(\frac{w}{g}\right)v_1^2 + w(h+x) + \left(\frac{1}{2}\right)kx_1^2 = \left(\frac{1}{2}\right)\left(\frac{w}{g}\right)v_2^2 + \left(\frac{1}{2}\right)kx_2^2$$

$$\left(\frac{1}{2}\right)\left(\frac{5}{32.2}\right)10^2 + 5(3+x) + \left(\frac{1}{2}\right)(10)(12)\left(\frac{3}{12}\right)^2 =$$

$$0 + \left(\frac{1}{2}\right)(10)(12)\left(x+\frac{3}{12}\right)^2$$

$$7.76 + 15 + 5x = 60x^2 + 30x$$

Solving gives the additional compression of the spring.

$$x = -0.859 \text{ ft } (-10.3 \text{ in}), 0.442 \text{ ft } (5.30 \text{ in})$$

Since the compression of the spring must be a positive number, $x = 5.30$ in is the solution.

Power and Efficiency

Power is the work per unit of time and may be written as

$$\text{Power} = \frac{W}{\Delta t}$$

Power = Fv Linear power

Power = $T\omega$ Torsional or rotational power

where F = force, v = linear velocity, T = torsional moment, and ω = rotational velocity.

For torsional power, suppose a drill using 1,500 watts (1500 Nm/sec) is turning the drill bit at 300 revolutions/minute. The torsional moment is

$$T = \frac{\text{Power}}{(2\pi f)} = \frac{1,500}{\left(\frac{2\pi 300}{60}\right)} = 47.7 \text{ Nm}$$

where f = frequency = 300 rev/min/60 sec/min = 5 rev/sec.

Also, suppose a 2 horsepower (one horsepower = 6,600 in-lb/sec) drill is turning the drill bit at 300 rev/min. The torsional moment is

$$T = \frac{\text{Power}}{(2\pi f)} = 2 \times 6,600 \text{ in-lb/sec} / \left(\frac{2\pi 300}{60 \sec} \right)$$

$$T = 420 \text{ in-lb}$$

The **efficiency** of a process is the amount of output power or work obtained from the system compared to the amount of input or energy. The efficiency of an ideal system is unity.

$$\text{Efficiency} = \frac{\text{output}}{\text{input}}$$

Often problems involving power require the conversion of work or power from one system of units to another such as mechanical, thermal, or electrical.

Suppose that water is dropped into a mine 5,000 feet deep (h) at 20 gallons per minute (Q) for the purposes of cooling the air. With water weighing 8.33 lb/gal (γ) the mechanical energy is

$$\text{Power} = Q\,\gamma h$$
$$\text{Power} = 20 \text{ gal/min} \times 8.33 \text{ lb/gal} \times 5,000 \text{ feet}$$
$$\text{Power} = 833,000 \text{ ft-lb/min}$$

This amount of mechanical power would produce 833,000/33,000 = 25.24 horsepower, 25.24 × 0.746 = 18.83 kilowatts, or 25.24 × 42.44 = 1,071 Btu/min. When a generator at an assumed efficiency of 90 percent is used to retrieve the electrical power, we would get 0.90 × 18.83 = 17 kilowatts. The energy that is not retrieved first as electrical energy is lost as heat. The energy required to pump the water back out of the mine would again increase the air temperature.

An automobile having a mass of 1,200 kilograms and traveling at 15 m/sec has

$$\text{Energy} = \left(\frac{1}{2} \right) mv^2 = \left(\frac{1}{2} \right) 1,200 \times 15^2 = 135,000 \text{ N-m}$$

If this conversion were really 100 percent efficient, then a 100 watt (100 Nm/sec) light bulb would burn for 135,000/100/60 = 22.5 min.

Impulse and Momentum

The **momentum** of an object is the product of its mass (m) and its velocity (v). Momentum is conserved, thus the sum of momentum of a system is always equal to the same constant total. The total momentum of the objects before a collision is equal to the sum of the momentums of the objects after a collision.

$$m_1 v_1 + m_2 v_2 = m_1' v_1' + m_2' v_2'$$

Impulse is the change in momentum or the integral of the force (F) acting on the object with respect to time (dt).

$$\text{Impulse} = \int F dt \text{ (linear impulse)}$$

$$\text{Impulse} = \int T dt \text{ (angular impulse)}$$

Integrating $F = ma = \dfrac{mdv}{dt}$ as follows

$$\int F dt = \int m dv$$

results in the impulse-momentum equation

$$mv_1 = \int F dt = mv_2$$

where mv_1 = momentum at position one, mv_2 = momentum at position two, and $\int F dt$ = impulse between position one and position two.

Determine the time required to stop the sliding automobile considered previously.

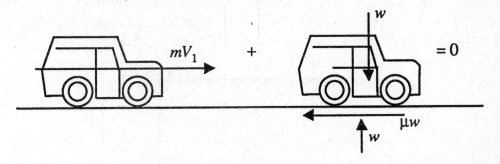

Figure 41. Force diagram on sliding automobile

$$mv_1 = \int F dt = mv_2$$

$$\left(\frac{w}{g}\right)v + \mu w t = 0$$

Solving gives $t = \dfrac{v}{(\mu g)}$

The beauty of the impulse-momentum approach is more evident when the force is not constant. The impulse is the area under the F/t curve such as shown below. Solving such problems by either the force-acceleration method or the energy method would require more effort.

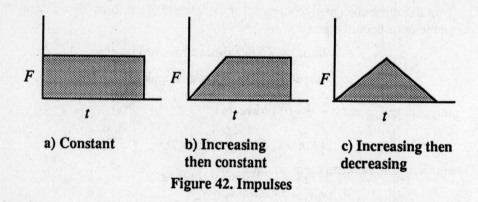

a) Constant   b) Increasing   c) Increasing then
                   then constant   decreasing

Figure 42. Impulses

Both impulse and momentum are vector quantities. For this reason, two equations are required for problems involving two weights as considered previously. Briefly consider the previous problem as follows.

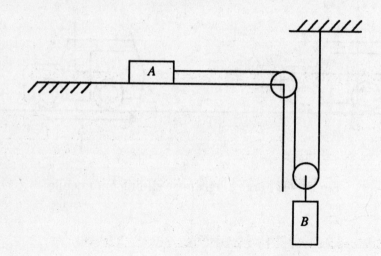

Figure 43. Schematic of block example

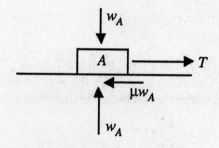

Figure 44. Forces on block A

Weight A $m_A v_{A1} + (-\mu w_A + T)t = m_A v_{A2}$

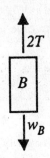

Figure 45. Forces on block B

Weight B $m_B v_{B1} + (w_B - 2T)t = m_B v_{B2}$

Using $v_A = 2v_B$ and the two momentum equations, the velocity at time two can be determined.

Impact

When two objects A and B collide, momentum is conserved. That is

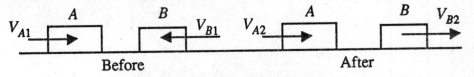

Before   After

Figure 46. Impact of two objects

$$m_A v_{A1} + m_B v_{B1} = m_A v_{A2} + m_B v_{B2}$$

The velocities are related by the **coefficient of restitution**

$$e = \frac{v_{B2} - v_{A2}}{v_{A1} - v_{B1}}$$

The coefficient of restitution varies from $0 \leq e \leq 1.0$. Two unknown velocities may be solved by using the two equations.

Special cases of impact are perfectly elastic impact ($e = 1.0$), perfectly plastic impact ($e = 0$), impact with one mass infinite, and oblique impact.

With **perfectly elastic impact** ($e = 1$), the kinetic energy

$$KE = \left(\frac{1}{2}\right)mv^2$$

is conserved.

With **perfectly plastic impact** ($e = 0$) the two objects stick together after impact and have the same final velocity. The impact equation is

$$m_A v_{A1} + m_B v_{B1} = (m_A + m_B)v'.$$

Two clay objects would normally involve plastic impact.

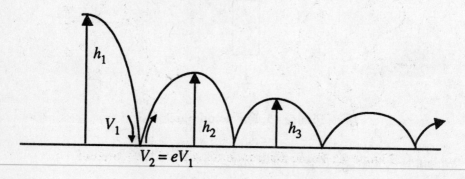

Figure 47. Ball striking the floor

When **one mass is infinite** such as a ball striking a floor or wall, the momentum equation is unnecessary and the velocity is related by

$$v_2 = e v_1.$$

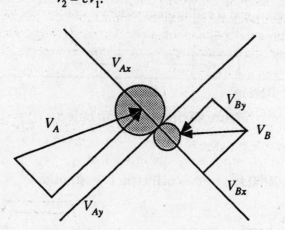

Figure 48. Oblique impact

With **oblique impact**, the momentum along the line of impact is determined in the same way as linear impact. The momentum and velocity transverse to the line of impact is not affected when the friction between the surfaces is neglected.

Two-Dimensional Rigid Body Motion

Kinetics of two-dimensional rigid body motion involves two force-acceleration equations (x and y components) and one moment equation about some point.

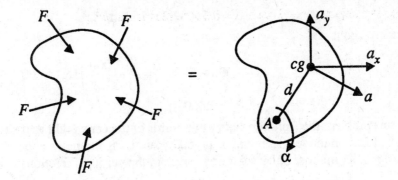

Figure 49. Force diagram during rigid body motion

$$F_x = ma_x$$
$$F_y = ma_y$$
$$M_A = I_A\alpha = I_{cg}\alpha + m(a)d$$

where F_x and F_y = force components, m = mass, a_x and a_y = linear acceleration components of the center of gravity (cg), M_A = moment of forces about point A, I_A = mass inertia about point A, I_{cg} = mass inertia about the center of gravity, a = component of linear acceleration normal to a line connecting points A and cg, α = angular acceleration, and d = distance from point A to the center of gravity.

Rigid body motion can be divided into three cases: translation only, rotation only, and general plane motion.

Determine the forces at the front and rear tires of the automobile as it slides to a stop. Assume a 3,000 lb car and a coefficient of friction of 0.5. Wheel base and center of gravity are shown in the sketch.

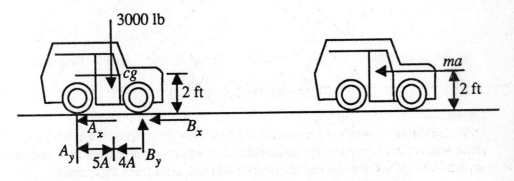

Figure 50. Forces on sliding automobile

Linear acceleration $\qquad F = ma,$

$$\mu w = \left(\frac{w}{g}\right)a,$$

$$a = \mu g$$

$$a = 0.5 \times 32.2 = 16.1 \text{ ft/sec}^2$$

Moment about rear $\qquad M_A = mad$

$$3{,}000 \times 5 - 9B_y = -\left(\frac{3{,}000}{32.2}\right)16.1 \times 2.0$$

$$B_y = 2{,}000 \text{ lb}$$

For this problem the front tires carry two-thirds of the car weight as it slides to a stop. The force on the rear tires can be found by summing moments about the front tires in a similar manner or by summing forces in the vertical direction.

Rotation

Problems having rotation constrained about a fixed point require the mass moment of inertia. One equation is necessary to find the angular acceleration.

$$M_A = I_A\,\alpha = I_{cg}\,\alpha + m(a)d$$

where point A is often the center of gravity. The linear acceleration components are required to find the forces at the point of rotation.

Rotation about a fixed point could involve such objects as a spinning disk, a beam hinged at one end, a plate hinged about one corner, or a wheel rolling on a large axle.

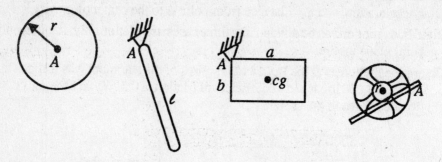

a) Spinning disk   b) Hinged beam   c) Hinged plate   d) Rolling wheel

Figure 51. Rotation of objects

Inertia

Newton's First Law of Motion states that if an object is at rest, its inertia will act upon it to keep it at rest. Likewise, an object in motion will remain in motion. The mass inertia of an object about its center of gravity (*cg*) is dependent upon its shape and size.

Mass inertia of the objects sketched previously are:

Spinning disk $\qquad I_{cg} = \left(\frac{1}{2}\right)mr^2$

Beam about cg $\qquad I_{cg} = \left(\dfrac{1}{12}\right)ml^2$

Plate about cg $\qquad I_{cg} = \left(\dfrac{1}{12}\right)m(a^2 + b^2)$

Wheel about cg $\qquad I_{cg}' = mk^2$

where m = mass and $r, l, a,$ and b = object dimensions, and k = radius of gyration of

the wheel. The radius of gyration of any object is $k = \sqrt{\dfrac{I}{m}}$.

The transfer axis theorem must be used to find the mass inertia about a point some distance d from the center of gravity. The transfer axis theorem states

$$I_A = I_{cg} + md^2$$

where d = distance from the center of gravity to point A.

Thus the mass inertia for the beam, plate, and wheel results in

Beam $\qquad I_A = \left(\dfrac{1}{12}\right)ml^2 + m\left(\dfrac{1}{2}\right)^2 = \left(\dfrac{1}{3}\right)ml^2$

Plate $\qquad I_A = \left(\dfrac{1}{12}\right)m(a^2 + b^2) + m\left[\left(\dfrac{a}{2}\right)^2 + \left(\dfrac{b}{2}\right)^2\right] = \left(\dfrac{1}{3}\right)m(a^2 + b^2)$

Wheel $\qquad I_A = mk^2 + mr^2$

The beams of the frame weigh 150 lb/ft. Find the mass moment of inertia about the center of gravity and point A.

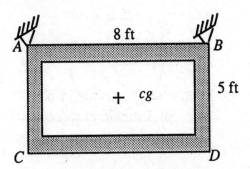

Figure 52. Hanging frame

The mass inertia of each beam about their respective centers of gravity is

$$I_{AB} = I_{CD} = \left(\dfrac{1}{12}\right)ml^2 = \left(\dfrac{1}{12}\right)\left(8 \times \dfrac{150}{32.2}\right)8^2 = 199 \text{ ft-lb-sec}^2$$

$$I_{AC} = I_{BD} = \left(\dfrac{1}{12}\right)\left(5 \times \dfrac{150}{32.2}\right)5^2 = 48.5 \text{ ft-lb-sec}^2 \text{ or } 48.5 \text{ slug-ft}^2$$

Combining the results and using the transfer axis theorem gives

$$I_{cg} = 2 \times 199 + 2 \times 48.5 + 2\left(8 \times \frac{150}{32.2}\right)2.5^2 + 2\left(5 \times \frac{150}{32.2}\right)4^2$$

$I_{cg} = 1{,}706$ ft-1b-sec^2

Transferring the mass inertia to point A results in

$I_A = 1{,}706 + (26 \times 150/32.2)(2.5^2 + 4^2) = 4{,}400$ ft-1b-sec^2

General Plane Motion

Plane motion could have both linear and angular acceleration components. The equations stated previously are

$$F_x = ma_x$$
$$F_y = ma_y$$
$$M_A = I_A \alpha = I_{cg} \alpha + m(a)d$$

Suppose a spherical ball is released from rest. Determine the angular acceleration.

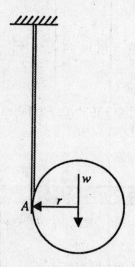

Figure 53. Rotation of spherical ball

The angular acceleration is

$$M_A = I_{cg} \alpha + m(a)d$$

$$wr = \left(\frac{1}{5}\right)\left(\frac{w}{g}\right)r^2\alpha + \left(\frac{w}{g}\right)(r\alpha)r = \left(\frac{6}{5}\right)\left(\frac{w}{g}\right)r^2\alpha$$

$$\alpha = \left(\frac{5}{6}\right)\left(\frac{g}{r}\right)$$

The tensile force is

$$F_y = ma_y = mr\alpha$$

$$w - T = \left(\frac{w}{g}\right)\left(\frac{5}{6}\right)\left(\frac{g}{r}\right)r = \left(\frac{5}{6}\right)w$$

$$T = \frac{w}{6}$$

As a second example, suppose that both a pipe and cylinder are rolling without sliding on a sloping surface. Assume that the mass and outside radius of both objects are the same. Determine which will roll the fastest.

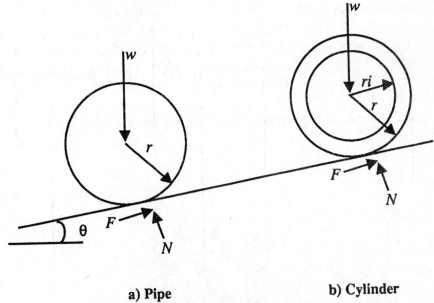

a) Pipe   b) Cylinder

Figure 54. Rolling objects and associated forces

Since the cylinder and pipe are rolling without sliding, moments are summed at the contact point for each object. The cylinder has an angular acceleration of

$$M_A = I_{cg}\,\alpha + m(a)d$$

$$wr\sin\theta = \left[\left(\frac{1}{2}\right)\left(\frac{w}{g}\right)r^2\right]\alpha + \left(\frac{w}{g}\right)(r\alpha)r = \left(\frac{3}{2}\right)\left(\frac{w}{g}\right)r^2\alpha$$

$$\alpha = \left(\frac{2g\sin\theta}{3r}\right)$$

The pipe has an angular acceleration of

$$wr\sin\theta = \left[\left(\frac{1}{2}\right)\left(\frac{w}{g}\right)\left(r^2 + r_i^2\right)\right]\alpha + \left(\frac{w}{g}\right)(r\alpha)r$$

$$wr\sin\theta = \left[\left(\frac{1}{2}\right)\left(\frac{w}{g}\right)\left(3r^2 + r_i^2\right)\right]\alpha$$

$$\alpha = \frac{(2gr\sin\theta)}{\left(3r^2 + r_i^2\right)}$$

Comparing the angular acceleration of the cylinder and pipe, it is noted that the cylinder will accelerate and roll faster than the pipe. The mass inertia of the pipe is larger than a cylinder having the same outside radius and mass.

In the previous problem, both cylinder and pipe rolled without sliding. The question is, at what angle will a round object roll and slide simultaneously? A rectangular block will begin to slide on a sloping surface when $\tan\theta = \mu$. Therefore, when the coefficient of friction is 0.60, a rectangular object will begin to slide at an angle at or exceeding 31°.

Determine the slope angle that a sphere will roll and slide simultaneously.

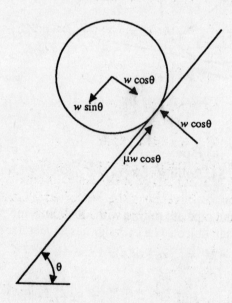

Figure 55. Rolling sphere

When the sphere is sliding, the angular acceleration is

$$M_{cg} = I_{cg}\alpha$$

$$\mu wr\cos\theta = \left(\frac{2}{5}\right)\left(\frac{w}{g}\right)r^2\alpha$$

$$\alpha = \left(\frac{5}{2}\right)\mu g\cos\theta\,/\,r$$

When the sphere is just beginning to slide $a_{cg} = r\alpha$, and the linear acceleration is

$$F = ma$$

$$wr\sin\theta - w\cos\theta = \left(\frac{w}{g}\right)r\left[\left(\frac{5}{2}\right)\mu g\cos\theta\,/\,r\right]$$

Solving

$$\tan\theta = \frac{7\mu}{2}$$

When the coefficient of friction is 0.6, a block will begin to slide at a slope of 31° and a sphere will begin to roll and slide at an angle of 65°. In general, the cylinder, pipe, or sphere will begin to roll and slide simultaneously at an angle roughly twice the angle that a block will begin to slide. If no friction exists, the sphere will slide down the slope without rolling; this is known as **constrained motion**.

Energy Methods for Rigid Body Motion

Rigid body problems can be solved by examining the energy relationships of the system.

$$KE_1 + \text{Work}_{1-2} = KE_2$$

where

$$KE = \left(\frac{1}{2}\right)m(v_{cg})^2 + \left(\frac{1}{2}\right)I_{cg}\omega^2 = \left(\frac{1}{2}\right)I_A\omega^2$$

and

$$W_{1-2} = \int F\,ds + \int M\,d\theta$$

Variables are KE = kinetic energy, v_{cg} = linear velocity of the center of gravity, ω = rotational velocity, I_{cg} = mass inertia at the center of gravity, I_A = mass inertia at point A, F = component of force along the line of movement, ds = increment of movement, M = moment, and $d\theta$ = increment of rotation.

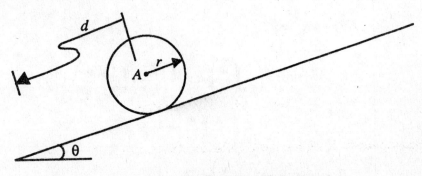

Figure 56. Rolling cylinder

Determine the rotational velocity of a cylinder after it has rolled a distance of d down a slope. Assume that the initial velocity = 0, and the cylinder rolls without sliding.

$$KE_1 + W_{1-2} = KE_2$$

$$0 + wd\sin\theta = \left(\frac{1}{2}\right)m(v_{cg})^2 + \left(\frac{1}{2}\right)I_{cg}\omega^2$$

$$wd\sin\theta = \left(\frac{1}{2}\right)\left(\frac{w}{g}\right)(r\omega)^2 + \left(\frac{1}{2}\right)\left[\left(\frac{1}{2}\right)\left(\frac{w}{g}\right)r^2\right]\omega^2$$

$$wd\sin\theta = \left(\frac{3}{4}\right)\left(\frac{w}{g}\right)r^2\omega^2$$

Solving $$\omega = \sqrt{\frac{(4gd\sin\theta)}{(3r^2)}}$$

For a second example, determine the rotational velocity of cylinder A when it has traveled a distance d. Assume no sliding, weight and radius of both cylinders are the same, and the initial velocity = 0. The relationship of the rotational velocities is $\omega_B = 2\omega_A$.

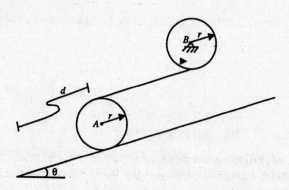

Figure 57. Two rolling cylinders

$$KE_1 + W_{1-2} = KE_2$$

$$0 + wd\sin\theta = \left(\frac{1}{2}\right)\left(\frac{w}{g}\right)(r\omega_A)^2 + \left(\frac{1}{2}\right)\left[\left(\frac{1}{2}\right)\left(\frac{w}{g}\right)r^2\right]\omega_A^2$$

$$+ \left(\frac{1}{2}\right)\left[\left(\frac{1}{2}\right)\left(\frac{w}{g}\right)r^2\right]\omega_B^2$$

$$wd\sin\theta = \left(\frac{7}{4}\right)\left(\frac{w}{g}\right)r^2\omega_A^2$$

Solving $\qquad \omega_A = \sqrt{\dfrac{(4gd\sin\theta)}{(7r^2)}}$

Mechanical Vibration

Simple harmonic motion results from a single mass vibrating about a position of equilibrium. The focus of this discussion is on problems that have a single degree of freedom, no damping, no forcing function, and a linear spring.

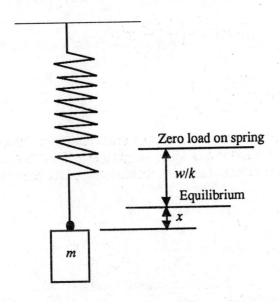

Figure 58. Simple harmonic motion

Suppose a mass is displaced a distance x from the position of equilibrium in the vertical plane. The unbalanced force and the force-acceleration relation give

$$F = ma$$

$$-kx = \frac{md^2x}{dt^2}$$

where k = spring constant, m = mass, x = displacement, and $\dfrac{d^2x}{dt^2}$ = acceleration of the mass. The differential equation is

$$\frac{md^2x}{dt^2} + kx = 0.$$

The solution of the differential equation gives

Position $$x = x_m \sin\left[\sqrt{\frac{k}{m}}\,t + \theta\right]$$

Velocity $$v = \frac{dx}{dt} = x_m \sqrt{\frac{k}{m}}\,\cos\left[\sqrt{\frac{k}{m}}\,t + \theta\right]$$

Acceleration $$a = \frac{d^2x}{dt^2} = -x_m\left(\frac{k}{m}\right)\sin\left[\sqrt{\frac{k}{m}}\,t + \theta\right]$$

Variables are x_m = maximum amplitude, θ = phase angle, t = time, and $\sqrt{\frac{k}{m}}$ = angular velocity. Other quantities are:

Maximum values $$x = x_m,\, v = x_m\sqrt{\frac{k}{m}},\, a = -x_m\left(\frac{k}{m}\right)$$

Period $$T = \frac{2\pi}{\sqrt{\frac{k}{m}}}$$

Frequency $$f = \frac{1}{T} = \frac{\sqrt{\frac{k}{m}}}{2\pi}$$

Consider two examples. Maximum displacement = 0.050 m, k_1 = 5,000 N/m, k_2 = 4,000 N/m, and m = 60 kg.

For springs in parallel, the spring force is

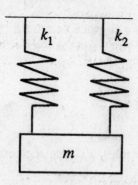

Figure 59. Springs in parallel

Assume the mass is moved a distance of one. For springs in parallel, the spring force is

$$k = k_1 + k_2 = 5,000 + 4,000 = 9,000 \text{ N/m}$$

Maximum values are $x = 0.05$ m, $v = 0.61$ m/sec, $a = -7.50$ m/sec^2. Maximum velocity occurs at zero displacement, and maximum acceleration occurs at maximum displacement. Period is 0.51 sec per cycle, and frequency is 1.95 cycles per sec.

Springs in series

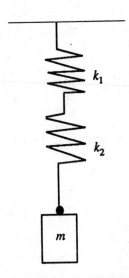

Figure 60. Springs in series

Assume a load = 1 is applied to the springs. Total displacement resulting from both springs is

$$\text{Displacement} = \frac{1}{k} = \frac{1}{k_1} + \frac{1}{k_2}$$

The spring constant of two springs in series is $\dfrac{1}{k} = \dfrac{1}{4,000} + \dfrac{1}{5,000}$ and $k = 2,222$

N/m. Maximum values are $x_m = 0.05$ m, $v = 0.30$ m/sec, $a = -1.85$ m/sec^2. Period is 1.03 sec per cycle and frequency is 0.97 cycles per second.

REVIEW PROBLEMS

PROBLEM 1

An electric motor is used to lift a 1,600 lb block at a rate of 10 fpm. How much electric power (watts) must be supplied to the motor, if the motor lifting the block is 60 percent efficient? The weight W is 1,600 lb.

SOLUTION: The power requirement of the motor is the power exerted in lifting the block. The power exerted by the motor is the same as the power exerted by the force F in lifting the block. This is

$$P = Fv.$$

Since the block is moving at a constant velocity,

$$F = 1600 \text{ lb}$$

so that the power exerted by the motor is

$$P = Fv = (1600)\left(\frac{10}{60}\right) = 267 \text{ ft lb/sec}$$

or, in terms of horsepower,

$$P = \frac{267}{550} = 0.485 \text{ hp.}$$

This is the power required to lift the block at the velocity required, and this is the power that the motor has to produce. However, since the motor is only 60 percent efficient, the power input to the motor must be greater. Thus,

$$P_{in} = \frac{P_{out}}{\varepsilon}$$

or

$$P_{in} = \frac{0.485}{0.60} = 0.81 \text{ hp.}$$

and the power input in terms of the electric power supplied to the motor is

$$P_{in} = 0.81 \times 746 = 605 \text{ watts.}$$

PROBLEM 2

Two similar cars, A and B, are connected rigidly together and have a combined mass of 4 kg. Car C has a mass of 1 kg initially, A and B have a speed of 5 m/sec and C is at rest as shown in the figure.

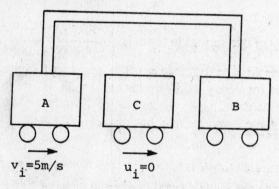

Figure 61. Cars rigidly connected

Assuming a perfectly inelastic collision between A and C, what is the final speed of the system?

SOLUTION:

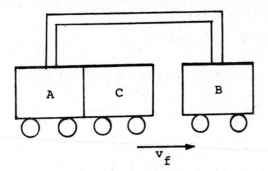

Figure 62. Inelastic collision

A perfectly inelastic collision means that the two colliding bodies stick together and move with the same velocity after the collision, as shown in the figure. From the Principle of Conservation of Linear Momentum, we may write

Total Momentum Before Collision = Total Momentum After Collision

Thus,

$$4(v_i) + 0 = (4 + 1)(v_f).$$

Solving for the final velocity, v_f:

$$v_f = \frac{4}{5}v_i = \frac{4}{5}(5 \text{ m / sec}) = 4 \text{ m / sec}.$$

PROBLEM 3

Two masses, $m_1 = a$ kg and $m_2 = b$ kg have velocities $u_1 = b$ m/sec in the $+x$ direction and $u_2 = a$ m/sec in the $+y$ direction. They collide and stick together. What is the measure of the angle to the horizontal?

SOLUTION: The total x and y components of linear momentum must be conserved after the collision. The mass of the body resulting after the collision is

$$m = m_1 + m_2$$

and the velocity v is inclined at angle to the x-axis. We know that the total momentum vector is unchanged, and we can write down the x and y components of momentum.

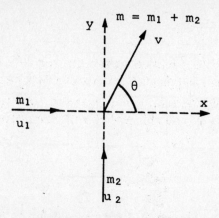

Figure 63. Components of momentum

	INITIAL MOMENTUM	FINAL MOMENTUM
x component	$m_1 u_1$	$(m_1 + m_2)v \cos \theta$
y component	$m_2 u_2$	$(m_1 + m_2)v \sin \theta$

Thus

$$m_1 u_1 = (m_1 + m_2)v \cos \theta$$
$$m_2 u_2 = (m_1 + m_2)v \sin \theta$$

$$\therefore \tan \theta = \frac{m_2 u_2}{m_1 u_1}$$

$$= \frac{b \times a}{a \times b} = 1$$

Hence $\theta = 45°$.

PROBLEM 4

A boat travels directly upstream in a river, moving with constant but unknown speed v with repsect to the water. At the start of this trip upstream, a bottle is dropped over the side. After 15 minutes the boat turns around and heads downstream. It catches up with the bottle when the bottle has drifted one mile downstream from the point at which it was dropped into the water. What is the current in the stream?

SOLUTION: Consider a coordinate system at rest with respect to the water. Then the water is at rest and it is the banks which appear to move upstream. The bottle is at rest with respect to the water. From the point of view of this coordinate system, it is just as though the boat were moving at speed v in a perfectly still pond. We can see that the return trip downstream must also take 15 minutes. Once it is known that the round trip takes half an hour, we can see that the current in the river must be 2 miles per hour since the bottle moves one miles in a half hour.

PROBLEM 5

If the coefficient of sliding friction for steel on ice is 0.05, what force is required to keep a man weighing 150 pounds moving at constant speed along the ice?

SOLUTION: To keep the man moving at constant velocity, we must oppose the force of friction tending to retard his motion with an equal but opposite force (see diagram).

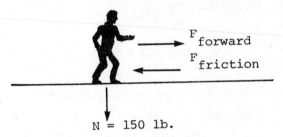

$$N = 150 \text{ lb.}$$

Figure 64. Constant velocity on ice

The force of friction is given by:
$$F = \mu_{\text{kinetic}} N$$
By Newton's Third Law
$$F_{\text{forward}} = F_{\text{friction}}$$
Therefore
$$F_{\text{forward}} = \mu_{\text{kinetic}} N$$
$$F_{\text{forward}} = (.05)(150 \text{ lb}) = 7.5 \text{ lb.}$$

PROBLEM 6

In Figure 65, a man applies a force F of 880N to a belt, wrapped four times around a pipe PQ, to draw water from a well. Given that the belt is just about to slip, calculate a) the coefficient of static friction between the belt and the pipe, b) F, if the belt is wrapped twice around the pipe.

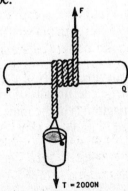

Figure 65. Belt around a pipe

SOLUTION: The equation relating the tensions at the beginning and end of the belt wrapped around the pipe is given as

$$\ln\frac{T}{F} = \mu_s\theta \tag{1}$$

where μ_s is the coefficient of static friction and θ is the angle subtended by the belt when in contact with the pipe. θ is given in radians and is multiplied by the number of turns that the belt is wrapped on the pipe.

Thus

$$\theta = 4(2\pi \text{ rad}) = 25.1 \text{ rad.}$$

From equation (1)

$$\mu_s = \ln\left(\frac{T}{F}\right)\left(\frac{1}{25.1 \text{ rad}}\right)$$

$$= \left[\ln\left(\frac{2000\text{N}}{880\text{N}}\right)\right]\left(\frac{1}{25.1 \text{ rad}}\right)$$

$$\mu_s = 0.033$$

b) From equation (1)

$$\frac{T}{F} = e^{\mu_s\theta}$$

or

$$F = \frac{T}{e^{\mu_s\theta}} \tag{2}$$

Now $\theta = 2$ turns $(2\pi \text{ rad}) = 12.6$ rad.

Calculating F from equation (2) gives

$$F = \frac{2000\text{N}}{e^{(0.033)(12.6)}}$$

$$= \frac{2000}{1.52} = 1315.8\text{N}$$

PROBLEM 7

A skid weighing 500N and carrying 9500N of paper, is being pulled up onto a loading dock. The static and kinetic coefficients of friction are $m_s = 0.30$ and $m_k = 0.20$, respectively. What is the magnitude of the force F (1) when the skid begins to move upward; (2) while the skid is moving; (3) to prevent the skid from sliding downward?

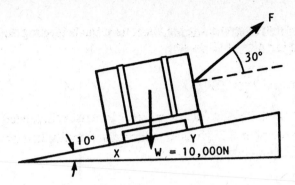

Figure 66. Loading a skid

SOLUTION: This problem may be solved by resolving forces into normal and tangential compounds with respect to the incline, to find values of the frictional force. However, this procedure actually does more than necessary. A more direct, geometrical procedure is possible. It is based on the fact that the combined normal and frictional forces, that is, the reaction of the incline on the skid, is directed at an angle ϕ defined by the friction coefficient μ and the equation $\tan \phi = \mu$.

This magnitude of the combined weight W of the skid and paper is

$$W = 500N + 9500N = 10kN$$

The free body diagram for part (1) is shown in Figure 67.

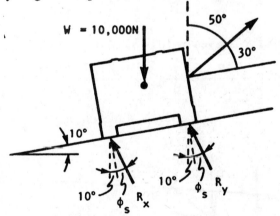

Figure 67. Free body diagram

Noting that R_x and R_y have the same direction (same angle of friction ϕ_s), draw a force triangle including the weight W, the force F, and the sum $R = R_x + R_y$. The triangle is closed since there is no acceleration. The direction of R will be different for each part of this problem. For part (1)

$$\tan \phi_s = \mu_s = 0.30$$
$$\phi_s = 16.7°$$

This angle is measured from the normal, as shown, so the actual direction of R is given by $10° + 16.7° = 26.7°$. The force triangle, then, is

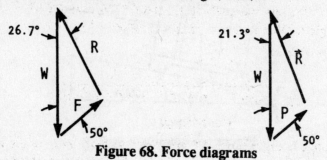

Figure 68. Force diagrams

Using the law of sines,

$$\frac{F}{\sin 26.7°} = \frac{W}{\sin\left[180° - (50° + 26.7°)\right]}$$

$$F = .478 \text{ kN, direction shown}$$

The situation is similar for part (2) except now the kinetic coefficient must be used. Therefore,

$$\tan\phi_k = \mu_k = 0.20$$

$$\phi_k = 11.3°.$$

And the direction of $\vec{R}$ is given by $10° + 11.3° = 21.3°$. Consequently

$$\frac{F}{\sin 21.3°} = \frac{W}{\sin\left[180° - (50° + 21.3°)\right]}$$

$$F = .329 \text{kN, direction shown}$$

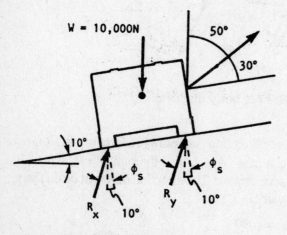

Figure 69. Motion of the skid

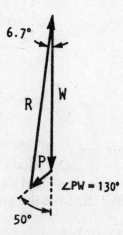

Figure 70. Force triangle

For part (3), consider the frictional force to be opposite the downward motion of the skid, so that the sense of R_x and R_y has changed considerably (see Figure 69). Again, $\phi_s = 16.7°$, but now the resulting direction of R is given by $16.7° - 10° = 6.7°$. The force triangle for this situation is shown in Figure 70.

Using the law of sines,

$$\frac{F}{\sin 6.7°} = \frac{W}{\sin[180° - (130° + 6.7°)]}$$

$$F = 1.701\text{kN}$$

However, notice in Figure 70 that F is directed from the head of the vector W to the tail of the vector R, as was done in Figure 68, and is now pointing downward. In other words, it would take a push of more than 1.7501kN along the direction of F to start the skid moving. Thus, the answer to (3) is that no force F is needed to keep the skid from sliding down; it will not slide down under its own weight.

PROBLEM 8

A small pump is mounted on a concrete slab which, in turn, rests on four concrete posts. Two of the posts have settled slightly, so that the slab must be leveled. The foreman decides to raise the low side with two large wedges located as drawn on the figure. Fortunately, there is a concrete floor on which the bottom wedge can be fastened. If μ_s between the two wedges is 0.4 and between the wedge and the concrete 0.6, and if the motor slab transmits a 1000 lb vertical force to the wedge, determine the force P required for lifting the slab.

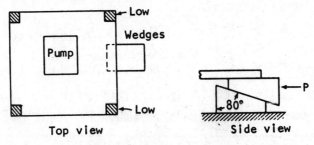

Figure 71. Pump mounted on a concrete slab

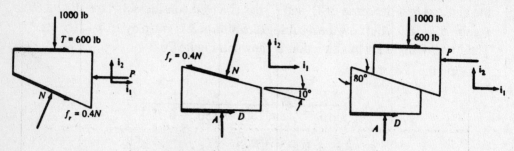

Figure 72. Free body diagrams

SOLUTION: The free body diagrams are shown in Figure 72. Notice that part (c), showing the two wedge system, has only the external forces shown. The forces between the wedges are equal and opposite and cancel out by Newton's Third Law. If the sum of all the external forces are set equal to zero, the system is in equilibrium. That is, nothing moves. Any greater force will start lifting the slab.

$$\overset{+\downarrow}{}\sum F_{i_1} = 0; \qquad 600 + D - P = 0 \tag{1}$$

$$\overset{+\uparrow}{}\sum F_{i_2} = 0; \qquad -1000 + A = 0$$

so $\qquad\qquad\qquad A = 1000 \text{ lb.}$

The bottom wedge, as described, is fastened to the floor. The sum of the forces must be equal to zero throughout. This means that the force D is entirely reactive and does not depend on any coefficient of friction or the normal force A. The equililbrium equations are, from diagram (2b),

$$\overset{+\downarrow}{}\sum F_{i_1} = 0; \qquad D - 0.4 \cos 10° - N \sin 10° = 0, \tag{2}$$

$$\overset{+\uparrow}{}\sum F_{i_2} = 0; \qquad A - N \cos 10° + 0.4N \sin 10° = 0$$

or, $\qquad\qquad 1000 - N \cos 10° + 0.4N \sin 10° = 0. \tag{3}$

Solving equations (2) and (3) simultaneously for D yields

$$D = 1000 \frac{0.4 \cos 10° + \sin 10°}{\cos 10° - 0.4 \sin 10°}$$
$$= 621 \text{ lb.}$$

Putting this values into equation (1) yields

$$P = 1221 \text{ lb.}$$

The force P needed by this method is greater than the 1000 lb needed for a direct lift.

CHAPTER SEVEN

Mechanics of Materials

Chapter 7

MECHANICS OF MATERIALS

The study of Mechanics of Materials covers two important areas: 1) Dynamics, which covers bodies in motion, and 2) Statics, which deals with bodies of matter at rest.

For a basic review of Statics, see Chapter 5. We will continue the discussion of Statics here. When a body is held at rest and exposed to external forces (such as gravity), the tendency of the body is to change shape or to be deformed. Stress can be defined as a body's internal resistance to these external forces. This chapter will cover various aspects of stress and the resultant deformation—strain. This will include its application to bodies such as beams and columns.

STRESS AND STRAIN

Remember that **stress** is thought of in terms of unit stress and total stress. The total resistance to an external force (expressed in pounds, kips, or newtons) or other force dimension is called total stress. The resistance over a unit area of a body is called unit stress.

Unit stress is expressed in units of pounds per square inch or kips per square inch or other force per unit area. In this review, stress refers to unit stress. The various types of stress are: 1) direct or simple stress, which can be in the form of tension, compression, or shear; 2) indirect stress, which applies to bending or torsion; and 3) combined stress which is any combination of stresses 1) and 2).

To review, direct stress (simple) is developed under direct loading, such as simple tension and compression. When the load is applied (applied force) axially (axial loading), we have simple tension and compression. When equal, opposite, and parallel forces cause two surfaces to slide relative to each other, simple shear occurs.

We can calculate the magnitude of simple stress from the formula:

$$\text{Stress, } \sigma = \frac{F(\text{Force})}{A(\text{Area})},$$

where:

F = External force causing stress (i.e., lb)

A = Area on which stress is acting (in^2)

σ = Average unit stress (i.e., psi).

While there are several types of stress/strain responses of materials when subjected to one or more loads, for our purposes, we will refer to the most useful behavioral response, which is a **linear elastic response**. Linear elastic response is characterized by stress proportional to strain, independent of time, and strain recoverable upon load removal. This response is called **Hooke's Law**.

If a beam is undergoing simple tension or compression and has the same cross-sectional area throughout, then the stress will be the same throughout the length of the body, as shown in Figure 1.

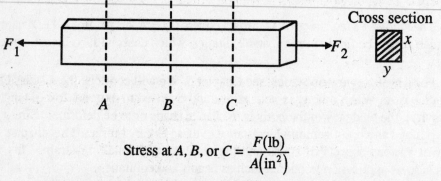

$$\text{Stress at } A, B, \text{ or } C = \frac{F(\text{lb})}{A(\text{in}^2)}$$

Figure 1. Stress on a beam of uniform cross-sectional area

However, if the cross section varies in total area as in Figure 2, then the largest stress will be carried on the smallest area. This is because the load must be carried by all areas of the body—otherwise failure will occur.

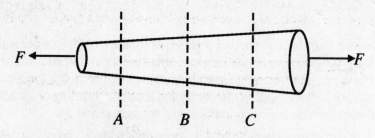

$$\sigma_C < \sigma_B < \sigma_A; \text{ Area } A < \text{Area } B < \text{Area } C$$

Figure 2. Stress on a body of varying cross-sectional area

Note that in Figures 1 and 2, the areas resisting the external forces are **perpendicular** to the direction of the application of those forces. It is also important to note that we assume no bending occurs in the beam under compression. Therefore, the force is applied through the centroidal axis of the straight member.

As stated earlier, the resultant deformation of a rigid body is called **strain** and can be measured from the formula:

$$\text{Unit Strain } (\varepsilon) = \frac{\text{New length} - \text{Original length}}{\text{Original length}}.$$

Strain is always related to these simple stresses—tension, compression, and shear. Remember that the deformation is always measured in the direction of the applied force.

The elongation of an axially loaded body experiencing normal stress is:

$$\delta = \frac{FL_o}{EA}$$

where L_o is the original length of the body and E is the modulus of elasticity.

SHEAR

In the **Stress and Strain** section, we examined the resultant stress when force is applied to the area of a rigid body in a perpendicular direction, as in Figure 3.

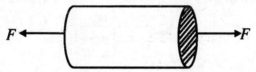

Figure 3. Force applied in a perpendicular direction

However, as mentioned, when equal opposite and parallel forces cause two surfaces to slide relative to each other, simple **shear** occurs. In other words, when the resisting area is **parallel** to the applied force (external), simple shear develops.

In Figure 4, the two pieces of wood are held together by a simple wooden dowel.

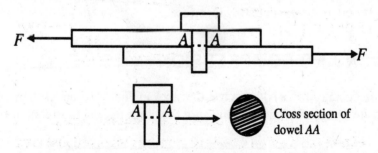

Figure 4. Resisting area is parallel to the applied force

The circular area, *AA*, of the dowel is subject to shear strain by the force, *F*. In this case, the magnitude of the stress is determined by:

$$\text{Stress, } \sigma = \frac{F(\text{Force})}{A(\text{Area})}.$$

Of course, if the dowel were to fail, it would shear across the surface, A, as shown in Figure 5.

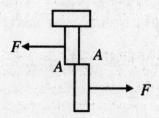

Figure 5. Failure across surface A

Another form of shear would occur in an operation such as hole punching, as in Figure 6.

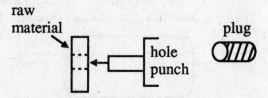

Figure 6. Hole punching

In this case, the shear area is taken to be the cylindrical surface of the plug, and the shear stress is still calculated by:

$$\sigma = \frac{F}{A},$$

except A is now the outer cylindrical surface.

TENSION AND COMPRESSION

Tension occurs when a body such as the block in Figure 7 is acted on by two equal and opposite forces, F. Tension will cause the block to tend to stretch or be pulled apart.

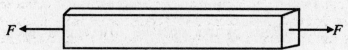

Figure 7. Block in tension

If the forces act in the opposite direction from those in Figure 7, then **compression** occurs. These compressive forces tend to crush or shorten the body.

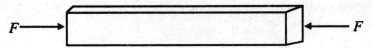

Figure 8. A block in compression

Remember that, although the forces acting on a body are equal and occurring in equilibrium, tension or compression is always present. These forces are acting in an axial direction.

It is important to distinguish between axial tension and axial compression. In axial tension, the length of a long body subjected to axial tension will not normally be a factor. However, in axial compression, length plays an important role. Obviously a long length of bar subjected to axial compression will experience both compression and buckling, as shown in Figure 9a. However, a short length of bar in axial compression will only experience compression, not buckling, as shown in Figure 9b.

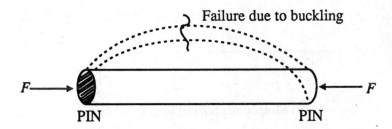

a) Compression and buckling on a long bar

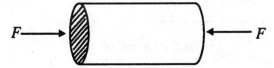

b) Compression only on a short bar

Figure 9. Axial compression

Bodies that are acted on by only axial forces or resultant axial forces are called two-force members.

BEAMS

A member that bends elastically in order to resist transverse forces and loads is called a **beam**. To examine our beams, we'll agree that the value of unknown forces or moments can be discovered if we apply the following condition of static equilibrium:

(1) The sum of all forces acting on the beam in the x and y direction is zero:

$$\sum F_x = 0, \ \sum F_y = 0$$

(2) The sum of all moments acting on the beam is zero:

$$\Sigma M = 0.$$

These conditions apply to a statically determinate beam only. Examples of statically determinate beams are:

(1) A simple supported beam with supports at each end and forces applied in between, Figure 10a,

(2) A cantilever beam supported at one end (i.e., by a wall or clamp) with force(s) applied along its length, Figure 10b,

(3) An overhanging beam with the force(s) applied between the end of the beam and the support point, Figure 10c.

In all cases, we assume that the forces are known values, and the resistance (or reaction at the supports) and the wall moments are unknown.

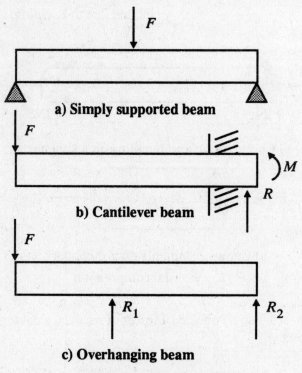

a) Simply supported beam

b) Cantilever beam

c) Overhanging beam

Figure 10. Beam supports

Let's look at a simply supported beam when a load is applied, as in Figure 11. First, remember that the sum of all forces equals zero, and the sum of all moments equals zero.

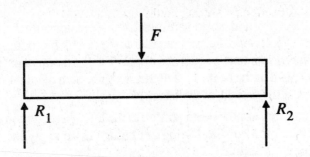

Figure 11. Load applied to a simply supported beam

Under loading, assuming that the elastic limit of the material is not exceeded, the bar will bend as shown in Figure 12.

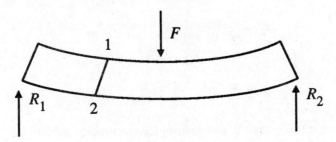

Figure 12. Beam bends upon loading

As we can see, some of the bar is in compression and some of it is in tension. In addition to bending, certain sections of the bar are trying to slip past one another in shear. This shearing tendency must be resisted by the material fiber.

If we cut the beam, along section 1 – 2, we see the acting forces more clearly.

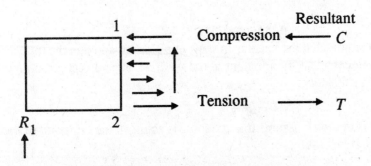

Figure 13. Section 1 – 2 of loaded beam

Because the greatest forces of compression and tension lie along the top and bottom edges of the beam, a resultant arrow for total compression and total tension will be placed at the outer edges. All forces in the xy plane equal zero:

$$C - T = 0; \; C = T$$

The resultant compression and tension are equal, but act in opposite directions. This effect is a couple (equal, opposite, and parallel forces) which produces a resisting moment. This resisting moment is equal and opposite to the moment produced by R_1, around the opposite end of the beam $1-2$. Otherwise, the sum of all moments would not be zero. (Neglect the weight of the beam.)

For ease of discussion, use the beam portion to the left of a section in determining the solutions to beam problems. Referring to Figure 11, we apply the condition to calculate the unknown reactions:

$$\Sigma F_y = 0$$

therefore

$$R_1 + R_2 - F = 0, \; R_1 + R_2 = F.$$

Because F is centrally located on the beam:

$$R_1 = \frac{F}{2}$$

and

$$R_2 = \frac{F}{2}.$$

In cases where the load is uniformly distributed over the entire length, the result is the same. A uniform load is usually indicated by weight per linear measurement. The total uniform load may be calculated by multiplying the uniform load by the total length of the beam.

For cantilever beams, such as the one shown in Figure 14, the wall must carry the load on the beam, and the resistance at the wall is equal to the load itself, whether the load is uniform or concentrated:

$$F - R = 0$$

therefore

$$R = F.$$

Because we know that the beam is in static equilibrium and that the total moment is zero, we can calculate the moment at the wall, M. Recall that a moment is force multiplied by distance:

$$F\ell - M = 0$$

For a uniform load, assume that the force is acting at the midpoint of the load on the beam. In this case:

$$F\left(\frac{\ell}{2}\right) - M = 0$$

$$F\left(\frac{\ell}{2}\right) = M.$$

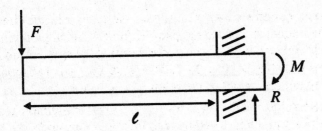

Figure 14. Cantilever beam

Knowing the shear force at a section of a beam can also be helpful in its design. The shear force is simply the sum of all forces acting vertically, to the left of that particular section that we are examining. To find shear force, it is helpful to plot a shear force diagram. All external forces are plotted along the beam. Upward forces considered positive and downward forces are given a negative sign. In other words, if the force acts upward on the left side of a cross section of interest or downward on the right side, that is considered **positive shear**. If the force acts in the opposite direction, that is **negative shear**. Cases of positve and negative shear are illustrated in Figure 15.

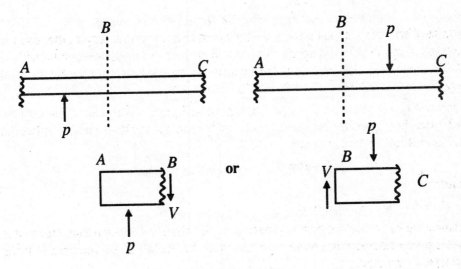

a) Positive shear

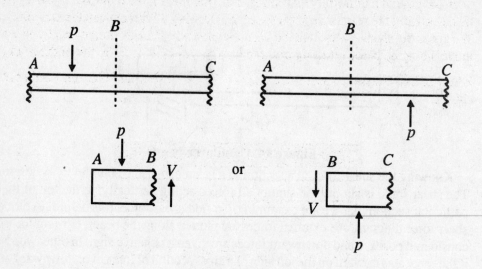

b) Negative shear

Figure 15. Shear stresses

EXAMPLE: Draw the shear force diagram for the system as shown.

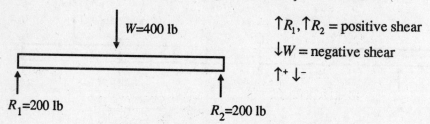

$$\uparrow R_1, \uparrow R_2 = \text{positive shear}$$
$$\downarrow W = \text{negative shear}$$
$$\uparrow^+ \downarrow^-$$

Figure 16. Stresses applied to beam

SOLUTION: The shear force diagram would be:

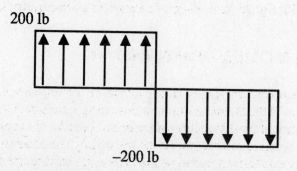

Figure 17. Shear force diagram

Of course, directly at each support, the shear forces are zero.

In the case of a uniformly distributed load, the shear force diagram is developed by calculating the resistance at each end and drawing a line between the endpoints. We know the shear forces acting on each section will change incrementally as we move along the beam resulting in a line passing through the midpoint of the beam.

EXAMPLE: Draw the shear force diagram for the beam under uniform loading as shown.

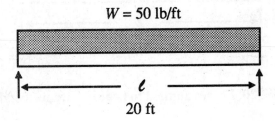

$W = 50$ lb/ft

ℓ

20 ft

Figure 18. Beam under uniform loading

SOLUTION:

$$\text{Total WT} = 1,000 \text{ lb}$$
$$R_1 + R_2 = 1,000 \text{ lb}$$
$$R_1 = R_2 = 500 \text{ lb}$$

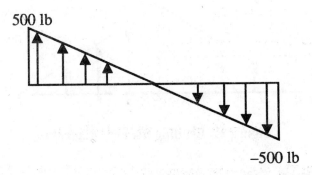

500 lb

−500 lb

Figure 19. Shear force diagram for beam under uniform loading

BENDING MOMENT DIAGRAM

A **bending moment diagram** is used to illustrate bending moment changes along a beam. The sum of the moments due to forces acting to the left of a section will give the bending moment for that section of the beam. In order to maintain equilibrium, these external bending moments must be resisted by the beam material. Bending occurs as the beam material undergoes stress and resists these bending moments.

EXAMPLE: Draw the bending moment diagram for the beam as shown. The 2,000 lb force is applied at the midpoint of the beam.

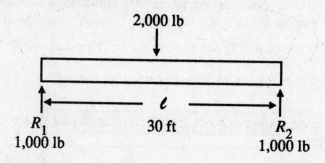

Figure 20. Forces applied to beam to induce bending

SOLUTION: Obviously the bending moment is greatest at the center of the beam:

$$\sum M = 0,$$
$$M = (15 \text{ ft}) (1{,}000 \text{ lb})$$
$$= 15{,}000 \text{ ft-lb}$$

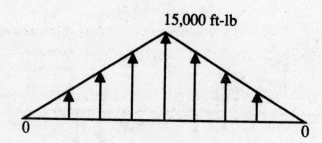

Figure 21. Bending moment diagram

If the effect of the bending moment causes a concave upward shape, this is called positive bending moment, as illustrated in Figure 22.

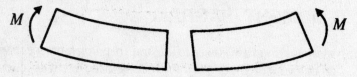

Figure 22. Positive bending moment

If the effect is a downward convex bending movement, this is called a **negative bending movement**, as shown in Figure 23.

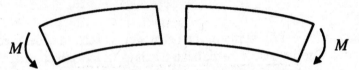

Figure 23. Negative bending moment

Figure 24 summarizes the shear and bending moment diagrams for two common types of loading.

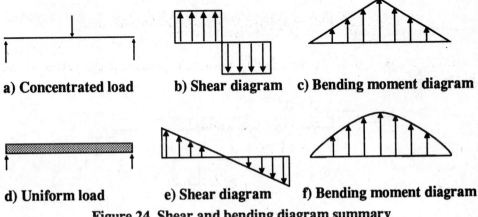

a) Concentrated load   **b) Shear diagram**   **c) Bending moment diagram**

d) Uniform load   **e) Shear diagram**   **f) Bending moment diagram**

Figure 24. Shear and bending diagram summary

Beams which are loaded with both concentrated and distributed loads, as in the following example, necessarily have more complicated bending moment and shear diagrams.

The maximum stress along a cross-section of a beam is along the outer edges, and the stress is zero along the neutral axis, as shown in Figure 25.

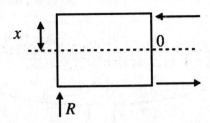

Figure 25. Stresses on the cross-section of a beam

The maximum fiber stress, σ_F, is:

$$\sigma_F = \frac{M}{Z}$$

where Z, the section modulus is:

$$Z = \frac{I_c}{C}.$$

The centroidal moment of inertia of the beam's cross-section is denoted I_c in the equation above. The distance from the neutral axis to the top or bottom surface (most distant from the neutral axis), is the quantity c. Note that the moment of inertia is generally listed in the table form for common objects.

TORSION

When two equal and opposite twisting moments, which exist in parallel planes, act upon a shaft, then the shaft is said to be in **torsion**.

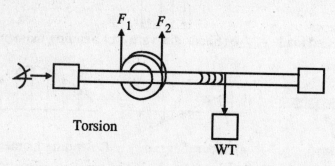

Torsion

Figure 26. Torsion

A shaft in torsion may be rotating or it may be stationary. If it is rotating, it will do so in a uniform manner, such as a hoist. In the above figure, looking down the shaft from the left-hand end, if F_2 is greater than F_1, the shaft will tend to rotate in a counterclockwise direction. Because of the weight attached to the shaft, the shaft will also tend to rotate in a clockwise direction. These opposite twisting movements put the shaft in **torsion**.

The twisting moment, or torque, has units of in-lb or N-m. Referring to Figure 26 above, if F_2 is 500 lb and F_1 is 200 lb and the pulley radius is 8 in, then the torque (T) is:

$$T = (500\ lb - 200\ lb) \times 8\ in$$
$$= 2{,}400\ in\text{-}lb.$$

If we assume that the system is in equilibrium and that the radius of the shaft is 2 in, then the weight (WT) equals 1,200 lb. In a shaft with several pulleys on it, the torque available along the length of the shaft will vary according to each pulley size.

Torsion shearing stress occurs when any cross-sectional area of the shaft tends to shear across the adjacent face due to the torsion applied to the body. The resistance of the material fibers per unit area is torsional shearing stress. Examine the torque wrench on the lug nut of the wheel shown in Figure 27.

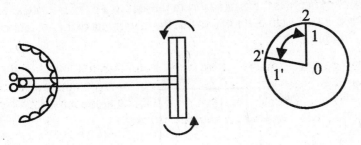

a) Rotating wrench   **b) Cross-section of shaft**

Figure 27. Torque wrench

The elongation 1-1' is proportional to the distance from the center of the shaft. Therefore, 2-2' is greater than 1-1' in deformation. If the shear stress at 2 is 10,000 lb/in² and the radius from $0 \rightarrow 2$ is 2 in, then the shearing stress at a point halfway between 0 and 2 would equal half that amount or 5,000 lb/in².

If we define σ_s as the maximum shearing stress, due to torsion, at the outer fiber of a uniform, solid circular member (in psi), then to find the maximum safe shearing stress use:

$$\sigma_s = \frac{16T}{\pi d^3},$$

for a hollow circular shaft, where d_o is the outer diameter and d_i is the inner diameter:

$$\sigma_s = \frac{16Td_o}{\pi\left(d_o^{\,4} - d_i^{\,4}\right)}.$$

COLUMNS

A body or member, which is in compression (axial loading), is referred to as a **column**. Compression of a short beam may be determined using the familiar equation F/A. For a relatively long column, of course, material defects, misalignment, and unknown initial stresses will affect the behavior of the material and the maximum load. Localized buckling or sudden bending are typical sources of failure in columns.

Because of the previously mentioned factors, maximum safe axial loads are found using several column design factors that determine the shape and size of the column. For our purposes in this discussion, both ends of the column are pinned or fixed.

Column length is a very important factor in determining maximum load. Also, cross-sectional area and moment of inertia can be important in determining column strength. However, because the moment of inertia can be found from the cross-sectional area, we combine these two factors into the radius of gyration (r).

The radius of gyration is the distance from the axis to a point in a plane where all the area is imagined to be concentrated so that the movement of inertia is unchanged, as illustrated in Figure 28.

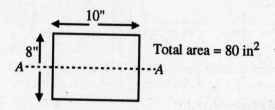

a) True area

or

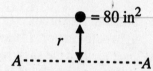

b) Idealized schematic of area

Figure 28. Radius of gyration

Then the radius of gyration, *r*, equals:

$$r = \sqrt{\frac{I}{A}},$$

where *I* is the moment of inertia, and *A* is the total area as shown in the figure. Therefore, along with end conditions, factors governing column strength include 1) length, and 2) radius of gyration. The ratio of length to radius of gyration is the slenderness ratio:

$$\text{Slenderness Ratio} = \frac{\ell}{r}.$$

There are generally three types of compression bodies — short, intermediate, and long—depending on their slenderness ratio. Typical slenderness ratios range from 80 to 120. In short compression bodies, the slenderness ratio $\frac{\ell}{r}$ has no effect on load performance, and the ultimate stress of the material is found from the equation:

$$\sigma = \frac{F}{A}.$$

In long columns, the length is of primary importance because slender columns can fail in buckling. Their strength can be defined by the Euler formula:

$$\sigma = \frac{F}{A} = \frac{\pi^2 E}{\left(\frac{\ell}{r}\right)^2},$$

where E is the modulus of elasticity, in units of lb/in² (or appropriate equivalent units), and the column ends are pinned. The effective (or critical) slenderness ratio takes other end mount configurations into account using the end restraint coefficient C as follows:

$$\text{effective slenderness ratio} = \frac{C\ell}{r}.$$

End restraint coefficients are listed in Figure 29 for various end mount configurations. Note that for equivalent cases, the end mount configuration in Figure 29a is the strongest followed by the configuration in Figure 29d, Figure 29c, and Figure 29b.

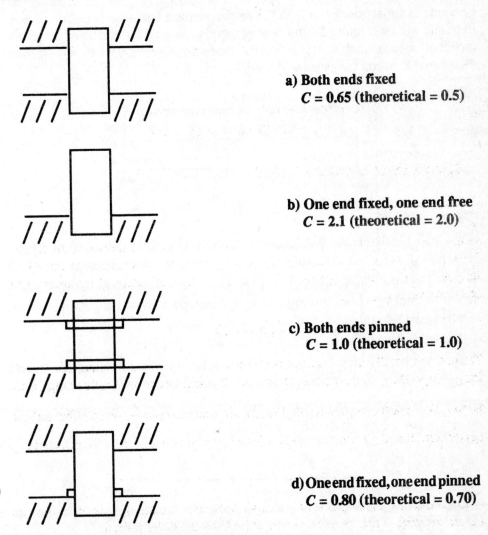

a) **Both ends fixed**
$C = 0.65$ (theoretical = 0.5)

b) **One end fixed, one end free**
$C = 2.1$ (theoretical = 2.0)

c) **Both ends pinned**
$C = 1.0$ (theoretical = 1.0)

d) **One end fixed, one end pinned**
$C = 0.80$ (theoretical = 0.70)

Figure 29. End restraint coefficients, C

COMBINED STRESSES

In the earlier sections of this chapter, it was shown that direct axial loading or bending can induce tensile (or compressive) stresses. In addition, it was shown that torsion or direct shearing forces can cause shear stress. When the same kind of stresses act simultaneously on a specific area, these stresses may be added to produce a **combined** effect. If these stresses act along the same line (**collinear**), then their resultant may be found **algebraically**. These stresses must be added **vectorially** if they do not act along the same line. For further information on resolving concurrent force systems into resultants, consult the chapter on Statics.

In situations of combined tension or compression, we will demonstrate the principle of **superposition**. For this purpose, combined axial and bending stresses in beams will be examined. This method can also be used for combined direct and torsional shear situations. Examine the beam in Figure 30, with a uniformly distributed load and subjected to the axial tensile force *F*.

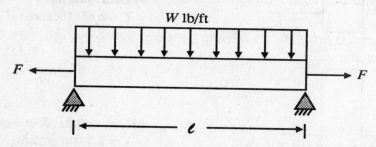

Figure 30. Uniform load distributed on a beam subjected to axial tensile forces

To determine the combined stresses at any given point, first determine the stresses due individually to axial loading and to the uniform load and superimpose their effects to get the combined result.

The tensile stress due to the axial force *F* is:

$$\sigma_1 = \frac{F}{A},$$

where *A* is the cross section of beam and it is assumed that σ_1 acts equally at all points on the cross-sections, as shown in the next figure.

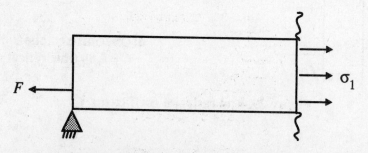

Figure 31. Tensile stress on beam

The uniform load W causes the beam to bend, which induces a stress at each cross section. The stresses vary from maximum compression at the top of the beam to maximum tension at the bottom of the beam, as shown in Figure 32.

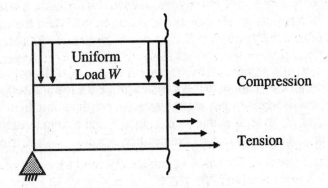

Figure 32. Tensile and compressive stresses due to uniform load W

As discussed in an earlier section of this chapter, the bending stress (maximum fiber stress) is:

$$\sigma_2 = \frac{M}{\left(\dfrac{I_c}{C}\right)}.$$

For design purposes, use the maximum values of σ_2 located at the extreme outer fibers of the beam. Also, note that for a uniformly loaded beam, on a simple span, the bending moment is greatest at the midpoint of the beam, which is also where the greatest stress values are located.

Axial and bending stresses are combined in Figure 33.

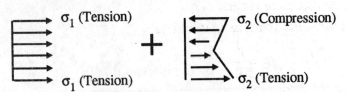

a) Addition of axial and bending stresses

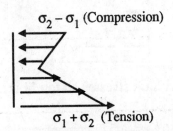

b) Resultant stress

Figure 33. Combining stresses

Algebraically:

$$\sigma_{Total} = \sigma_1 \pm \sigma_2$$

or

$$\sigma_{Total} = \frac{F}{A} \pm \frac{M}{\left(\frac{I_c}{C}\right)}.$$

To determine sign convention, in this case, examine the top and bottom fibers of the beam to determine whether they are in tension or compression. In the case of the bottom fibers, they are in tension due to axial load F and also in tension due to the bending moment so it is proper to add both resultant stresses, as shown in Figure 34.

$$\sigma_{Bottom} = \frac{F}{A} + \frac{M}{\left(\frac{I_c}{C}\right)}$$

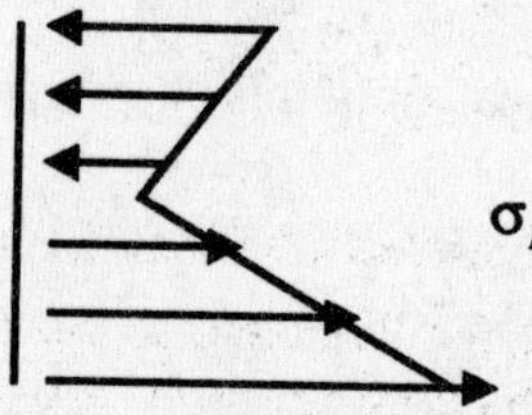

$$\sigma_{Bottom} = \sigma_1 + \sigma_2 (\text{Tension})$$

Figure 34. Resultant stresses on bottom fibers of beam

In the case of the top fibers, they are in tension due to axial load F, but in compression due to the bending moment, so it is necessary to subtract one stress value from the other, as in Figure 35.

$$\sigma_{Top} = \frac{F}{A} - \frac{M}{\left(\frac{I_c}{C}\right)}$$

$$\sigma_{Top} = \sigma_2 - \sigma_1 (\text{Compression})$$

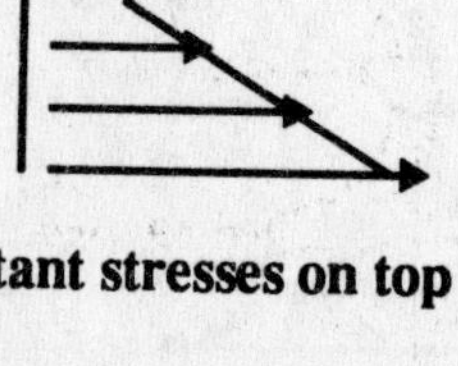

Figure 35. Resultant stresses on top fibers of beam

Remember to always identify tension or compression at the beam fibers to help maintain proper sign convention.

THIN-WALLED PRESSURE VESSELS

When a pressure vessel has a diameter at least ten times the thickness of its wall, it is considered a **thin-walled pressure vessel**, such as a boiler drum. The forces tending to rupture the vessel act perpendicular to the surface of the vessel. These forces tend to rupture the vessel along the seam parallel to the element of the shell longitudinally and along a seam corresponding to the vessel's circumference transversely.

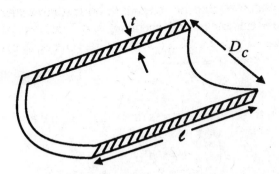

Figure 36. Longitudinal section of a thin-walled pressure vessel

Although D_c is shown as outer diameter, as this section deals with thin-walled vessels; outer or inner diameter may be used in equations.

As shown in Figure 37 below, the pressure is distributed uniformly along the curved surface. Only components acting perpendicular to the longitudinal surface tend to cause failure.

a) Uniform pressure　　　　**b) Vertical and horizontal components of the uniform pressure**

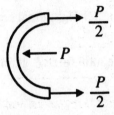

c) Effect of all horizontal pressure components

Figure 37. Pressure distributed in a thin-walled pressure vessel

Notice how the vertical components of the force p will cancel themselves out because they act in equal and opposite directions (Figure 37b). If force P is the total effect of all horizontal components of p acting against the curved surface (Figure 37c), then we calculate P as follows:

P = the pressure p × the area of the curved surface on the longitudinal plane, or

$$P = p D_c \ell.$$

Because the equations in this section refer to *thin*-walled vessels, it is usually not necessary to specify inner or outer diameter when using D_c. It is the resistance of the material that holds the pressure force in equilibrium in the longitudinal section. If the resistance is shared equally by both sections, then:

$$P = \frac{P}{2} + \frac{P}{2}$$

as shown in Figure 37c.

Then, based on force equal to stress times the area, the total resistive force is:

$$2\left(\frac{P}{2}\right) = (2\ell t)\sigma_t,$$

$$P = 2\ell t \sigma_t,$$

where

σ_t = Tangential stress in material, psi

$2\ell t$ = Total area of resistance, in^2.

Equating resistive and pressure forces:

$$(2\ell t)\sigma_1 = pD_c\ell,$$

$$\sigma_t = \frac{pD_c}{2t},$$

where

σ_t = Average tangential stress, psi

p = Internal pressure in cylinder

D_c = Cylinder diameter, in

t = Thickness of plate, in

To determine the force along each longitudinal seam of the vessel, refer back to:

$$P = pD_c\ell.$$

If each seam carries half of the total force P, then:

$$F = \frac{pD_c\ell}{2}.$$

where

F = Force on the longitudinal seam ℓ

p = Internal pressure in cylinder, psi

D_c = Cylinder diameter, in

ℓ = Length of seam, in

The total pressure acting in the longitudinal plane is:

$$P = pa,$$

$$P = p\frac{\pi D_c^2}{4}.$$

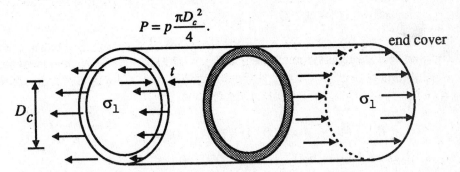

Figure 38. Longitudinal stresses σ_ℓ in a thin-walled pressure vessel

The pressure force must be balanced by resistive forces in the thin ring of metal on the transverse section which is approximately equal to its circumference $\pi D_c \times$ thickness t:

$$P = (\pi D_c t)\sigma_\ell.$$

Equating pressure force and resistive force to determine the longitudinal stress σ_ℓ:

$$(\pi D_c t)\sigma_\ell = p\left(\frac{\pi D_c^2}{4}\right)$$

$$\sigma_\ell = \frac{pD_c}{4t}.$$

REVIEW PROBLEMS

PROBLEM 1

Construct the shear force and bending moment diagrams for the beam shown in Figure 39.

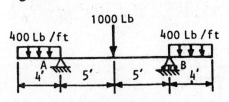

Figure 39. Shear force and bending moment diagram

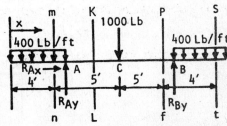

Figure 40. Loading diagram

SOLUTION: The loading diagram is shown in Figure 40. Reactions at the supports are calculated to be:

$$\Sigma F_x = 0 \qquad R_{Ax} = 0$$

$$\Sigma M_A = 0 + 400(4)(10+2) - R_{By}(10) + 1,000\,(5) - 400(4)(2) = 0$$

$$R_{By} = \frac{19,200 + 5,000 - 3,200}{10} = 2,100\ \text{lb}.$$

Because the loading is symmetric $R_{Ay} = R_{By} = 2,100\ \text{lb}$. Check whether the beam is in equilibrium:

$$\Sigma F_y = 0$$

$$R_{Ay} + R_{By} - (400)(4) - (400)(4) - 1,000 = 0$$

$$2,100 + 2,100 - 1,600 - 1,600 - 1,000 = 0$$

$$0 = 0.$$

Beam is in vertical equilibrium and calculated values of R_{Ay} and R_{By} are correct.

The free-body diagram for a portion of the beam to the left of section *mn* (refer to Figure 40) is shown in Figure 41.

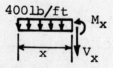

Figure 41. Force diagram left of section mn

$$0 < x \le 4$$

$$V_x + 400x = 0$$

$$V_x = -400x$$

$$M_x + 400(x)\left(\frac{x}{2}\right) = 0$$

$$M_x = -400\frac{x^2}{2} = -200x^2$$

The free-body diagram for a portion of the beam to the left of section *KL* (refer to Figure 40) is shown in Figure 42.

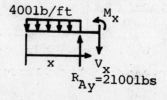

Figure 42. Force diagram left of section KL

$$4 < x \le 9$$
$$V_x - 2{,}100 + 400(4) = 0$$
$$V_x = 500 \text{ lb}$$
$$M_x + 400(4)(x - 4 + 2) - 2{,}100(x - 4) = 0$$
$$M_x = 2{,}100x - 8{,}400 - 1{,}600x + 3{,}200$$
$$M_x = 500x - 5{,}200$$

Figure 43 illustrates a portion of the beam to the left of section *Pf* (refer to Figure 40).

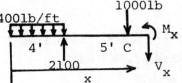

Figure 43. Force diagram left of section *Pf*

$$9 < x \le 14$$
$$V_x + 1{,}000 - 2{,}100 + 400(4) = 0$$
$$V_x = 2{,}100 - 1{,}000 - 1{,}600 = -500 \text{ lb}$$
$$M_x + 1{,}000(x - 9) - 2{,}100(x - 4) + 400(4)(x - 4 + 2) = 0$$
$$M_x = -1{,}000x + 9{,}000 + 2{,}100x - 8{,}400 - 1{,}600x + 3{,}200$$
$$M_x = -500x + 3{,}800$$

The portion of the beam to the right of section *St* (refer to Figure 40) is shown in Figure 44.

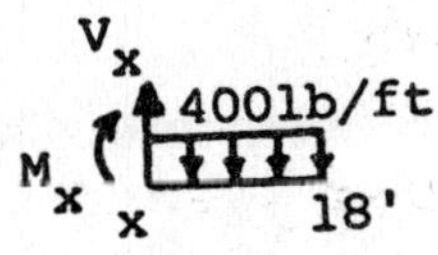

Figure 44. Force diagram of section *St*

$$V_x - 400(18 - x) = 0 \qquad\qquad M_x + 400(18 - x)(18 - x)/2 = 0$$
$$V_x = -400x + 7{,}200 \qquad\qquad\quad M_x = -200(18 - x)^2$$

Multiplied out:

$$M_x = -200x^2 + 7{,}200x - 64{,}800$$

This section of the beam, $14 < x \le 18$, can also be analyzed by looking at the loads to the left of the cut.

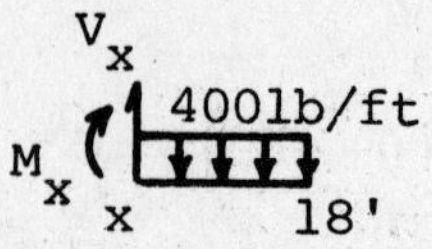

V_x

M_x ⟳ 400lb/ft

x   18'

Figure 45. Summary force diagram

$$V_x + 400(x - 14) - 2,100 + 1,000 - 2,100 + 400(4) = 0$$
$$V_x = -400x + 7,200$$

Summing the moments:

$$M_x + 400(x - 14)(x - 14)/2 - 2,100(x - 14) + 1,000(x - 9)$$
$$- 2,100(x - 4) + 400(4)(x - 4/2) = 0$$
$$M_x = -200x^2 + 7,200x - 64,800.$$

The shear force and bending moment diagrams are then plotted by substituting for x in the various expressions for V_x and M_x (Figures 46 and 47).

FIG-MS

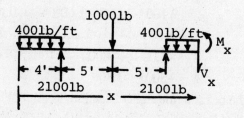

Figure 46. Shear force diagram

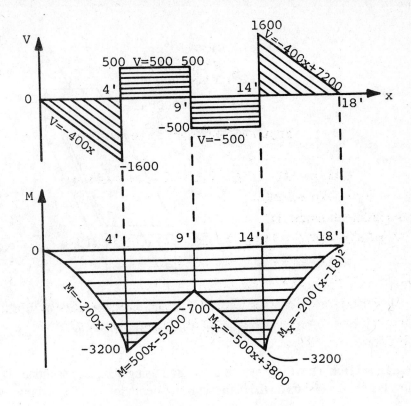

Figure 47. Bending moment diagram

PROBLEM 2

A 5-foot structural steel bar $\frac{1}{2}$ by 4 inches in cross section is to support an axial tensile load with allowable normal and shearing stresses of 18,000 psi and 8,000 psi, respectively, and with an allowable elongation of 0.035 in. Determine the maximum permissible load.

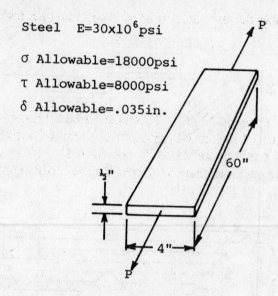

Steel   E=30x10^{6}psi

σ Allowable=18000psi

τ Allowable=8000psi

δ Allowable=.035in.

60"

1½"

4"

Figure 48. Stresses on steel bar

SOLUTION: This force can be calculated using shear stress alone. The maximum shear stress occurs on the plane with an orientation of 45°, as shown in Figure 49.

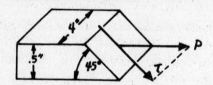

Figure 49. Maximum shear stress on beam

$$\tau = \frac{F}{A} = \frac{P\sin 45°}{(.5"/\sin 45°)4"} = \frac{.707P}{(.707)4"}$$

$$\tau = \frac{P}{4}$$

$$P = 4\tau = 4(8,000)$$

$$P = 32,000 \text{ lb.}$$

It still remains for us to determine whether the extension for this load exceeds the allowable elongation:

$$\delta = \frac{PL}{AE} = \frac{(32,000 \text{ lb})(60\text{in})}{(2 \text{ in}^2)(30\times10^6 \text{ lb/in}^2)} = .032 \text{ in}$$

This is within the allowable range of elongation, therefore:

$$P_{max} = 32,000 \text{ lb.}$$

PROBLEM 3

The beam of Figure 50 has a rectangular cross section 4 inches wide 12 inches deep and a span of 12 feet. It is subjected to a uniformly distributed load of 100 lb per foot and concentrated loads of 600 lb and 1,200 lb at 2 feet and 7 feet, respectively, from the left support. Determine the maximum fiber stress at the center of the beam.

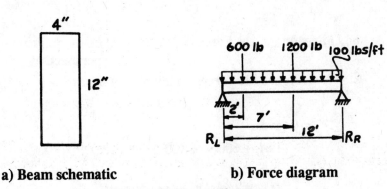

a) Beam schematic   b) Force diagram

Figure 50. Beam diagram

SOLUTION: The reactions are evaluated from a free-body diagram of the entire beam and moment equations with respect to each of the two reactions. By this method, the reactions are evaluated independently, and summation of forces may be used to check the results. $\Sigma M = 0$ about the right support. Therefore:

$$0 = -12 R_L + (100)(12)\left(\frac{12}{2}\right) + (1,200)(5) + 600(10)$$

$$12 R_L = 19,200$$
$$R_L = 1,600 \text{ lb.}$$

$\Sigma M = 0$ about the left support. Therefore:

$$0 = 12 R_R + 600(2) + 100\left(\frac{12}{2}\right)(12) + 1,200(7)$$

$$12 R_R = 16,800$$
$$R_R = 1,400 \text{ lb.}$$

Check: $\Sigma F_y = 0$

$$600 + 1,200 + 100 \times 12 - 1,600 - 1,400 = 0$$
$$0 = 0.$$

The left reaction, R_L, is 1,600 lb and the right reaction, R_R, is 1,400 lb, both upward.

Figure 51 is a free-body diagram of the left half of the beam.

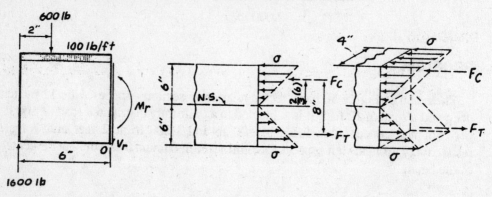

a) Free body diagram   b) Force diagram   c) Force diagram in 3-d perspective

Figure 51. Forces on left half of beam

The equation $\sum M_r = 0$ gives:

$$+ M_r + 600(4) + 100(6)(3) - 1{,}600(6) = 0,$$

from which

$$M_r = 5{,}400 \text{ ft-lb or } 64{,}800 \text{ in-lb.}$$

The moment, M_r, is the resultant of the flexural stresses indicated in Figure 2b and c in which F_C and F_T are the resultants of the compressive and tensile stresses, respectively. The resultants F_C and F_T (which act through the centroids of the wedge-shaped stress distribution diagram of Figure 2c) are located through the centroids of the triangular stress distribution diagrams of Figure 2b. The magnitude of the couple M_r is:

$$M_r = F_C(8)$$

or

$$M_r = F_T(8).$$

The stress, σ is:

$$\sigma = \frac{F}{A}$$

$$= \frac{F_c}{(6)(4)\left(\frac{1}{2}\right)},$$

therefore:

$$F_c = 12\sigma$$
$$M_r = F_c(4)$$
$$64{,}800 = (12\sigma)(8)$$

$$\sigma_{max} = 675 \text{ psi } C \text{ at top and } T \text{ at bottom.}$$

PROBLEM 4

Shown in Figure 52 is a simply supported beam with a triangular loading. What is the deflection curve of this beam?

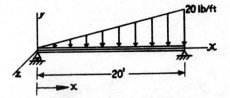

Figure 52. Beam with triangular loading

SOLUTION: This problem will be solved by integrating the fourth order differential equation of the curvature due to loading

$$EI\frac{d^4y}{dx^4} = w$$

The equation for the loading is

$$w = \frac{20\,\text{lb}/\text{ft}\,(x)}{20\,\text{ft}} = x\,\text{lb}/\text{ft}^2$$

Integrating the loading gives the shear

$$EI\frac{d^3y}{dx^3} = \frac{x^2}{2} + c_1$$

Integrating the shear gives the moment

$$EI\frac{d^2y}{dx^2} = \frac{x^3}{6} + c_1x + c_2$$

The moments at the ends of the beam are both zero. At $x = 0$

$$EI(0) = \frac{(0)_3}{6} + c_1(0) + c_2$$

$$c_2 = 0$$

At $x = 20$

$$EI(0) = \frac{(20)_3}{6 + c_1(20)}$$

$$c_1 = \frac{-400}{6}$$

Substituting the moment equation becomes

$$EI \frac{d^2y}{dx^2} = \frac{x^3}{6} - \frac{400x}{6}$$

Integrating the moment gives the slope

$$EI \frac{dy}{dx} = \frac{x^4}{24} - \frac{400x^2}{12} + c_3$$

Integrating the slope gives the deflection

$$EIy = \frac{x^5}{120} - \frac{400x^3}{36 + c_3x + c_4}$$

The deflections at the ends of the beam are both zero. At $x = 0$

$$EI(0) = \frac{(0)^5}{120} - \frac{400(0)^3}{36 + c_3(0) + c_4}$$

$$c_4 = 0$$

At $x = 20$

$$EI(0) = \frac{(20)^5}{120} - \frac{500(20)^3}{36 + c_3(20)}$$

$$c_3 = \frac{88,888.9 - 26,666.7}{20} = 3,110$$

Substituting yields

$$EIy = \frac{x^5}{120} - \frac{400x^3}{36 + 3,110x}$$

Thus the equation of the deflection curve of the beam is

$$y = \frac{1}{EI} \left(\frac{x^5}{120} - \frac{400x^3}{36 + 3,110x} \right)$$

PROBLEM 5

Two small lathes are driven by the same motor through a $\frac{1}{2}$ inch diameter steel shaft, as shown in Figure 53. Determine the maximum shear stress in the shaft due to twisting.

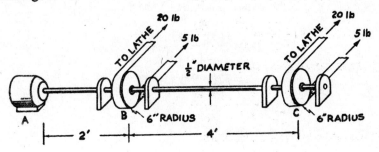

a) Lathes diagram

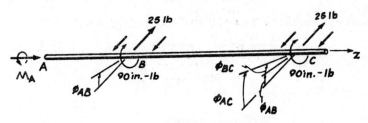

b) Force diagram

Figure 53. Shear stress on shaft

SOLUTION: We begin the analysis by idealizing the situation as shown in Figure 53b. Here we represent each pulley loading by its static equivalent of a force of 25 lb through the axis of the shaft and a couple about the z axis of $6(20 - 5) = 90$ in-lb. Because each pulley is supported by a pair of immediately adjacent bearings, we make the idealization that the 25-lb transverse forces are balanced by the bearing reactions in such a way that there is negligible shear force and bending moment transmitted beyond the bearings. In this case it is only necessary for the motor to supply a torque M_A, as shown.

Establishing moment equilibrium, we have, since all moment vectors are parallel to z, $\sum M_A = 0$ if:

$$M_A - 90 - 90 = 0$$
$$M_A = 180 \text{ in-lb.}$$

The twisting moments in sections AB and BC of the shaft are then clearly

$$M_{AB} = 180 \text{ in-lb} \qquad M_{BC} = 90 \text{ in-lb}.$$

The maximum shear stress occurs at the outside of the shaft in section *AB*. Using equation:

$$\sigma_s = \frac{16T}{\pi d^3},$$

where the torque (*T*) here is the twisting moment in section *AB*:

$$\sigma_s = \frac{(16)(180 \text{ in-lb})}{\pi(.5 \text{ in})^3}$$

$$= 7333.9 \text{ psi}.$$

PROBLEM 6

Determine the required section modulus *Z* for a beam *AB* to support the distributed load shown in Figure 54 if $q = 4,000$ lb per foot and the allowable bending stress σ_w = 16,000 psi. Neglect the weight of the beam.

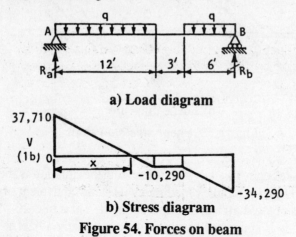

a) Load diagram

b) Stress diagram

Figure 54. Forces on beam

SOLUTION: 1) To locate the section of maximum bending moment, it is helpful to construct a shear force diagram, shown in Figure 54. The reactions at the supports are:

$$R_a = 37,710 \text{ lb} \qquad R_b = 34,290 \text{ lb}$$

and the distance *x* defining the point of zero shear is given by the equation:

$$R_a - qx = 0$$

from which $x = \dfrac{R_a}{q} = 9.43$ ft. At this distance from support *A*, the bending moment is a maximum:

$$M_{max} = R_a x - \frac{qx^2}{2}$$

$$= (37,710 \text{ lb})(9.43 \text{ ft}) - \frac{(4,000 \text{ lb / ft})(9.43 \text{ ft})^2}{2}$$

$$= 177,775.5 \text{ ft-lb.}$$

The required section modules, from $\sigma = \dfrac{M}{Z}$ is:

$$Z = \frac{M}{\sigma}$$

$$= \frac{(177,755.5 \text{ ft-lb})(12 \text{ in / ft})}{(1,600 \text{ lb / in}^2)}$$

$$= 133.3 \text{ in}^3.$$

PROBLEM 7

A 12-inch-by-16-inch wooden cantilever beam weighing 50 lb per foot carries an upward concentrated force of 4,000 lb at the end, as shown in Figure 55. Determine the maximum bending stresses at a section 6 feet from the free end.

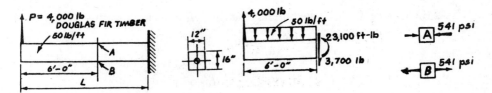

a) Beam schematic   b) Cross-section c) Free-body diagram d) Summary of forces

Figure 55. Cantilever beam

SOLUTION: A free-body diagram for a 6-ft segment of the beam is shown in Figure 55c. The weight, 50 lb/ft is considered as a uniformly distributed external load and the beam is then considered massless. To keep this segment in equilibrium requires a shear of 4,000 – 50(6) = 3,700 lb and a bending moment of 4,000(6) – 50(6)3 = 23,100 ft-lb at the cut section. Both these quantities are shown with their proper sense in Figure 55. By inspecting the cross-sectional area, the distance from the neutral axis to the extreme fibers is seen to be 8 in., hence, $c = 8$ in. This is applicable to both the tension and the compression fiber.

Therefore,
$$I_{zz} = \frac{bh^3}{12} = \frac{12(16)^3}{12} = 4{,}096 \text{ in.}^4$$

$$\sigma_{max} = \frac{Mc}{I} = \frac{23{,}100(12)8}{4{,}095} = \pm 541 \text{ psi}.$$

From the sense of the bending moment shown in Figure 55c, the top fibers of the beam are seen to be in compression, and the bottom ones in tension. In the answer given, the positive sign applies to the tensile stress, and the negative sign applies to the compressive stress. Both of these stresses decrease at a linear rate toward the neutral axis where the bending stress is zero. The normal stresses acting on infinitesimal elements at A and B are shown in Figure 55d.

PROBLEM 8

A beam with overhanging ends (Figure 56) carries a uniform load of intensity $q = 10$ kip/ft on each overhang. Assuming that the beam is a $W30 \times 172$ section with $E = 30 \times 10^6$ psi, determine the maximum normal stress in the beam.

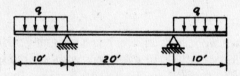

Figure 56. Overhanging beam

SOLUTION: A $W30 \times 172$ section has the following properties:

nominal size	$30" \times 15"$
depth of section	29.88 in
area	50.67 in^2
I	7,891.5 in^4

Maximum normal stress occurs at a point where moment is maximum. This is at the midpoint of the beam.

In order to find the moment, the reaction forces need to be determined. The beam is symmetrically loaded; therefore, the reactions are equal.

By vertical equilibrium:

$$\Sigma F_y = 0,$$

$$(2)(10 \text{ ft})(10 \text{ kip/ft}) - 2R = 0$$

$$R = 100 \text{ kip}.$$

The moment at the midpoint of the beam by moment equilibrium is:

$$\Sigma M = 0$$

$$M_{max} + (100 \text{ kip})(10 \text{ ft}) - (10 \text{ kip/ft})(15 \text{ ft})(10 \text{ ft}) = 0$$

$$M_{max} = 500 \text{ kip-ft} = 6000 \text{ kip-in},$$

and σ_{max} is:

$$\sigma_{max} = \frac{M_{max}c}{I}$$

$$= \frac{(600 \text{ kip} - \text{in})(29.88 \text{ in}^2)}{7891.5 \text{ in}^4}$$

$$= 11.3 \text{ ksi.}$$

PROBLEM 9

Member *GF* of the pin-connected truss in Figure 57 has a cross-sectional area of 2.4 sq. in. 1) Determine the axial stress in *GF*. The lengths shown are center to center of pins. 2) If member *CF* of the pin-connected truss is a $\frac{7}{8}$-in diameter rod, determine the axial stress in *CF*. The lengths shown are center to center of pins.

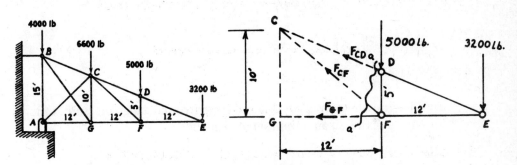

Figure 57. Truss   Figure 58. Force diagram of truss

SOLUTION: In general, to determine the forces in the truss members, the reactions for the whole structure should be calculated. Since the truss in this problem is a cantilever truss, the member forces can be found without finding the reaction.

1) To find the axial force in GF, make a section cut *a-a* as shown in Figure 58. By taking the moment about point C, F_{CD} and F_{CF} will not contribute since their lines of action go through C. Therefore, the only unknown is F_{GF}:

$$F_{GF}(10) + 5,000(12) + 3,200(24) = 0$$
$$F_{GF} = -13,680 = 13,680 \text{ lb, compression.}$$

The axial stress in *GF* is:

$$\sigma_{GF} = \frac{F_{GF}}{A_{GF}} = \frac{13,680}{2.4} = 5,700 \text{ psi } C.$$

2) To find the axial force in *CF*, again determine the moment: (The vertical component of F_{CF} passes through the point D and contributes no moment.)

The horizontal component of F_{CF} is:

$$F_{CF_x} = F_{CF}\left(\frac{12}{\sqrt{12^2 + 10^2}}\right)$$

$$\Sigma M_D = F_{CF}\left(\frac{12}{\sqrt{12^2 + 10^2}}\right)(5) - 13,680(5) + (3,200)(12) = 0$$

$$F_{CF} = 7,810 \text{ lb tension.}$$

Axial stress in *CF*:

$$\sigma_{CF} = \frac{F_{CF}}{A_{CF}}$$

where

$$A_{CF} = \frac{\pi d^2}{4} = \frac{\pi(7/8)^2}{4} = .601 \text{ in}^2$$

$$\sigma_{CF} = \frac{7,810}{.6} = 12,988 \text{ psi } T.$$

PROBLEM 10

A thin-walled pressure vessel is subjected to an internal gauge pressure. Determine the normal stresses on an infinitesimal element.

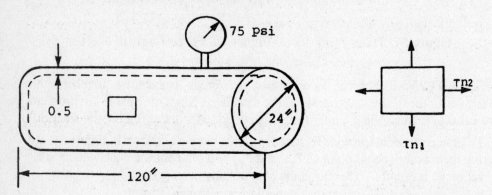

Figure 59. Pressure vessel   **Figure 60. Free body diagram**

SOLUTION: The equations for the normal stresses $(\sigma_n)_1$ and $(\sigma_n)_2$ for a thin cylinder acted on by internal pressure p, at a point far from the ends of the cylinder are:

$$(\sigma_n)_2 = \frac{pr}{t} \tag{1}$$

and

$$(\sigma_n)_1 = \frac{pr}{2t}. \tag{2}$$

The internal pressure is 75 psi above atmospheric pressure. This is the same as the net internal pressure, or the equivalent internal pressure if the external pressure were zero. Since this is the pressure used to derive (1) and (2), we can use it without modification for this problem.

From (1)

$$(\sigma_n)_2 = \frac{(75\,\text{psi})(12\,\text{in})}{(1/2\,\text{in})} = 1{,}800\,\text{psi} \tag{3}$$

From (2)

$$(\sigma_n)_1 = \frac{(75\,\text{psi})(12\,\text{in})}{2(1/2\,\text{in})} = 900\,\text{psi} \tag{4}$$

It is seen from (3) and (4) that both components of normal stress are much larger in value than the pressure that caused them. This can be accounted for by the fact that the internal pressure acts over a large area to create certain forces that must be balanced by forces in the cylinder acting over small areas. This produces rather large stresses in the cylinder.

Also note that the axial stress $(\sigma_n)_1$ is half as large as the tangential stress $(\sigma_n)_2$.

Equation (1) and (2) should be used only where $t \ll r$ and away from the ends of the cylinder.

PROBLEM 11

The steel shaft of Figure 61a is in equilibrium under the torques shown. Determine the maximum shearing stress in the shaft.

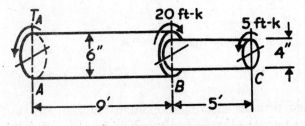

a) Shaft in equilibrium

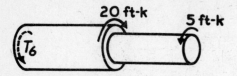

**b) Free body diagram of shaft
cut by plane**

**c) Free body diagram of
4 in section**

Figure 61. Steel shaft

SOLUTION: In general, free-body diagrams should be drawn in order to evaluate the resisting torque correctly. Such diagrams are shown in Figure 61b and 61c, where in 61b, the shaft is cut by any transverse plane through the 6-inch segment, and T_6 is the resisting torque on this section. Similarly, in Figure 61c, the plane is passed through the 4-inch section, and T_4 is the resisting torque on this section. The location of the maximum shearing stress is not apparent; hence, the stress must be checked at both sections. Thus, from Figure 61b:

$$\Sigma M = 0, \ T_6 = 20 - 5 = 15 \text{ ft-kips} = 15,000(12) \text{ in-lb},$$

and

$$\sigma_{max} = \frac{T_6 c}{I} = \frac{15,000(12)(3)}{(\pi/2)(3^4)} = \frac{40,000}{3\pi} \text{psi}.$$

From Figure 61c,

$$\Sigma M = 0,$$
$$T_4 = 5 \text{ ft-kips} = 5,000(12) \text{ in-lb},$$

and

$$\sigma_{max} = \frac{T_4 c}{I}$$

$$= \frac{5,000(12)(2)}{(\pi/2)(2^4)} = \frac{15,000}{\pi} \text{psi};$$

$$\frac{15,000}{\pi} > \frac{40,000}{3\pi}$$

therefore, the maximum shearing stress is

$$\sigma_{max} = \frac{15,000}{\pi} = 4,774.6 \text{ psi}$$

This stress is less than the shearing proportional limit of any steel; hence, the torsion formula applies. Note that had the larger torque been carried by the smaller section, the maximum stress would obviously occur in the small section and only one stress determination would have been required.

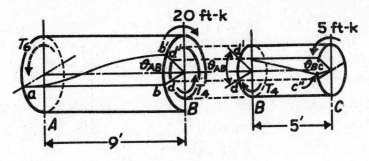

d) Segments *AB* and *BC*

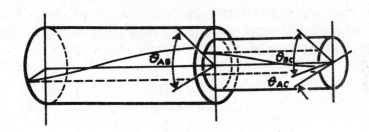

e) Resultant distortion

Figure 62. Torque diagrams

As an aid to visualizing the distortion of the shaft, the segments *AB* and *BC* and the torques acting on them are drawn in Figure 62d and 62e with the distortions greatly exaggerated. As the resultant torque of 15 ft-kips twists segment *AB* through the angle θ_{AB}, points *b* and *d* move to *b'* and *d'* respectively, and segment *BC* may be considered as a rigid body rotating through the same angle, point *c* moving to *c'*, after which the torque of 5 ft-kips acting on *BC* twists this part of the shaft back through the angle θ_{BC}, point *c'* moving back to *c"*. The resultant distortion for the entire shaft is shown in Figure 62e.

PROBLEM 12

Calculate the internal forces and moments acting at sections 1 and 2 in the structure shown.

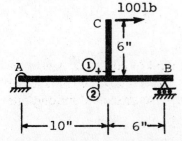

Figure 63. Forces on structure

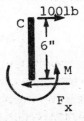

Figure 64. Forces on structure part 1

279

SOLUTION: Consider a cut at section 1 and draw a free-body diagram indicating the internal forces and moments as shown in Figure 64.

For equilibrium

$$\sum F = 0$$

$$\therefore F_x - 100 = 0$$

$$\therefore F_x = 100 \text{ lbs}$$

$\sum M = 0$ about the cut

$$\therefore M - 6 \times 100 = 0$$

$$\therefore M = 600 \text{ in-lb}$$

To get the internal forces and moment at section 2, first determine the reactions at the supports. See Figure 65.

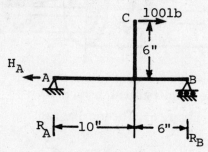

Figure 65. Reactions at structure supports

$F = 0$ (horizontal forces)

$$\therefore -H_A + 100 = 0$$

$$\therefore H_A = 100 \text{ lb}$$

$\sum M = 0$ at support $A +$

$$\therefore 100 \times 6 - 16 R_B = 0$$

$$\therefore R_B = 37.5 \text{ lb}$$

$\sum M = 0$ at support $B +$

$$100 \times 6 + 16 R_A = 0$$

$$\therefore R_A = 37.5 \text{ lb}$$

Now consider a cut through section 2, and draw a free-body diagram as shown in Figure 66.

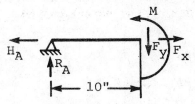

Figure 66. Free body diagram

$\sum F_y = 0$

$$\therefore R_A - F_y = 0$$
$$\therefore F_y = R_A = -37.5 \text{ lb}$$

$\sum F_x = 0$

$$\therefore F_x - H_A = 0$$
$$\therefore F_x = H_A = 100 \text{ lb}$$

$\sum M = 0$ about the cut $+$

$$-M + 10 R_A = 0$$
$$\therefore M = -10 \times 37.5$$
$$= 375 \text{ in-lb clockwise}$$

CHAPTER EIGHT

Chemistry

Chapter 8

CHEMISTRY

This chapter is a brief review of general chemistry, reviewing such fundamental concepts as nomenclature and equations to redox reactions and acid-base equilibria. Since being able to recognize and identify chemicals is of primary importance, we will start there.

NOMENCLATURE

A chemical formula is a representation of a compound in terms of its constituent atoms and their relative numbers. Chemical formulas may be empirical or molecular. The **empirical formula** of any compound gives the *relative number* of atoms in each element in the compound. It is the simplest formula of a material compound that can be derived solely from its components. The molecular formula of a substance indicates the *actual number* of atoms in a molecule of a substance. To determine the molecular formula of a compound, one must calculate the empirical formula and then extrapolate to the molecular formula via molecular weight. It is easiest to remember that a molecular formula is simply a multiple of an empirical formula.

Now, let's look at naming inorganic compounds (nomenclature for organic compounds is discussed in the Organic Chemistry section). Binary compounds consist of two elements. The formula of a binary compound identifies the two elements present and ends in "-ide" (e.g., NaCl = sodium chloride). If a metal is bonded to a nonmetal, the metal is named first. If a metal element has only two possible oxidation states, use the suffix "-ous" for the lower state and "-ic" for the higher state. For example,

$$FeCl_2 = \text{ferrous chloride [iron (II) chloride]}$$
$$FeCl_3 = \text{ferric chloride [iron (III) chloride]}.$$

When naming binary covalent compounds formed between two nonmetals, a third system is preferred. The numbers of each atom in the molecule are specified by a Greek prefix: di-, tri-, tetra-, penta-, and so on. For example, N_2O_5 = dinitrogen pentoxide.

Ternary (three element) compounds are usually made up of an element and a radical. To name these compounds, each component making up the compound is identified. The positive component is written first and the negative one second. For example, in $CaCO_3$, Ca is the element, and CO_3 is the radical anion: $CaCO_3$ (Ca^{+2} + CO_3^{-2}) = calcium carbonate.

Binary acids use the prefix "hydro-" in front of the stem (or full name) of the nonmetallic element and place the suffix "-ic" at the end (e.g., HCl = hydrochloric acid).

Ternary acids are slightly more complicated. They may be examined along with salts containing polyatomic anions. Both oxy-acids and oxy-ions are named according to their oxygen content. Anions with more oxygen atoms end in "-ate," while their counterparts end in "-ite." Acids are defined almost the same way. Acids with more oxygen atoms end in "-ic,"and acids with fewer end in "-ous." If there are more than two types of anions or acids, prefixes are also added. These naming conventions are summarized in table form below.

TABLE 1
SUFFIX NOMENCLATURE

Oxidation State	Anion		Acid		Example	Name
+1	hypo-	-ite	hypo-	-ous	HClO	hypochlorous acid
+3		-ite		-ous	$HClO_2$	chlorous acid
+5		-ate		-ic	$HClO_3$	chloric acid
+7	per-	-ate	per-	-ic	$HClO_4$	perchlorous acid

A simple rule of thumb is "-ate/-ic...-ite/-ous."

Partial neutralization of an acid that is capable of furnishing more than one H+ per acid molecule produces salts that are called acid salts. When only one acid salt is formed, the salt can be named by adding the prefix "bi-" to the name of the anion of the acid (e.g., $NaHSO_4$ = sodium bisulfate). The salt can also be named by specifying the presence of H. For example, Na_2HPO_4 = sodium hydrogen phosphate or disodium hydrogen phosphate. In writing formulas, there are a couple of general observations to keep in mind:

(1) Metals have positive oxidation numbers while nonmetals (and most common radical ions except the ammonium ion) have negative oxidation numbers.

(2) A radical ion is a group of elements that remain bonded as a group even when involved in the formation of compounds.

Some basic rules for writing formulas are:

(1) Represent the symbols of the components using the positive part first and the negative part second.

$$(NH_4)(SO_4) = \text{ammonium sulfate}$$

(2) Indicate the respective oxidation number above and to the right of each symbol.

$$(NH_4)^{+1}(SO_4)^{-2}$$

(3) Write the subscript number equal to the oxidation number of the other element or radical, and omit the subscript "1" as well as any + or – signs.

$$(NH_4)^{+1}(SO_4)^{-2} = (NH_4)_2(SO_4)$$

(4) As a general rule, the subscript numbers are reduced to their lowest terms. Hence, Ca_2O_2 becomes CaO. There are some exceptions. Hydrogen peroxide H_2O_2 and acetylene C_2H_2 are two.

EQUATIONS

Now that simple inorganic nomenclature has been established, let's look at equations. An **equation** is a chemical sentence. It gives information about the states of the reactants and products, their relative molar proportions, and the possible directions of the reaction. The formula of each substance is followed by an indication of state (e.g., s, l, g, aq). Remember that state *does* depend on temperature and pressure, so when writing an equation, be sure to check for these values. If none are given, assume standard temperature and pressure (STP) values (1 atm, 273 K).

Aqueous solutions may contain strong electrolytes (which dissociate into ions readily), weak electrolytes (which dissociate to a small extent), or nonelectrolytes (which do not dissociate at all). If a substance is a strong electrolyte, it is written in its dissociated form. For example, sodium chloride would be written in aqueous solution as $Na^+(aq) + Cl^-(aq)$. Strong electrolytes include salts (a metal ion with a nonmetal or polyatomic ion) and strong acids and bases (HCl, NaOH). Weak electrolytes and nonelectrolytes are generally written in undissociated form. These include weak acids and bases, most nonpolar substances, and many organic materials.

Any substances in the system whose quantities do not change during the reaction are not included in the equation. Since their quantities and states remain constant, they do not affect the net reaction. For example,

$$Ag^+(aq) + NO_3^-(aq) + Na^+(aq) + Cl^-(aq) \rightleftharpoons AgCl(s) + Na^+(aq) + NO_3^-(aq)$$

should be written

$$Ag^+(aq) + Cl^-(aq) \rightleftharpoons AgCl(s).$$

The bottom equation is called a net equation because it only involves the net reacting components of the system. This is usually the part of the reaction in which we are interested. Finally, equations are not complete until they are properly balanced. The law of conservation of mass must be observed when writing an equation. No quantities of elements can suddenly appear or disappear; the same number of atoms must be on either side of the equation. For

$$2H_2(g) + O_2(g) \rightleftharpoons 2H_2O(l),$$

there are four H atoms on the left and four H atoms on the right. Oxygen is balanced the same way. In general, a properly balanced equation is the starting point in solving many chemical problems.

EXAMPLE: Balance the following by filling in the missing species and proper coefficient:

(a) $NaOH +$ _____ $\rightarrow NaHSO_4 + HOH$,

(b) $PCl_3 + 3\ HOH \rightarrow$ _____ $+ 3HCl$,

(c) $CH_4 +$ _____ $\rightarrow CCl_4 + 4HCl$.

SOLUTION: To balance chemical equations you must remember that ALL atoms (and charges) must be accounted for. The use of coefficients in front of compounds is a means to this end. Thus,

(a) $NaOH +$ _____ $\rightarrow NaHSO_4 + HOH$

On the right side of the equation, you have 1 Na, 3 H's, 5 O's and 1 S. This same number of elements must appear on the left side. However, on the left side, there exists only 1 Na, 1 O, and 1 H. You are missing 2 H's, 1 S, and 4 O's. The missing species is H_2SO_4, sulfuric acid. You could have anticipated this since a strong base ($NaOH$) reacting with a strong acid yields a salt ($NaHSO_4$) and water. The point is, however, that H_2SO_4 balances the equation by supplying all the missing atoms.

(b) $PCl_3 + 3HOH \rightarrow$ _____ $+ 3HCl$.

Here, the left side has 1 P, 3 Cl's, 6 H's, and 3 O's. The right has 3 H's and 3 Cl's. You are missing 1 P, 3 O's and 3 hydrogens. Therefore, $P(OH)_3$ is formed.

(c) $CH_4 +$ _____ $\rightarrow CCl_4 + 4HCl$

Here, there are 1 C, 8 Cl's, and 4 H's on the right and 1 C and 4 H's on the left. The missing compound, therefore, contains 8 Cl's and thus is $4\ Cl_2$. One knows that it is $4\ Cl_2$ rather than Cl_8 or $8\ Cl$ because elemental chlorine gas is a diatomic or 2 atom molecule.

STOICHIOMETRY

The next fundamental review is that of stoichometry. **Stoichiometry** is really chemical arithmetic. Basically, it involves measuring relative amounts and proportions of reactants and products in chemical reactions. To perform such measurements, a few concepts must first be outlined.

One **mole** of a substance is the standard chemical unit for "amount." It contains Avogadro's number of particles (atoms, molecules, ions, electrons, etc.), approximately 6.02×10^{23}. The definition of the mole is

$$\text{number of moles} = \frac{\text{mass}}{\text{molecular weight}} = \frac{g}{g/mol}.$$

The gram-atomic weight of any element is defined as the mass, in grams, which contains one mole of atoms of that element. For example, approximately 12.0 g of carbon, 16.0 g of oxygen, and 32.1 g of sulfur each contain one mole of atoms.

The molecular weight (formula weight) of a molecule or compound is determined by the addition of its component atomic weights. For example, look at the formula weight (F.W.) of $CaCO_3$:

F.W. $CaCO_3$ = 1(40) + 1(12) + 3(16) = 100 g/mol.

The formula for molecular weight is given below:

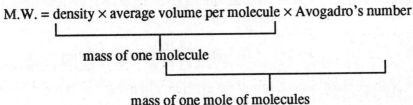

M.W. = density × average volume per molecule × Avogadro's number

mass of one molecule

mass of one mole of molecules

Finally, equivalent weights are the amounts of substances that react completely with one another in chemical reactions. In *electrolytic* reactions, the equivalent weight is defined as that weight which either receives or donates one mole of electrons at an electrode. For *oxidation-reduction* reactions, an equivalent is defined as the quantity of a substance that either gains or loses one mole of electrons. In *acid-base* reactions, an equivalent of an acid is defined as the quantity of acid that supplies one mole of H^+. An equivalent of base supplies one mole of OH^-.

It is important to note that a given substance may have any of several equivalent weights, depending on the particular reaction in which it is involved. For example, for Fe^{3+}:

$Fe^{3+} + e^- = Fe^{2+}$ one equivalent weight per mole; eq wt = 56 g/eq

$Fe^{3+} + 3e^- = Fe^0$ three equivalent weights per mole; eq wt = (1/3)(56) = 18.7 g/eq

Stoichiometric calculations utilize reaction equations. Since stoichiometry involves examining relative amounts and/or proportions of reactants and products, it is very important that the equations be properly balanced. The coefficients in chemical equations provide the ratios in which moles of one substance react with moles of another. Although the coefficients may represent numbers of molecules, they also represent the number of moles of reactants and products required. For the equation, $C_2H_4 + 3O_2 \rightleftharpoons 2CO_2 + 2H_2O$, the number of moles of O_2 consumed is always equal to three times the number of moles of C_2H_4 that react.

Usually, reactants and products are not set up in a neat economical package containing correct stoichiometric ratios. Some reagents may be in excess of their needed ratios and will not be completely consumed in the reaction. Chemical reactions can only proceed as far as their least abundant reactant will allow (similar to the strong chain/weakest link idea). The reactant is called **limiting reagent**. It is always consumed first in the chemical reaction, and it determines the amount of product that will be formed unless equilibrium is reached first (see the section on

Equilibrium). To perform limiting reagent calculations, you must modify your calculations according to the amounts of reagents present. Calculate the number of moles of each reagent and compare them to the number of actual moles needed (according to the stoichiometric ratios and actual amounts of reagents you have present). Whatever amount of reactant present that is less than the amount required is the limiting reagent (and will determine the quantities of product formed).

This brings us to "yields" of reactions. The **theoretical yield** of a given product is the maximum yield that can be obtained from a given reaction if the reaction goes to completion (rather than to equilibrium). For example:

$$50 \text{ g O}_2 \times \frac{18 \text{ g H}_2\text{O}}{16 \text{ g O}} = 56.25 \text{ g H}_2\text{O} \text{ [water is } \left(\frac{16}{18}\right)\text{oxygen } \left(\frac{2}{18}\right)\text{hydrogen]}$$

The **percentage yield** is a measure of the efficiency of a reaction. It is defined as

$$\text{percentage yield} = \frac{\text{actual yield}}{\text{theoretical yield}} \times 100\%.$$

We may also look at **percentage compositions**, the percentage of the total mass contributed by each element:

$$\text{percentage composition} = \frac{\text{mass of element in compound}}{\text{mass of compound}} \times 100\%$$

Remember that at "STP," standard temperature and pressure, 273 K and 1 atm, 1 mole of any ideal gas occupies 22.4 liters. In other words, the molar volume of an ideal gas is 22.4 liters/mol. Density can be converted to molecular weight using this relationship:

$$\text{M.W.} = (\text{density})(\text{molar volume})$$
$$(\text{g/mol}) = (\text{g/l})(\text{l/mol})$$

STATES OF MATTER

Matter occupies space and possesses mass. It is found in three states or phases: solid, liquid, and gas. A solid has both definite volume and shape. A liquid has definite volume, but takes the shape of its container. A gas has neither definite shape nor definite volume. Usually, temperature and pressure are the primary factors that determine the state of a substance. We will first review solids.

The properties of solids are as follows:

(1) They retain their shape and volume when transferred from one container to another.
(2) They are virtually incompressible.
(3) They exhibit extremely slow rates of diffusion.

In a solid, the attractive forces between the atoms, molecules, or ions are relatively strong. The particles are held in a rigid structural array, wherein they exhibit only vibrational motion. When solids are heated at certain pressures, some vaporize without passing through the liquid phase. This is called sublimation. The heat required to change one mole of a solid completely to vapor is called the molar heat of sublimation, ΔH_{sub}. Note that $\Delta H_{sub} = \Delta H_{fus} + \Delta H_{vap}$.

Next, we move to liquids. A liquid is composed of molecules constantly moving in a random fashion; however, attractive forces between molecules prevent it from being a gas. These attractive forces hold the molecules close together, so that increasing the pressure has little effect on the volume. As a result, liquids are virtually incompressible. Also, changes in temperature cause only small volume changes. Liquids diffuse more slowly than gases, but their rates increase with increasing temperature. The heat of vaporization of a substance is the number of calories required to convert 1g of liquid to 1g of vapor without a change in temperature. The reverse process is called the heat of condensation, and involves the removal of the same amount of heat energy. The heat needed to vaporize 1 mole of a substance is called the molar heat of vaporization or the molar enthalpy of vaporization, ΔH_{vap}. Again, note that $\Delta H_{vap} = \Delta H_{vapor} - \Delta H_{liquid}$. We can also look at the change from a solid to a liquid, fusion. The number of calories needed to change 1g of a solid substance (at the melting point) to 1g of liquid (at the melting point) is called the heat of fusion. Note that $\Delta H_{fus} = \Delta H_{liquid} - \Delta H_{solid}$.

When the rate of evaporation equals the rate of condensation, the system is in equilibrium. Vapor pressure is the pressure exerted by gas molecules in equilibrium with their liquid. This pressure increases with increasing temperature. To determine the vapor pressure of an ideal solution at a particular temperature, we can use **Raoult's law**. This states that the vapor pressure is equal to the mole fraction of a solvent in the liquid phase multiplied by the vapor pressure of the pure solvent at the same temperature:

$$P_{solution} = X_{solvent} \, P^0_{solvent}$$

or

$$P_A = X_A \, P^0_A$$

where P_A is the vapor pressure with solute added, P^0_A is the vapor pressure of pure A, and X_A is the mole fraction of A in the solution (A is assumed to be nonvolatile).

EXAMPLE: The vapor pressures of pure benzene and toluene at 60°C are 385 and 139 Torr, respectively. Calculate (a) the partial pressures of benzene and toluene, (b) the total vapor pressure of the solution, and (c) the mole fraction of toluene in the vapor above a solution with 0.60 mole fraction toluene.

SOLUTION: The vapor pressure of benzene over solutions of benzene and toluene is directly proportional to the mole fraction of benzene in the solution. The vapor pressure of pure benzene is the proportionality constant. This is analogous to the vapor pressure of toluene. This is known as Raoult's law. It may be written as

$$P_1 = X_1 \, P°_1$$
$$P_2 = X_2 \, P°_2$$

where 1 and 2 refer to components 1 and 2, P_1 and P_2 represent the partial vapor pressure above the solution, $P°_1$ and $P°_2$ are the vapor pressures of pure components, and X_1 and X_2 are their mole fractions. Solutions are called ideal if they obey Raoult's law.

The mole fraction of a component in the vapor is equal to its pressure fraction in the vapor. The total vapor pressure is the sum of the vapor's component partial pressures.

To solve this problem one must
- (1) calculate the partial pressures of benzene and toluene using Raoult's law;
- (2) find the total vapor pressure of the solution by adding the partial pressures; and
- (3) find the mole fraction of toluene in the vapor.

One knows the mole fraction of toluene in the solution is 0.60 and, thus, one also knows the mole fraction of benzene is $(1 - 0.60)$ or 0.40. Using Raoult's law:

$$P°_{benzene} = 385 \text{ Torr} \quad P°_{toluene} = 139$$

a) $P_{benzene} = (0.40)\,(385 \text{ Torr}) = 154.0 \text{ Torr}$

$P_{toluene} = (0.60)\,(139 \text{ Torr}) = 83.4 \text{ Torr}$

b) $P_{total} = 154.0 + 83.4 = 237.4 \text{ Torr}$

c) The mole fraction of toluene in the vapor =

$$X_{toluene}, \text{vap} = \frac{P_{toluene}}{P_{toluene} + P_{benzene}} = \frac{83.4}{237.4} = 0.351.$$

This brings us to gases. Gases, remember, have no fixed volume, so they assume the volumes of their containers. Gases are measured in terms of pressure, which is defined as "force per unit area." Standard atmospheric pressure is used in many problems and can be expressed in several ways:

$$14.7 \text{ psi} = 760 \text{ mm Hg} = 760 \text{ torr} = 1 \text{ atm}.$$

A familiar equation used in gas calculations is the ideal gas equation (or ideal equation of state):

$$PV = nRT,$$

where P is pressure, V is volume, T is temperature, n is number of moles, and R is the gas constant.

Molecules of the hypothetical ideal gas have no attraction for one another and have no intrinsic volume; they are "point masses." Several equations are derived from the ideal gas law. They are shown in Table 2.

TABLE 2 GAS LAW CORRELATIONS		
Law	**Definition (k = constant)**	**Form**
Boyle's	$T = k, PV = k, V \propto \dfrac{1}{P}$	$P_1 V_1 = P_2 V_2$
Charles'	$P = k, \dfrac{V}{T} = k, V \propto T$	$\dfrac{V_1}{T_1} = \dfrac{V_2}{T_2}$
Gay-Lussac	$V = k, \dfrac{P}{T} = k, P \propto T$	$\dfrac{P_1}{T_1} = \dfrac{P_2}{T_2}$
Combined Gas	$n = k, V \propto \dfrac{1}{P} \propto T$	$\dfrac{P_1 V_1}{T_1} = \dfrac{P_2 V_2}{T_2}$
Avogadro's	if $P_1 = P_2$ and $T_1 = T_2$, $V \propto n$	$\dfrac{V_2}{V_1} = \dfrac{n_2}{n_1}$

EXAMPLE: What pressure is required to compress 5 liters of gas at 1 atm pressure to 1 liter at a constant temperature?

SOLUTION: In solving this problem, one uses Boyle's Law: The volume of a given mass of gas at a constant temperature varies inversely with the pressure. This means that, for a given gas, the pressure and the volume are proportional, at a constant temperature, and their product equals a constant.

$$P \times V = k$$

where P is the pressure, V is the volume and k is a constant. From this one can propose the following equation

$$P_1 V_1 = P_2 V_2,$$

where P_1 is the original pressure, V_1 is the original volume, P_2 is the new pressure, and V_2 is the new volume.

In this problem, one is asked to find the new pressure and is given the original pressure and volume and the new volume.

$$P_1 V_1 = P_2 V_2$$

$P_1 = 1 \text{ atm}$

$V_1 = 5 \text{ liters}$

$$1 \text{ atm} \times 5 \text{ liters} = P_2 \times 1 \text{ liter}$$

$P_2 = ?$

$$\frac{1 \text{ atm} \times 5 \text{ liters}}{1 \text{ liter}} = P_2$$

$V_2 = 1 \text{ liter}$

$$5 \text{ atm} = P_2$$

Sometimes we have mixtures of gases. The pressure exerted by each gas in a mixture is called its partial pressure. The total pressure exerted by a mixture is equal to the sum of the partial pressures of the gases in the mixture. This is **Dalton's law of partial pressures**:

$$P_T = P_a + P_b + P_c \dots$$

The ideal gas obeys exactly the mathematical statement of the ideal gas law; however, most gases are not ideal. Real gases act in a less ideal way, especially under condition of increased pressure and/or decreased temperature. Real gas behavior approaches that of ideal gases as the gas pressure becomes very low.

Last, we'll look at effusion and diffusion. **Effusion** is the process in which a gas escapes from one chamber of a vessel by passing through a very small opening or orifice. **Graham's law of effusion** states that the mean molecule velocity and, hence, the rate of effusion (ρ) is inversely proportional to the square root of the density of the gas:

$$\text{rate of effusion} \propto \sqrt{\frac{1}{\rho}},$$

and

$$\frac{\text{rate of effusion (A)}}{\text{rate of effusion (B)}} = \sqrt{\frac{\rho_B}{\rho_A}} = \sqrt{\frac{MW_B}{MW_A}}$$

where M.W. is the molecular weight of each gas, and where temperature is the same for both gases.

EXAMPLE: Two gases, HBr and CH_4, have molecular weights 81 and 16, respectively. The HBr effuses through a certain small opening at the rate of 4 ml/sec. At what rate will the CH_4 effuse through the same opening?

SOLUTION: The comparative rates or speeds of effusion of gases are inversely proportional to the square roots of their molecular weights. This is written

$$\frac{\text{rate}_1}{\text{rate}_2} = \frac{\sqrt{MW_2}}{\sqrt{MW_1}}$$

For this case $\dfrac{\text{rate}_{HBr}}{\text{rate}_{CH_4}} = \dfrac{\sqrt{MW_{CH_4}}}{\sqrt{MW_{HBr}}}$

One is given the $\text{rate}_{HBr}, MW_{CH_4}$, and MW_{HBr} and asked to find rate CH_4.

Solving for rate CH_4:

$$\frac{rate_{HBr}}{rate_{CH_4}} = \frac{\sqrt{MW_{CH_4}}}{\sqrt{MW_{HBr}}}$$

$rate_{HBr} = 4 \ ml/sec$

$rate_{CH_4} = ?$

$$\frac{4 \ ml/sec}{rate_{CH_4}} = \frac{\sqrt{16}}{\sqrt{81}}$$

$MW_{CH_4} = 16$

$MW_{HBr} = 81$

$$rate_{CH_4} = \frac{4 \ ml/sec \times \sqrt{81}}{\sqrt{16}}$$

$$= \frac{4 \ ml/sec \times 9}{4} = 9 \ ml/sec$$

Mixing of molecules of different gases by random motion and collision until the mixture becomes homogeneous is called **diffusion**. Graham's law of diffusion states that the relative rates at which gases will diffuse will be inversely proportional to the square roots of their respective densities or molecular weights:

$$rate \propto \frac{1}{mass} \ (\text{where, again, } T_1 = T_2)$$

and

$$\frac{rate \ 1}{rate \ 2} = \frac{\sqrt{MW_2}}{\sqrt{MW_1}} \left(or \ \frac{r_1}{r_2} = \frac{\sqrt{d_2}}{\sqrt{d_1}} \right)$$

SOLUTIONS

Often in chemistry, we work with solutions; therefore, it is important that we understand some concepts of solution chemistry. A **solution** is a homogeneous mixture of substances. They may be gaseous, liquid, or solid. **Concentration** quantifies the amount of solute in a solution. Some concentration units are defined in Table 3:

Unit	Definition	Formula
TABLE 3 CONCENTRATION UNITS		
Mole fraction	$\dfrac{\text{\# moles of one component of solution}}{\text{total \# moles of all components in solution}}$	$X_A = \dfrac{n_A}{n_A + n_B + n_C + \ldots}$
Mole percent	Mole fraction expressed as %	$(X_A)(100\%)$
Weight fraction	$\dfrac{\text{g one component of solution}}{\text{total g of all components in solution}}$	$W_A = \dfrac{g_A}{g_A + g_B + g_C + \ldots}$
Weight percent	Weight fraction expressed as %	$(W_A)(100\%)$
Molarity	Moles of solute per liter of solution	$M = \dfrac{n \text{ solute}}{\ell \text{ solution}}$
Molality	Moles of solute per kg of solvent	$m = \dfrac{n \text{ solute}}{\text{kg solvent}}$
Normality	# equivalents of solute per liter of solution	$N = \dfrac{\text{equiv solute}}{\ell \text{ of solution}}$

EXAMPLE: Calculate the molarity of a solution containing 10.0 grams of sulfuric acid in 500 ml of solution. (MW of H_2SO_4 = 98.1)

SOLUTION: The molarity of a compound in a solution is defined as the number of moles of the compound in one liter of the solution. In this problem, one is told that there are 10.0 grams of H_2SO_4 present. One should first calculate the number of moles that 10.0 g represents. This can be done by dividing 10.0 g by the molecular weight of H_2SO_4.

$$\text{number of moles} = \frac{\text{amount present in grams}}{\text{molecular weight}}$$

$$\text{number of moles of } H_2SO_4 = \frac{10.0 \text{ g}}{98.1 \text{ g / mole}} = 0.102 \text{ moles}$$

Since molarity is defined as the number of moles in one liter of solution, and since, one is told that there is 0.102 moles in 500 ml ($\frac{1}{2}$ of a liter), one should multiply the number of moles present by 2. This determines the number of moles in H_2SO_4 present in 1,000 ml.

$$\text{Number of moles in 1,000 ml} = 2 \times 0.102 = 0.204$$

Because molarity is defined as the number of moles in 1 liter, the molarity (M) here is 0.204 M.

Now that concentration units have been defined, we can look at **solubility**, the concentration of dissolved solute in a saturated solution. A **saturated** solution is one in which solid solute is in equilibrium with dissolved solute, and no more solute will go "into solution." **Unsaturated** solutions contain less solute than required for saturation. **Supersaturated** solutions contain more solute than required for saturation (this is metastable state). Solubility may be affected by changes in temperature or pressure, but the exact change depends on the solution. The solubility of most *solids* in liquids increases with increasing temperature. For *gases* in liquids, the solubility usually decreases with increasing temperature. To analyze the change in solubility due to change in temperature, we may use the following equation:

$$\log \frac{K_2}{K_1} = \frac{-\Delta H^O}{2.303R}\left(\frac{1}{T_2} - \frac{1}{T_1}\right)$$

where K_2 = solubility constant at T_2, K_1 = solubility constant at T_1 (see the section on Equilibrium for a detailed definition of the solubility constant) and ΔH^O = enthalpy change when the solute dissolves in the solvent at standard conditions. A positive ΔH^O indicates that solubility increases with increasing temperature. Pressure has very little effect on the solubility of liquids or solids in liquid solvents; however, the solubility of gases in liquid (or solid) solvents always increases with increasing pressure.

Some solids are only sparingly soluble in solutions. The equilibrium constant for sparingly soluble compounds is termed the solubility product, K_{sp}. When performing calculations on sparingly soluble compounds, it is important to have a balanced equation for the dissolution of the compound. For example, let's look at the reaction

$$Ag_2CrO_4(s) \rightleftharpoons 2Ag^+(aq) + CrO_4^{-2}(aq), K_{sp} = 8.5 \times 10^{-8}.$$

The solubility of Ag_2CrO_4 at room temperature in moles/liter is determined as follows:

$$Ag_2CrO_4(s) \rightleftharpoons 2Ag^+(aq) + CrO_4^{-2}(aq)$$

start	a	0	0
equil	a-x	2x	x

$K_{sp} = [Ag^+]^2[CrO_4^{-2}] = 8.5 \times 10^{-8}$ (Ag_2CrO_4 is not included, because it is a solid)

Now, synthesize this information to determine "x":

$$(2x)^2(x) = 8.5 \times 10^{-8}$$
$$2x^3 = 8.5 \times 10^{-8}$$
$$x = 3.5 \times 10^{-3}$$

recalling that the solubility product is equal to the product of the concentrations of the ions, each raised to the power of its coefficient.

Aqueous solutions are a very important part of chemistry. Ionic compounds generally dissolve readily in water, but there are some exceptions.

TABLE 4
SOLUBLE IONS

Ion	Solubility	Exceptions
Nitrates, NO_3^-	All soluble	none
Chlorates, ClO_3^-	"	"
Acetates, CH^3COO^-	"	"
Sulfates, SO_4^{-2}	Soluble	$BaSO_4$, $SrSO_4$, $PbSO_4$ are insoluble $CaSO_4$, Ag_2SO_4, $HgSO_4$ are slightly insoluble
Chlorides, Cl^-	"	AgCl (insoluble), Hg_2Cl_2 (insoluble), $PbCl_2$ (slightly soluble)
Bromides, Br^-	"	AgBr (insoluble), Hg_2Br_2 (insoluble), $PbBr_2$ (soluble)
Iodides, I^-	"	AgI (insoluble), Ag_2I_2 (insoluble), PbI_2 (soluble)
Hydroxides, OH^-	Insoluble	Group IA metals (Li, Na, …) are soluble $Sr(OH)_2$, $Ca(OH)_2$ are slightly soluble
Carbonates, CO_3^{-2}	Insoluble	Group IA metals, $(NH_4)_2(CO_3)$ are soluble
Phosphates, PO_4^{-3}	"	Group IA metals, $(NH_4)_3(PO_3)$ are soluble
Sulfides, S^{-2}	"	Group IA, IIA metals, $(NH_4)_2S$ are soluble
Common salts of NH_4^+, Na^+, K^+	Almost all soluble	

We looked at vapor pressure when we reviewed states of matter. Vapor pressure is an example of a **colligative property**, one that depends on the number of solute particles in the solution. The colligative property law states that the freezing point, boiling point, and vapor pressure of a solution differ from those of the pure solvent by amounts directly proportional to the molal concentration of the solute. The vapor pressure of an aqueous solution is always lowered by the addition of more solute (which also causes the boiling point to be raised—boiling point elevation). The freezing point is always lowered by the addition of solute. The formulas are:

$$\text{freezing point depression} \quad \Delta T_f = -K_f m$$

$$\text{boiling point elevation} \quad \Delta T_b = K_b m$$

where $\Delta T = T_{solution} - T_{pure\ solvent}$, and m = molality of solute. K_f and K_b represent constants.

EXAMPLE: The freezing point constant of toluene is 3.33°C per mole per 1,000 g. Calculate the freezing point of a solution prepared by dissolving 0.4 mole of solute in 500 g of toluene. The freezing point of toluene is –95.0°C.

SOLUTION: The freezing point constant is defined as the number of degrees the freezing point will be lowered per 1,000 g of solvent per mole of solute present. The freezing point depression is related to this constant by the following equation:

freezing point depression = molality of solute × freezing point constant

The molality is defined as the number of moles per 1,000 g of solvent. Here one is given that 0.4 moles of solute are added to 500 g of solvent; therefore; there will be 0.8 moles in 1,000 g.

$$\frac{0.4 \text{ moles}}{500 \text{ g}} = \frac{0.8 \text{ moles}}{1,000 \text{ g}}$$

The molality of the solute is thus 0.8 m. One can now find the freezing point depression. The freezing point constant for toluene is 3.33°.

freezing point depression = molality × 3.33°

$$= 0.8 \times 3.33° = 2.66°$$

The freezing point of toluene is thus lowered by 2.66°.

freezing point of solution = (–95°C) – 2.66° = –97.66°C.

Another colligative property is osmotic pressure. Osmosis is the diffusion of a solvent through a semipermeable membrane into a more concentrated solution. The osmotic pressure of a solution is the minimum pressure that must be applied to the solution to prevent the flow of solvent from pure solvent into the solution. The formula for osmotic pressure in very dilute solutions is

$$\Pi = cRT,$$

where Π is osmotic pressure, c is concentration in molality or molarity, R is the gas constant, and T is temperature in Kelvins.

PERIODICITY

Now that we're familiar with compounds, let's familiarize ourselves with the periodic table and its elements. **Periodic law** states that chemical and physical properties of elements are periodic functions of their atomic numbers. The periodic table provides a systematic representation of these chemical and physical properties. Vertical columns are called groups, and each contain a family of elements possessing similar chemical properties. The horizontal rows in the table are called periods. The elements lying in the two rows just below the main part of the table are called the inner transition elements. In the first of these rows are elements 58 through 71, called the lanthanides or rare earth elements. The second row consists of elements 90 through 103, the actinides. Group IA elements are the alkali metals; they have a single S electron in the outer orbital. Group IIA elements are the alkaline earth metals; they have two S electrons in the outer orbital. Group VIIA elements are the halogens, and the Group VIIIA elements are the noble gases. The Group B elements are the semimetals and transition elements. For the Group A elements, the column numbers indicate the number of filled valence shell electron orbitals (since the Group VIIIA noble gases have eight electrons in their valence shells, they are the most stable of all elements). Group B elements are harder to classify since they commonly possess several valence states.

The periodic table provides a lot of information, and various trends can be outlined in its design. The diagram below defines some of these trends or periodicity.

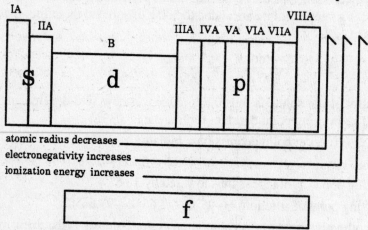

Figure 1. General characteristics of periodic table

The diagram above illustrates periodic trends in atomic size, electronegativity, and ionization energy. The electronegativity of an element is a number that measures the relative strength with which the atoms of the element attract valence electrons in a chemical bond. (Recall that ionic bonding occurs when electrons are transferred during the formation of a bond, whereas electrons are shared in covalent bonding.) Ionization energy is defined as the energy required to remove an electron from an isolated atom in its ground state.

EXAMPLE: Distinguish a metallic bond from an ionic bond and from a covalent bond.

SOLUTION: The best way to distinguish between these bonds is to define each and provide an illustrative example of each.

When an actual transfer of electrons results in the formation of a bond, it can be said that an ionic bond is present. For example,

$$2K° + \ddot{S}: \rightarrow 2K^+ + :\ddot{S}:^{2-} \rightarrow K_2S$$

| potassium atoms | sulfur atom | potassium ions (unlike ions due to transfer of electrons from potassium to sulfur) | sulfur ion | ionic bond due to the attraction of unlike ions |

When a chemical bond is the result of the sharing of electrons, a covalent bond is present. For example:

$$: \ddot{Br} \cdot \quad + \quad \cdot \ddot{F}: \quad \rightarrow \quad : \ddot{Br}: \ddot{F}:$$

These electrons are shared by both atoms.

A pure crystal of elemental metal consists of millions of atoms held together by metallic bonds. Metals possess electrons that can easily ionize, i.e., they can be easily freed from the individual metal atoms. This free state of electrons in metals binds all the atoms together in a crystal. The free electrons extend over all the atoms in the crystal and the bonds formed between the electrons and positive nucleus are electrostatic in nature. The electrons can be pictured as a "cloud" that surrounds and engulfs the metal atoms.

The periodic table also reveals the number of electrons in the valence shell of an atom. The "A" group numbers describe the number of electrons that are filling the valence shell of an atom in that column. This is not a foolproof rule, but it helps to generalize valence shell occupancies and vacancies when determining bonding or ion charges.

The table also follows the sequence for filling electron orbital subshells (s, p, d, f, etc.). Recall the mnemonic device for electron distribution.

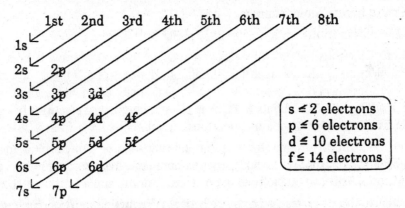

Figure 2. Electron distributions

Also, remember that the filling of orbitals follows several rules. Only two atoms may occupy an orbital at a time, and their spins must be paired, or opposite. If we look across the periodic table, we may see the trends of electron addition. For example:

	1s	2s	$2p_x$	$2p_y$	$2p_z$	
3_{Li}	⇅	↑				$1s^2 2s^1$
4_{Be}	⇅	⇅				$1s^2 2s^2$
6_C	⇅	⇅	↑	↑		$1s^2 2s^2 2p^2$
8_O	⇅	⇅	⇅	↑	↑	$1s^2 2s^2 2p^4$

Figure 3. Electron addition

Note: The number of electrons generally follows the atomic number of the element. Since the atomic number represents the number of protons in the element, a neutral atom would have the same number of electrons as protons. Also, note that Hund's rule is in effect—electrons fill all empty subshells within a level before they pair up in a subshell.

EXAMPLE: The faint light sometimes seen over marshland at night, the "will-o'-the-wisp," is believed to come about as a result of the burning of a compound of phosphorus (P) and hydrogen (H). What is the formula of this compound?

$$\begin{array}{c} H \\ {}^{\times}{}_{\bullet} \\ H\,{}^{\times}_{\bullet}\,P\,{}^{\times}_{\bullet}\,H \\ {}_{\bullet\bullet} \end{array}$$

SOLUTION: To find the formula of this compound, it is necessary to determine the valence of the elements from which it is composed. The valence of an element is the number of electrons that are involved in chemical bonding.

To find the valence of phosphorus and hydrogen, consider their atomic number and electronic configuration.

> Hydrogen: Atomic number = 1
>
> Electronic configuration = $1s^1$
>
> Phosphorus: Atomic number = 15
>
> Electronic configuration = $1s^2\, 2s^2\, 2p^6\, 3s^2\, 3p^3$

The outer electrons are $1s^1$ for hydrogen and $3p^3$ for phosphorus. It takes one additional electron to fill hydrogen's s orbital; its valence is one. It takes three more electrons to fill phosphorus' p orbital; its valence is three. Elements react with the purpose of filling all their orbitals with the maximum number of electrons by either a transfer of electrons or by sharing electrons.

It would take three hydrogen atoms to complete phosphorus' outer orbital. In turn, each electron of phosphorus would serve to complete the outer orbital of each hydrogen atom. This can be pictured in an electron-dot formula as shown above.

In this figure, the x's represent the outer electrons of hydrogen and dots represent the electrons in the outer shell of phosphorus. The formula of this compound is, thus, PH_3.

EQUILIBRIUM

Now let's look at a specific type of chemical state: equilibrium. When a system is in a state of equilibrium, both the forward and reverse reactions take place at the same rate. There is no spontaneous forward or reverse reaction favored, and the concentrations of reactants and products no longer change with time. If the system is disturbed by an outside stress (change in temperature, pressure, or concentration, for example), it adjusts to a new equilibrium. Chemical equilibrium provides useful information in areas such as kinetics, thermodynamics, and the compositions of chemical systems.

We work with chemical equilibrium by using K_{eq}, the equilibrium constant. It is usually written in the form of a fraction, with product values in the numerator and reactant values in the denominator. For the reaction $aA + bB \rightleftharpoons eE + fF$:

$$K_{eq} = \frac{[E]^e[F]^f}{[A]^a[B]^b},$$

where [A-F] indicates the thermodynamic activity. This relationship is known as the mass action equation, and

$$\frac{[E]^e[F]^f}{[A]^a[B]^b}$$

is the mass action expression. It is very important to note that pure solids and pure liquids are not included in the expression, since their concentrations are fairly constant at any given temperature. If the reaction involves solutions, we express concentration in terms of molarity, molality, etc. If the reaction involves gases, we express concentration as pressures. Thus, for the reaction

$$NH_3(g) + HCl(l) \rightleftharpoons NH_4^+(aq) + Cl^-(aq)$$

$$K_{eq} = \frac{[NH_4^+][Cl^-]}{P_{NH_3}}$$

It is also useful to note that the size of the K_{eq} value provides an indication of the direction of the reaction. If the K_{eq} value is large (e.g., 10^5), we know that there must be a large concentration of products. The reaction must favor the formation of products. If the K value is small, reactants are favored.

Kinetic information can be gathered by analyzing the mass action expression of a reversible reaction. The rate of an elementary chemical reaction is proportional to the concentrations of the reactants raised to powers equal to their stoichiometric coefficients in the balanced equation. For the equation, $aA + bB \rightleftharpoons eE + fF$:

$$\text{rate}_{forward} = k_f[A]^a[B]^b$$
$$\text{rate}_{reverse} = k_r[E]^e[F]^f$$

where k_f and k_r are the rate constants for the forward and reverse reactions, respectively. If we arrange this information into the form of the mass action expression, the relationship between K_{eq} and rate can be observed, as shown below:

$$K_{eq} = \frac{[E]^e[F]^f}{[A]^a[B]^b} = \frac{k_f}{k_r} \quad (\text{at equilibrium rate}_f = \text{rate}_r).$$

Chemical equilibrium also provides thermodynamic information. If we use the symbol, Q, to represent the mass action expression for a reaction at any given time, we can determine that reaction's favored direction. Q is written using the same concentrations and form as K_{eq}, so we may say that

when Q < K then products favored

when Q = K then reaction at equilibrium

when Q > K then reactants favored

Q may also be used in Gibb's free energy determinations:

$$\Delta G = \Delta G^0 + 2.303 \ RT \ \log Q,$$

where ΔG is the free energy change under conditions other thAn standard conditions, and ΔG^0 is the free energy under standard conditions. At equilibrium, $Q = K$, and the products and reactants have the same total Gibb's free energy such that $\Delta G = 0$. Therefore,

$$\Delta G^0 = -2.303 RT \ \log K_{eq} = -RT \ \ln K_{eq}.$$

In some situations, gases are also expressed in terms of concentration rather than pressure. We can differentiate between using concentration values and pressure values by using K_c and K_p, respectively. K_c and K_p are related by the ideal gas equation, $PV = nRT$. Since $P = \left(\dfrac{n}{V}\right) RT$, we can substitute $\dfrac{n}{V}$ (concentration) into the mass action expression:

$$K_p = \frac{P_E^e P_F^f}{P_A^a P_B^b} = \frac{[E]^e (RT)^e [F]^f (RT)^f}{[A]^a (RT)^a [B]^b (RT)^b} = \frac{[E]^e [F]^f}{[A]^a [B]^b} (RT)^{(e+f)-(a+b)}$$

Therefore, $K_p = K_c (RT)^{\Delta n}$, where Δn represents the change in the number of moles of gas upon going from reactants to products.

K_{eq} can also be used to determine the status of other equilibrium states of the reaction at the same temperature. For example, if a chemist performs the reaction

$$2NO \ (g) + Br_2 \ (g) \rightleftharpoons 2NOBr \ (g)$$

where $K_{eq} = 100$ and the initial pressure of the reactants is 1 atm, we have enough information to determine the quantity of NOBr formed. First, write the equilibrium expression from the balanced equation:

$$K_{eq} = \frac{[NOBr]^2}{[NO]^2 [Br_2]} = 100.$$

To find out how much NOBr is produced, we have to know how many moles of NO and Br_2 reacted. Then, we can find the number of grams produced. The equilibrium expression is based on the concentrations of reactants and products. We can express concentrations as moles per liter. This means that if the volume and concentration of NOBr is known, we can find moles (since mol/liter $\times$ liter $=$ moles).

If we express the concentration as $\dfrac{n}{V}$, the expression becomes

$$K = \frac{\left(\dfrac{n_{NOBr}}{V}\right)^2}{\left(\dfrac{n_{NO}}{V}\right)^2 \left(\dfrac{n_{Br_2}}{V}\right)}.$$

Let x = moles of NOBr formed. Then, x moles of NO and $\frac{x}{2}$ moles of Br_2 are consumed (since the coefficients of the reaction show a 2:2:1 ratio for $NOBr:NO:Br_2$). Now,

$$K = 100 = \frac{n_{NOBr}^2 V}{n_{NO}^2 n_{Br_2}} = \frac{x^2 V}{(2-x)^2 (1-0.5x)}$$

$$2NO + Br_2 \rightleftharpoons 2NOBr$$

start	2	1	0

equil	$2-x$	$1-0.5x$	x

If x moles of NOBr form and we started with 2 moles of NO, then, at equilibrium, we have $(2-x)$ moles of NO left. We started with only 1 mole of Br_2, of which $0.5x$ moles form NOBr; thus, we only have $1-0.5x$ moles of Br_2 left. To determine the quantity of NOBr formed, we now have to calculate the volume.

$$V = \frac{nRT}{P} = \frac{(3\text{mol})\left(0.821\frac{\text{liters} - \text{atm}}{\text{mol} - \text{K}}\right)(273\text{ K})}{1\text{ atm}} = 673 \text{ liters}$$

Now that V is known:

$$K = \frac{x^2 (673)}{(2-x)^2 (1-0.5x)} = 100$$

We solve for x either by factoring or by the quadratic equation,

$$x = \frac{b \pm \sqrt{b^2 - 4ac}}{2a},$$

and obtain $x = 0.923$ moles = moles of NOBr formed. Now, use the molecular weight of NOBr to determine the grams produced (101.53 g).

One last bit of useful information involves LeChatelier's principle. It states that when a system at equilibrium is disturbed by the application of a stress, it reacts to minimize the stress and attain a new equilibrium position.

TABLE 5 LECHATELIER'S PRINCIPLE	
Stress	**Result**
Increase in concentration of reactants	Increase in amount of product formed
Decrease in concentration of reactants	Decrease in amount of product formed
Increase in temperature	Exothermic reaction: shift to left
	Endothermic reaction: shift to right
Increase in pressure	Shift in position of equilibrium in direction of fewest number of moles of gaseous reactant or product
Catalyst	Speeds approach to equilibrium, but K_{eq} does not change
Addition of inert gas	Increases pressure, but K_{eq} is not affected

ACIDS AND BASES

Next, we come to a familiar topic in chemistry. There are several definitions of acids and bases. The Arrhenius theory states that acids are substances that ionize in water to give H^+ ions, and bases are substances that produce OH^- ions in water. The Bronsted-Lowry theory defines acids simply as proton donors and bases as proton acceptors. Finally, the Lewis theory defines an acid as an electron pair acceptor and a base as an electron pair donor.

Several factors influence the strength of acids. The greater the number of oxygens bound to the element "E" in the hydroxy compound, H_xEO_y, the stronger the acid. This is also a positive correlation with oxidation state of E. In addition, as we move down a group in the periodic table, the strength of the oxoacids decreases.

Acids and bases are often examined in terms of acid-base equilibria in aqueous solution. First, we can study the ionization of water and the concept of pH. For the equation:

$$H_2O + H_2O \rightleftharpoons H_3O^+ + OH^-, \quad K_w = [H_3O^+][OH^-] = 1.0 \times 10^{-14} \text{ at } 25°C.$$
$$\text{(or } K_w = [H^+][OH^-])$$

$[H_3O^+][OH^-]$ is the product of ionic concentrations, and K_w is the dissociation, or ionization, constant for water. When we think of acids, we often think of pH. pH is related to K_w as follows:

$$pH = -\log [H^+]$$
$$pOH = -\log [OH^-]$$
$$pK_w = pH + pOH = 14.0$$

In a neutral solution, pH = 7.0, the concentration of acids and bases are equal. In an acidic solution, pH is less than 7.0; in basic solutions, pH is greater than 7.0. Note that since K_w varies with temperature (as do all equilibrium constants), neutral pH may differ from 7.0 when the temperature differs from 25°C.

The dissociation of weak electrolytes (weak acids and bases) can be defined in the same way. For the equation:

$$A^- + H_2O \rightleftharpoons HA + OH^-, \quad K_b = \frac{[HA][OH^-]}{[A^-]},$$

where K_b is the base ionization constant. Here, A^- is acting as the base (conjugate base).

For the reaction:

$$HA + H_2O \rightleftharpoons A^- + H^- (H_3O^+), \quad K_a = \frac{[H^+][A^-]}{[HA]},$$

where K_a is the acid ionization constant, and HA is the acid. For any conjugate acid-base pair, the two constants are related to K_w as follows:

$$K_w = K_a K_b.$$

Thus,

$$K_w = [H^+][OH^-].$$

For polyprotic acids, there is more than one dissociation constant. For example:

$$H_2S \rightleftharpoons H^+ + HS^- \qquad K_{a1} = \frac{[H^+][HS^-]}{[H_2S]}$$

$$HS^- \rightleftharpoons H^+ + S^{-2} \qquad K_{a2} = \frac{[H^+][S^{-2}]}{[HS^-]}$$

Also, $K_{a1} \gg K_{a2}$, since it is easier to "pull off" the first H^+ than the second. The total K_a is

$$K_a = K_{a1} \times K_{a2} = \frac{[H^+]^2[S^{-2}]}{[H_2S]}.$$

Acids and bases are used extensively in buffer systems. **Buffer solutions** are equilibrium systems that resist changes in acidity, maintaining constant pH when acids or bases are added to them. The most effective pH range for any buffer is at or near the pH where the acid and salt concentrations are equal (that is pK_a). The pH for a buffer is given by

$$pH = pK_a + \log\frac{[A^-]}{[HA]} = pK_a + \log\frac{[base]}{[acid]},$$

which is obtained very simply from the equation for weak acid equilibrium.

Hydrolysis refers to the action of salts of weak acids or bases with water to form acidic or basic solutions. Some examples are shown below.

Salts of Weak Acids and Strong Bases: Anion Hydrolysis

For $C_2H_3O_2^- + H_2O \rightleftharpoons HC_2H_3O_2 + OH^-$ $\quad K_h = \dfrac{[HC_2H_3O_2][OH^-]}{[C_2H_3O_2^-]}$

$K_h = \dfrac{K_w}{K_a}$ $\qquad\qquad K_a = \dfrac{[H^+][C_2H_3O_2^-]}{[HC_2H_3O_2]}$

Salts of Strong Acids and Weak Bases: Cation Hydrolysis

For $NH_4^+ + H_2O \rightleftharpoons H_3O^+ + NH_3$ $\quad K = \dfrac{[H_3O^+][NH_3]}{[NH_4^+]}$

$K_h = \dfrac{K_w}{K_b}$ $\qquad\qquad K_1 = \dfrac{[NH_4^+][OH^-]}{[NH_3]}$

Hydrolysis of Salts of Polyprotic Acids

For $S^{-2} + H_2O \rightleftharpoons HS^- + OH^-$ $\quad K_{h1} = \dfrac{K_w}{K_{a2}} = \dfrac{[HS^-][OH^-]}{[S^{-2}]}$

$HS^- + H_2O \rightleftharpoons H_2S + OH^-$ $\quad K_{h2} = \dfrac{K_w}{K_{a1}} = \dfrac{[H_2S][OH^-]}{[HS^-]}$

In chemistry, titration is often used as an analytical tool. Titration is the process of determining the amount of a solution of known concentration that is required to react completely with a certain amount of a sample that is being analyzed. The solution of known concentration is called a standard solution, and the sample being analyzed is the unknown. The concentration unit, normality, is used rather than molarity. Recall that equivalents are used in normal concentration units, and here an equivalent is defined as a substance that releases one mole of either protons or hydroxyl ions. One may define equivalents as

$$equiv_A = V_A N_A \text{ and } equiv_B = V_B N_B,$$

where V is the volume of solution in liters, and N is the normality in equivalents per liter. At the equivalence point, $eq_A = eq_B$ and $V_A N_A = V_B N_B$. The equivalence point occurs when equal numbers of equivalents of acid and base have reacted. It is often useful to look at titration curves. For example, look at the following strong acid-strong base curve.

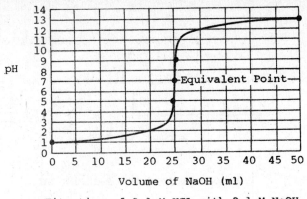

Titration of 0.1 M HCL with 0.1 M NaOH.

Figure 4. Titration of 0.1 M HCL with 0.1 M NaOH

At the equivalence point, the solution is neutral because neither of the ions of the salt solution undergoes hydrolysis.

One last term related to acid-base chemistry is the term amphoteric. An **amphoteric** substance is one that may act as an acid or a base. For example, aluminum hydroxide, $AL(OH)_3$, can donate H^+ or OH^- ions depending on the concentration of H^+ or OH^- ions from other compounds in the solution.

OXIDATION-REDUCTION

We will now review oxidation-reduction or "redox" reactions. **Oxidation** is defined as a reaction in which atoms or ions undergo an increase in oxidation state (they lose electrons to become more positive). **Reduction** is defined as a reaction in which atoms or ions undergo a decrease in oxidation state (they gain electrons and become more negative). The term "oxidation state" can be used interchangeably with the term "oxidation number." **Oxidation number** can be defined as the charge that an atom would have if both of the electrons in each bond were assigned to the more electronegative element. The following are basic rules for assigning oxidation numbers:

(1) The oxidation number of any element in its elemental form is zero.
(2) The oxidation number of any monatomic ion is equal to the charge on that ion.
(3) The sum of all the oxidation numbers of all of the atoms in a neutral compound is zero (or, more generally, the sum of the oxidation numbers of all the atoms in a given species is equal to the net charge on that species).

Recall that a redox reaction involves an oxidizing and a reducing agent. The oxidizing agent is reduced and the reducing agent is oxidized.

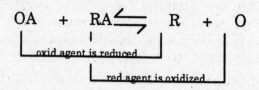

Figure 5. Oxidation and reduction equilibrium

Like any other chemical equation, a "redox" reaction equation must also be properly balanced. The transfer of electrons makes it a little tricky, but here are a couple of step-by-step methods:

The Oxidation-Number-Change-Method

(1) Assign oxidation numbers to each atom in the equation.
(2) Note which atoms change oxidation number and calculate the number of electrons transferred per atom during the reaction.
(3) Sometimes, more than one atom of an element that changes oxidation number is present. If so, calculate the number of electrons transferred per formula unit.
(4) Make the number of electrons gained equal to the number of electrons lost.
(5) Once the coefficients from step 4 have been attained, the remainder of the equation can be balanced by inspection. Add H^+ and H_2O (in acidic solution), or OH^- and H_2O (in basic solution), as required.

Sometimes, you don't know (or don't want to calculate) the oxidation numbers of each atom. You may use the ion-electron (or half-reaction) method to balance your equation. Here the reaction is split in half: the oxidizing part and the reducing part.

The Ion-Electron Method

(1) First, determine which of the substances present are involved in the redox reaction.
(2) Break the overall reaction into two half-reactions; one for the oxidation step and one for the reduction step. Note: The half-reactions are written in the same directions that they appear in the original equation.
(3) Balance the numbers and kinds of atoms on each side of the equation (make sure that there is the same number of each kind of atom on each side) for all species *except* H and O.
(4) Balance the number of O atoms first by adding H_2O as required; then, balance the H atoms (for acidic solutions, balance with H^+; for basic solutions, use OH^-).

(5) Balance the reactions electrically by adding electrons to either side. The total electric charge should be the same on the left and right sides.

(6) Multiply the two balanced half-reactions by the appropriate factors so that the same number of electrons is transferred in each.

(7) Add these half-reactions to obtain the balanced overall reaction (the electrons should cancel from both sides of the final equation).

Redox reactions are a very important part of electrochemistry, and we can gain a better understanding of them by looking at their electrochemical applications (see problem 9).

EXAMPLE: Balance the equation for the following reaction taking place in aqueous acid solution:

$$Cr_2O_7^{2-} + I_2 \rightarrow Cr^{3+} + IO_3^-$$

SOLUTION: The equation in this problem involves both an oxidation and a reduction reaction. It can be balanced by using the following rules: (1) Separate the net reaction into its two major components: the oxidation process (the loss of electrons) and the reduction process (the gain of electrons). For each of these reactions, balance the charges by adding H^+, if the reaction is occurring in an acidic medium, or OH^- in a basic medium. (2) Balance the oxygens by addition of H_2O. (3) Balance hydrogen atoms by addition of H. (4) Combine the two half-reactions, so that all charges from electron transfer cancel out. These rules are applied in the following example.

The net reaction is

$$Cr_2O_7^{2-} + I_2 \rightarrow Cr^{3+} + IO_3^-.$$

The oxidation reaction is

$$I_2^0 \rightarrow 2IO_3^- + 10e^-.$$

The I atom went from oxidation number of O in I_2 to +5 in IO_3^-, because O always has a −2 charge. You begin with I_2, therefore, 2 moles of IO_3^- must be produced and 10 electrons are lost, 5 from each I atom. Recall, the next step is to balance the charges. The right side has a total of 12 negative charges. Add 12 H$^+$'s to obtain

$$I_2 \rightarrow 2IO_3^- + 10e^- + 12H^+.$$

To balance the oxygen atoms, add $6H_2O$ to the left side, since there are 6 O's on the right, thus,

$$I_2 + 6H_2O \rightarrow 2IO_3^- + 10e^- + 12H^+.$$

Hydrogens are already balanced. There are 12 on each side. Proceed to the reduction reaction:

$$Cr_2O_7^{2-} + 6e^- \rightarrow 2Cr^{3+}.$$

Cr began with an oxidation state of +6 and went to +3. Since $2Cr^{3+}$ are produced, and you began with $Cr_2O_7^{2-}$, a total of 6 electrons are added to the left. Balancing charges: the left side has 8 negative charges and the right side has 6 positive charges.

If you add $14H^+$ to the left, they balance. Both sides now have a net $+3$ charge. The equation can now be written.

$$Cr_2O_7^{2-} + 6e^- + 14H^+ \rightarrow 2Cr^{3+}$$

To balance oxygen atoms, add $7H_2O$'s to the right. You obtain

$$Cr_2O_7^{2-} + 6e^- + 14H^+ \rightarrow 2Cr^{3+} + 7H_2O.$$

The hydrogens are also balanced, 14 on each side. The oxidation reaction becomes

$$I_2 + 6H_2O \rightarrow 2IO_3^- + 10e^- + 12H^+.$$

The reduction reaction is

$$Cr_2O_7^{2-} + 6e^- + 14H^+ \rightarrow 2Cr^{3+} + 7H_2O.$$

Combine these two in such a manner that the number of electrons used in the oxidation reaction is equal to the number used in the reduction. To do this, note that the oxidation reaction has $10e^-$ and the reduction $6e^-$. Both are a multiple of 30. Multiply the oxidation reaction by 3, and the reduction reaction by 5, obtaining

oxidation: $\quad 3I_2 + 18H_2O \rightarrow 6IO_3^- + 30e^- + 36H^+$

reduction: $\quad 5Cr_2O_7^{2-} + 30e^- + 70H^+ \rightarrow 10Cr^{3+} + 35H_2O$

Add these two half-reactions together.

$$3I_2 + 18H_2O \rightarrow 6IO_3^- + 30e^- + 36H^+$$
$$+ \quad 5Cr_2O_7^{2-} + 30e^- + 70H^+ \rightarrow 10Cr^{3+} + 35H_2O$$
$$\overline{3I_2 + 18H_2O + 5Cr_2O_7^{2-} + 30e^- + 70H^+ \rightarrow 10Cr^{3+} + 35H_2O + 30e^- + 36H^+}$$

Simplifying, you obtain

$$3I_2 + 5Cr_2O_7^{2-} + 34H^+ \rightarrow 6IO_3^- + 10Cr^{3+} + 17H_2O.$$

This is the balanced equation.

ELECTROCHEMISTRY

First, we will look at electrochemical cells. These include things like common flashlight batteries to larger salt solution cells. Cells are important chemical "translators." They can convert electrical energy into chemical energy, or they may convert chemical energy into electrical energy.

Reactions that do not occur spontaneously can be forced to take place by supplying energy through an external current. These are called electrolytic reactions. In **electrolytic cells**, electrical energy is converted to chemical energy.

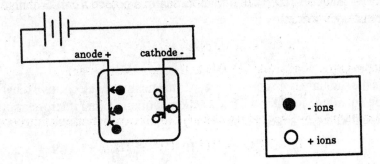

Figure 6. Simple battery

The other type of electrochemical cell is the galvanic (or voltaic cell). Flashlight batteries are of this type. In galvanic cells, chemical energy is converted to electrical energy (the opposite of electrolytic cells). Since the energy conversion process proceeds in the opposite direction in galvanic cells as compared to electrolytic cells, their anodes and cathodes are switched. In galvanic cells, the anode is negative, and the cathode is positive. Electrons flow from the negative electrode to the positive electrode.

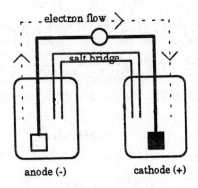

Figure 7. Salt bridge

The force driving the electrons is called the electromotive force (emf) and is measured in volts (V), where 1 V = 1 J/coulomb. The greater the tendency, or potential, of the two half-reactions to occur spontaneously, the greater will be the emf of the cell. The emf of the cell is called the cell potential, or E_{cell}. The cell potential for a Zn/Cu cell may be written as

$$E^0_{cell} = E^0_{Cu} - E^0_{Zn},$$

where the E^0's are the standard reduction potentials (these are generally provided in a table of reduction potentials). The overall standard cell potential is obtained by subtracting the smaller reduction potential from the larger one. A positive emf corresponds to a negative ΔG (and, therefore, to a spontaneous process). For a cell at concentrations and conditions other than standard, a potential can be calculated using the Nernst equation below:

$$E_{cell} = E^0_{cell} - \frac{0.059}{n} \log(Q),$$

where E^0_{cell} is the standard state cell voltage, n is the number of electrons exchanged in the reaction equation, and Q is the mass action quotient. Other interpretations of the Nernst equation are arranged below.

In terms of cell potential: $\quad E = E^0 - \frac{RT}{nF} \ln Q \quad E = E^0 - \frac{0.059}{n} \log(Q)$

In terms of Gibb's free energy: $\quad \Delta G = -nFE \quad\quad \Delta G = G^0 + 2.30RT \log (Q)$

In the above equations, F represents a faraday, the electrical charge on one mole of electrons. One faraday is approximately 96,500 coulombs per mole of electrons.

EXAMPLE: Calculate the voltage of the cell Fe; Fe^{+2} ∥ H^+; H_2 if the iron half cell is at standard conditions but the H^+ ion concentrations is .001M.

SOLUTION: The voltage (E) of a cell is found using the Nernst equation because it involves the use of concentration factors. It is stated

$$E = E^0 - \frac{RT}{nF} \ln Q,$$

where R is 8.314 joules per degree, F is 96,500 coulombs, n is the number of moles of electrons transferred, and Q is the concentration term. $T = 25°C$, by definition of standard conditions, in this equation (or 298K).

But to solve the problem we must first obtain E^0. This is done by writing down the appropriate half-reactions. Oxidation is the loss of electrons and reduction is the gain of electrons.

Reaction	Type	E^0_{red}
$Fe \rightarrow Fe^{+2} + 2e^-$	Oxidation	+.44
$2H^+ + 2e^- \rightarrow H_2$	Reduction	0

Next, take the algebraic sum of the E^0_{red} and E^0_{oxid}, which gives E^0.

$$E^0 = +.44 + 0 = +.44.$$

Now set up the concentration term.

$$\ln Q = \ln \frac{\left[H^+\right]^2}{\left[Fe^{+2}\right]} = \ln \frac{(.001)^2}{(1)}$$

Standard conditions always mean a concentration of 1M. Substituting all these terms into the Nernst equation, one calculates E to be

$$E = +.44 - \frac{.059}{2} \ln \frac{10^{-6}}{1} = .85 \text{volt}.$$

Now, we'll take quite a different turn. In electrochemistry, we can also include the behavior of electrons and photons. First, let's review some atomic structure theory. The ground state is the lowest energy state available to the atom. The excited state is any state of energy higher than that of the ground state. The formula for changes in energy, ΔE, is

$$\Delta E = E_{final} - E_{initial}.$$

When an electron moves from the ground state to an excited state, it absorbs energy. Conversely, when an electron moves from an excited state back down to the ground state, it emits energy. This exchange of energy is the basis for atomic spectra.

We can now look at the Bohr theory of the hydrogen atom. To the hydrogen atom, Bohr applied the concept that the electron can exist in only certain stable energy levels and that when the electronic state of the atom changes, it must absorb or emit exactly that amount of energy equal to the difference between the final and initial states:

$$\Delta E = E_a - E_b.$$

The equation below measures the energy difference between states a and b, where n = the (quantum) energy level, E = energy, e = charge on the electron, a_0 = Bohr radius, and z = atomic number:

$$E_a - E_b = \frac{z^2 e^2}{2a_0}\left[\frac{1}{n_a^2} - \frac{1}{n_b^2}\right].$$

The Rydberg-Ritz equation permits calculation of the spectral lines of hydrogen:

$$\frac{1}{\lambda} = R\left[\frac{1}{n_a^2} - \frac{1}{n_b^2}\right],$$

where R = 109,678 cm^{-1} (Rydberg constant), n_a and n_b are quantum numbers for states a and b, and λ is the wavelength of light emitted or absorbed.

EXAMPLE: The Rydberg-Ritz equation governing the spectral lines of hydrogen is $\frac{1}{\lambda} = R\left(\frac{1}{n_1^2} - \frac{1}{n_2^2}\right)$, where R is the Rydberg constant, n_1 indexes the series under consideration ($n_1 = 1$ for the Lyman series, $n_1 = 2$ for the Balmer series, $n_1 = 3$ for the Paschen series), $n_2 = n_1 + 1, n_1 + 2, n_1 + 3, ...$indexes the successive lines in a series, and λ is the wavelength of the line corresponding to index n_2. Thus, for the Lyman series, $n_1 = 1$ and the first two lines are 1215.56 Å ($n_2 = n_1 + 1 = 2$) and 1025.83 Å ($n_2 = n_1 + 2 = 3$). Using these two lines, calculate two separate values of the Rydberg constant. The actual value of this constant is R = 109,678 cm^{-1}.

SOLUTION: The first thing to do is to convert the wavelengths from Å to more manageable units, i.e., centimeters. Using the relationship 1 Å$=10^{-8}$ cm, the first two

Lyman lines are 1215.56 $\overset{\circ}{\text{A}}$ = 1215.56 × 10^{-8} cm for n_2 = 2, and 1025.83 $\overset{\circ}{\text{A}}$ = 1025.83 × 10^{-8} cm for n_2 = 3. Solving the Rydberg-Ritz equation for R, one obtains

$$R = \left(\lambda \left(\frac{1}{n_1^2} - \frac{1}{n_2^2} \right) \right)^{-1}.$$

For the first line,

$$R = \left(\lambda \left(\frac{1}{n_1^2} - \frac{1}{n_2^2} \right) \right)^{-1} = \left(1215.56 \times 10^{-8} \text{cm} \left(\frac{1}{1^2} - \frac{1}{2^2} \right) \right)^{-1}$$

$$= 109,689 \text{ cm}^{-1}$$

and for the second line,

$$R = \left(\lambda \left(\frac{1}{n_1^2} - \frac{1}{n_2^2} \right) \right)^{-1} = \left(1025.83 \times 10^{-8} \text{cm} \left(\frac{1}{1^2} - \frac{1}{3^2} \right) \right)^{-1}$$

$$= 109,667 \text{ cm}^{-1}$$

The first of these is 0.0100% greater than the true value, and the second is 0.0100% less than the true value.

Recall that light behaves as if it were composed of tiny packets, or quanta, of energy (now called photons). The energy of a photon is defined as

$$E_{photon} = h\upsilon,$$

where h is Planck's constant, and υ is the frequency of the light. The frequency can be further broken down to make the equation even more elementary:

$$E = \frac{hc}{\lambda},$$

where c is the speed of light, and λ is the wavelength of light.

The electron is restricted to specific energy levels in the atom; it is quantized. Specifically,

$$E = \frac{-A}{n^2},$$

where A = 2.18 × 10^{-11} erg, and n is the quantum number.

Finally, we'll look at components of atomic structure. The number of protons and neutrons in the nucleus is called the **mass number**, which corresponds to the isotopic atomic weight. The **atomic number** is the number of protons found in the nucleus.

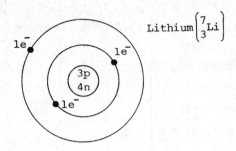

Figure 8. Lithium, $_3^7Li$ Electron(–), Proton(+), Neutron(0)

The electrons in the outermost shell are valence electrons. When these electrons are lost or partially lost (through sharing), the oxidation state is assigned a positive value for the element. If valence electrons are gained by an atom, its oxidation number is taken to be negative.

KINETICS

Now, we'll move on to another important area of chemistry, kinetics. **Kinetics** is the study of the rates of reactions. The measurement of reaction rate is based on the rate of appearance of a product or disappearance of a reactant. It is usually expressed in terms of a change in the concentration of one of the participants per unit time:

$$\text{rate of reaction} = \frac{\text{change in concentration}}{\text{time}} = \frac{\text{moles/liter}}{\text{sec}}.$$

For the general reaction $2AB \rightarrow A_2 + B_2$:

$$\text{average rate} = \frac{[AB]_{t2} - [AB]_{t1}}{t_2 - t_1} = -\frac{\Delta[AB]}{\Delta t},$$

where $[AB]_{t2}$ represents the concentration of AB at time t_2. There are a few important factors that control the rate of a reaction:

(1) The nature of the reactants and products (i.e., the nature of the transition state formed). Some elements and compounds, because of the bonds broken or formed, react more rapidly with each other than others do.

(2) The surface area exposed. Since most reactions depend on the reactants coming into contact with each other, increasing the surface area exposed proportionally increases the rate of the reaction.

(3) Concentrations. The reaction rate usually increases with increasing concentrations of reactants and products.

(4) Temperature. For most reactions, the reaction rate increases as temperature increases.

The Arrhenius equation relates rate to temperature and energy:

$$k = Ae^{\frac{-E_a}{RT}}$$

where k = Rate constant

A = Arrhenius constant

E_a = Activation energy

R = Universal gas constant

T = Temperature in Kelvins

EXAMPLE: What activation energy should a reaction have so that raising the temperature by 10°C at 0°C would triple the reaction rate?

SOLUTION: The activation energy is related to the temperature by the Arrhenius equation which is stated

$$k = Ae^{\frac{-E_a}{RT}}$$

where A is a constant characteristic of the reaction, e is the base of natural logarithms, E is the activation energy, R is the gas constant (8.314 J mol^{-1} deg^{-1}) and T is the absolute temperature. Taking the natural log of each side:

$$\ell nk = \ell nA - \frac{E}{RT}.$$

For a reaction that is 3 times as fast, the Arrhenius equation becomes

$$3k = Ae^{\frac{-E}{R(T+10°)}}$$

Taking the natural log:

$$\ell n3 + \ell nk = \ell nA - \frac{E}{R(T+10°)}$$

Subtracting the equation for the final state from the equation for the intital state:

$$\ell n\, k = \ell n\, A - \frac{E}{RT}$$

$$-\left(\ell n\, 3 + \ell n\, k = \ell n\, A - \frac{E}{R(T+10°)}\right)$$

$$-\ell n\, 3 = -\frac{E}{RT} + \frac{E}{R(T+10°)}$$

Solving for E:

$$\ell n\, 3 = \frac{E}{RT} + \frac{E}{R(T+10°)})$$

$$R = 8.314 \text{ J/mole °K}$$

$$T = 0 + 273 = 273$$

$$-\ell n\,3 = \frac{-E}{(8.314\ \text{J/mole}^\circ\text{K})}\,(273\text{K}) + \frac{E}{(8.314\ \text{J/mole}^\circ\text{K})}\,(283)\text{K}$$

$$-1.10 = \frac{-E}{(2269.72\ \text{J/mole}^\circ\text{K})} + \frac{E}{(2269.72\ \text{J/mole}^\circ\text{K})}$$

$$(2269.72\ \text{J/mole}\ ^\circ\text{K})(2352.86\ \text{J/mole}\ ^\circ\text{K}) \times (-1.10) =$$

$$\left(\frac{-E}{2269.72\ \text{J / mole}^\circ\text{K}} + \frac{E}{2352.86\ \text{J / mole}^\circ\text{K}}\right)$$

$$(2269.72\ \text{J/mole}\ ^\circ\text{K})\ (2352.86\ \text{J/mole}\ ^\circ\text{K})$$

$$-5.874 \times 10^6\ \text{J}^2/\text{mole}^2 = (-E)(2352.86\ \text{J/mole}) + E\,(2269.72\ \text{J/mole})$$

$$-5.874 \times 10^6\ \text{J}^2/\text{mole}^2 = 8.314 \times 10^1\ \text{J/mole} \times E$$

$$7.06 \times 10^4\ \text{J/mole} = E$$

When the temperature of the reaction mixture is very low or the activation energy is very large, k is small. If we take the natural log of both sides,

$$\ln k = \ln A - \frac{E_a}{RT}$$

$$(y = b + mx)$$

we can plot ln (k) versus $\dfrac{1}{T}$. The straight line plot will give a slope $= \dfrac{-E_a}{R}$ and an intercept with the ordinate = ln (A). For two different temperatures of the same reaction:

$$\ln\frac{k_2}{k_1} = -\frac{E_a}{R}\left(\frac{1}{T_2} - \frac{1}{T_1}\right) \quad \text{or} \quad \log\frac{k_2}{k_1} = -\frac{E_a}{2.303R}\left(\frac{1}{T_2} - \frac{1}{T_1}\right).$$

Activation energy E_a is the energy necessary to cause a reaction to occur. It is the difference in energy between the transition state (or activated complex) and the reactants. The diagram below illustrates the concept of activation energy.

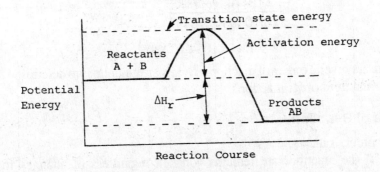

Figure 9. Activation energy

In an exothermic process, energy is released and ΔH_2 is negative. In an endothermic process, energy is absorbed and ΔH_2 is positive. For a reversible reaction, the energy liberated in the exothermic reaction is equal to the energy absorbed in the endothermic reaction (the energy of the reaction, ΔH_2, is also equal to the difference between the activation energies of the opposing reactions:

$$\Delta H_2 = E_{a\ reverse} - E_{a\ foward}$$

As mentioned earlier, the rate of an irreversible reaction is directly proportional to the concentration of the reactants raised to some power. If the reaction

$$A + B \rightarrow P$$

is an elementary reaction, the rate is proportional to

$$[A]^x[B]^y.$$

The order of the reaction with respect to A is x, and the order with respect to B is y. For this reaction, the rate law is

$$\text{rate} = k[A]^x[B]^y, \ k = \text{rate constant.}$$

EXAMPLE: Assume that one A molecule reacts with two B molecules in a one-step process to give AB_2. (a) Write a rate law for this reaction. (b) If the initial rate of formation of AB_2 is 2.0×10^{-5} M/sec and the initial concentrations of A and B are 0.30 M, what is the value of the specific rate constant?

SOLUTION: (a) The overall equation for this reaction is

$$A + 2B \rightarrow AB_2.$$

Since no other information is provided about the reaction, the rate law for the reaction is assumed to be written

$$\text{rate} = k\ [A][B]^2,$$

where k is the rate constant and [ ] indicates concentration.

(b) One can solve for k using the rate law when the rate, [A] and [B] are given as they are in this problem.

$$\text{Rate} = k\ [A][B]^2$$

$$2.0 \times 10^{-5} \text{ M/sec} = k(0.30 \text{ M})\ (0.30 \text{ M})^2$$

$$\frac{2.0 \times 10^{-5}\ M/\text{sec}}{(0.30\ M)(0.30\ M)} = k$$

$$k = 7.41 \times 10^{-4}\ M^{-2}\ \text{sec}^{-1}$$

There are several orders of reactions. Below are guidelines to determine zero, first, second, and third order reactions.

Orders of Reactions

Zero order: rate = k

  If the reaction rate remains constant regardless of changes in reactant concentration, it is a zero order reaction.

First order: rate = k[A]

If the reaction rate is doubled by doubling the concentration of the reactant, the order with respect to the reactant is 1.

Second order: rate = $k[A]^2$; rate = $k[2a]^2 = 4ka^2$

If the rate is increased by a factor of four when the concentration of a reactant is doubled, it is second order with respect to that component.

Third order: rate = $k[A]^3$; rate = $[2a]^3 = 8ka^3$

The rate of a third order reaction would undergo an eightfold increase when the concentration is doubled.

Another way to think about rate is to look at rates of appearance or disappearance of products or reactants. For the reaction:

$$aA + bB \rightarrow eE + fF$$

$$\frac{1}{a}\left(\frac{-\Delta A}{\Delta t}\right) = \frac{1}{b}\left(\frac{-\Delta B}{\Delta t}\right) = \frac{1}{e}\left(\frac{-\Delta E}{\Delta t}\right) = \frac{1}{f}\left(\frac{-\Delta F}{\Delta t}\right)$$

disappearance   appearance

METALS AND NONMETALS

Elements can be divided into three general categories: metals, semimetals, and nonmetals. We can further simplify the categories into metals and nonmetals. Metals comprise most of the elements on the periodic table, and most periodic tables have a line dividing the metals and nonmetals. Metals are excellent conductors of electricity; they possess a shiny, lustrous surface, are malleable, are usually solids (mercury is a liquid), and emit electrons when exposed to heat or short wavelength radiation. Metals rarely form diatomic molecules. Instead, they usually bond to nonmetals in an ionic bond (NaCl, LiOH, KBr). Gold is an interesting example of a metal. Gold is actually a noble metal because even concentrated acids do not attack it. Only a mixture of concentrated HCl and concentrated HNO^3 in a 3:1 ratio (aqua-regia) can dissolve gold (watch out for this in problem solving!).

The middle of the periodic table is primarily composed of semimetals and transition metals. In general, they are classified in terms of their conductivity. Semimetals are poor conductors of electricity but increase their conductivity with increasing temperature. These metals are electronically characterized as having an incomplete "d" subshell or may form ions with an incomplete "d" subshell (the elements Zn, Cd, and Hg do, however, have a complete "d" subshell and so are not included in the transition metal group). The metals in groups IB through VIIIB of the periodic table exhibit multiple valence states. These metals have two (sometimes one) "s" electrons in the outer orbit, and "d" electrons are added one orbit down with each successive element formed. These "d" electrons have significantly different reactivity than the "s" electrons in the outer orbit and lead to multiple valence states depending on whether or not they react.

Nonmetals are among the most abundant substances in the universe. Nonmetals include the noble gases, the halogens, and the elements C, N, P, O, S, and H. Nonmetals may bond to metals to form ionic salts, or they may bond with each other covalently to form various compounds. Also, bonding in nonmetals may be nonpolar or polar. Remember, electronegativity still follows periodic trends within the nonmetal groups! Now, let's look at some of these nonmetal groups.

The noble, or inert, gases (Group VIIIA) are so named because of their "reluctance" to bond with other atoms. Since their "s" and "p" orbitals are completed, they have a filled valence shell (an octet) and so are stable as atomic particles. Under ordinary conditions, they do not react with other molecules, and they exist naturally as monatomic gases.

The halogens (Group VIIA) are a very important group of nonmetals. They exist as diatomic molecules, but they also react readily with metals to form ionic salts. They may also react with oxygen to form various compounds. Halogens are very electronegative and form strong acids. In addition, they are good oxidizing agents; and each halogen can oxidize any of the halogens below it on the periodic table. (However, they cannot oxidize anything above them.) Their valence state is -1, and they have an electron configuration of ns^2np^5.

Hydrogen is, in a sense, a misfit element. It cannot truly be classified with any type of element group beyond its "nonmetallic" classification. Its electron configuration is $1s$, and it naturally exists as a diatomic molecule; however, it is one of the most common elements in the universe and exists in many other forms as well (water, organics, acids, bases, etc.). Hydrogen reacts with almost all of the elements to form hydrides, binary compounds consisting only of hydrogen and one other element (e.g., H_2O, HCl).

Carbon is one of the most abundant elements on the earth. It is the backbone element of all organic compounds, and forms most of the biological and polymeric compounds that we see everyday. The electron configuration of carbon is $2s^22p^2$, and it has only four valence electrons. This allows it to form many strong covalent bonds at its four sites.

The other nonmetals form vital compounds, and are also abundant throughout the earth. Nitrogen makes up most of the atmosphere and forms such compounds as ammonia, various oxides, and amino acids—the building blocks of proteins. Phosphorous is found in many biological molecules and makes up part of the helical structure of the DNA molecule. Finally, sulfur is present in various oxides and acids, but it is also prevalent within the earth's crust, rising in a gaseous form from the cracks in the surface.

ORGANIC CHEMISTRY

This brings us to organic chemistry. In general, organic chemistry focuses primarily on nonmetallic compounds, involving the chemistry of carbon and its compounds. Many carbon compounds are **hydrocarbons**, compounds that contain carbon and hydrogen. Carbon is a unique element in that it has four bonding sites.

As a result, it may form single, double, or triple bonds. It may also form ringlike structures or long hydrocarbon chains. Another phenomenon of organic molecules is that even though two compounds may have the same chemical formulas, they may not have the same chemical properties. Since the carbon atoms may bond to each other in any of several ways, several different structures may form. These are called isomers. Isomers may be structural or conformational; the molecules may differ in their actual physical makeup, or they may simply be rotated or inverted enough to prevent them from being the same molecule. For example, the formula C_5H_{12} can be drawn several ways:

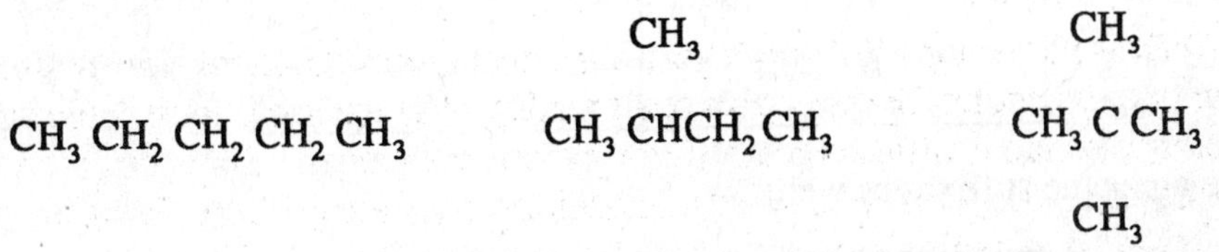

$$CH_3 \; CH_2 \; CH_2 \; CH_2 \; CH_3 \qquad\qquad CH_3 \; \overset{\displaystyle CH_3}{CH}CH_2 \; CH_3 \qquad\qquad CH_3 \; \overset{\displaystyle CH_3}{\underset{\displaystyle CH_3}{C}} \; CH_3$$

a) Pentane   b) 2-methyl-butane   c) 2,2-dimethyl-propane

Figure 10. Structures of C_5H_{12}

For compounds with double or triple bonds, isomers may be the result of a change in the position of the multiple bond:

$$CH_2 = CHCH_2CH_3 \qquad\qquad CH_3CH = CHCH_3$$

butene   2-butene

Or, they may simply have groups arranged in different ways:

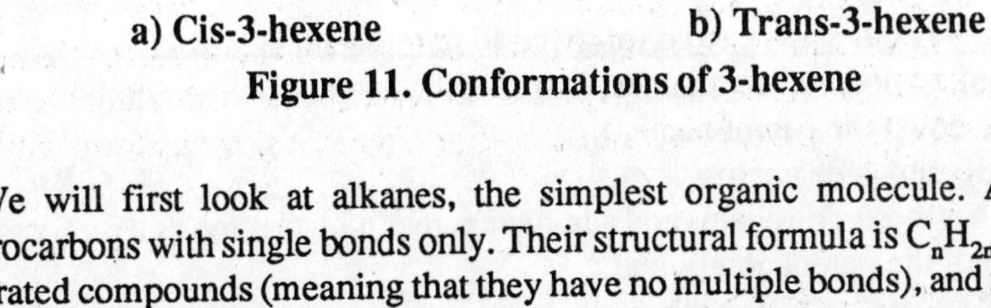

a) Cis-3-hexene   b) Trans-3-hexene

Figure 11. Conformations of 3-hexene

We will first look at alkanes, the simplest organic molecule. **Alkanes** are hydrocarbons with single bonds only. Their structural formula is C_nH_{2n+2}. They are saturated compounds (meaning that they have no multiple bonds), and may take the form of a straight chain or a ring. Nomenclature defines an organic molecule's nature and structure. Prefixes and suffixes indicate substituents, functional groups, and bonding type. Alkanes always end with the suffix "-ane." The prefix depends on the number of carbons in the longest continuous chain of the molecule.

Some prefixes are outlined below:

TABLE 6
CHAIN LENGTH PREFIXES

# of Carbons	Prefix
1	meth-
2	eth-
3	prop-
4	but-
5	pent-
6	hex-

Here are some suffixes as well:

TABLE 7
STURCTURAL SUFFIXES

Structure	Suffix
Single bond	-ane
Double bond	-ene
Triple bond	-yne

Alkanes are generally nonpolar, unreactive molecules. The boiling point, melting point, density, and viscosity of alkanes increase as the length of the carbon chain increases.

$$CH_3$$
$$|$$

EXAMPLE: Name the following alkanes: (a) CH_4 (b) CH_3CH_3 (c) $CH_3 — CH_2$
(d) $CH_3CH_2CH_2CH_3$

SOLUTION: Four steps can be followed in naming alkanes.
(1) In naming open-chain alkanes, first find the longest chain of carbon atoms.
(2) Write down the parent name.
(3) Identify any side chains.
(4) Number these side chains and add their names and location as a prefix to the name of the parent compound.

These steps are illustrated in the naming of the four compounds above.

(a)
$$\begin{array}{c} H \\ | \\ H-C-H \\ | \\ H \end{array}$$

This compound contains only one carbon atom and is the simplest of all the alkanes. It is called methane.

(b)
$$\begin{array}{c} H\ \ H \\ |\ \ \ | \\ H-C-C-H \\ |\ \ \ | \\ H\ \ H \end{array}$$

This is a two carbon alkane. A chain of two carbon atoms is given the root "eth." The parent names of alkanes are formed by adding the suffix "ane" to the root name of the longest carbon chain. This compound is called ethane.

(c)
$$H-C-----C-H$$

The longest chain in this compound is three carbons long. It is not significant that the chain is bent. The root used in naming three carbon chains is "prop." The name of this compound is propane.

(d)
$$H-C-C-C-C-H$$

The longest chain here is four carbons long. The root used in naming four carbon chains is "but." The name of this compound is butane.

Alkenes are unsaturated hydrocarbons with one or more carbon-carbon double bonds. They have the general formula C_nH_{2n}. Since alkenes have another bond, they have two fewer hydrogens than their alkanal counterparts. Alkenes are named in much the same way as alkanes (the location of the double bond must be specified; however, naming compounds can be complicated and tedious. This way, you will be able to recognize the compounds quickly). Alkenes with one to four C's are gases, with five to fifteen C's are liquids, and with sixteen or more C's are solids. Alkenes show relatively higher reactivity than alkanes, and many reactions focus on "attacking" the double bond. Addition reactions saturate the double bond and create alkanes.

Cleavage reactions break the double bond and create two new products. Dehydration, dehalogenation, dehydrohalogenation, and application of heat create alkenes from alkanes.

EXAMPLE: Write the chemical structures for each compound listed.

(a) 1-Hexene
(d) 1-Iodo-2-methyl-2-pentene
(b) 3-Methyl-1-butene
(e) 2-Chloro-3-methyl-2-hexene
(c) 2, 4-Hexadiene
(f) 6,6-Dibromo-5-methyl-5-ethyl-2,3-heptadiene

SOLUTION: (1) Look at the complete name of the compound and pick out the parent name. It is usually the last word of the complete name, and it denotes the longest continuous chain that contains the carbon-carbon double bond. (All of the structures are alkenes.)

(2) Write out the carbon skeleton that makes up the parent chain. Determine the number of double bonds present by examining the suffix of the parent name. For example, "ene" means one double bond, where as "diene" means two.

(3) Position the double bond (or bonds) in the carbon skeleton as specified by the number directly (usually) in front of the parent name. For example, if the compound is 2-pentene, one would write

$$C - C - C = C - C$$
$$5 \quad 4 \quad 3 \quad 2 \quad 1$$

(Recall, the position of the double bond is given by the number of the first doubly bonded carbon encountered when numbering from the end of the chain nearest the double bond.)

(4) Position the functional group substituents on the chain as specified by the number directly in front. For example, 3-methyl-2-pentene would be:

$$\overset{\displaystyle C}{\underset{\displaystyle 5 \quad 4 \quad 3 \quad 2 \quad 1}{C - C - C = C - C}}$$

The structures of the compound in (a) – (f) become:

(a) $CH_3CH_2CH_2CH_2CH = CH_2$

(b) $CH_2 = CH-CH-CH_3$
                          $|$
                        CH_3

(c) $CH_3-CH = CH-CH = CH - CH_3$

(d) 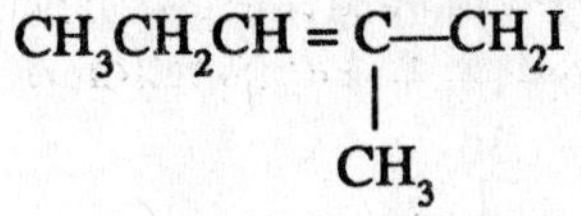 $CH_3CH_2CH = C-CH_2I$
                        |
                       CH_3

(e) 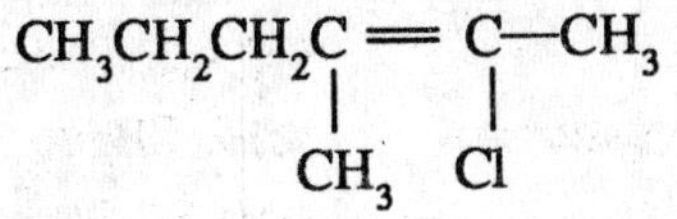 $CH_3CH_2CH_2C = C-CH_3$
                      | |
                     CH_3 Cl

                                    CH_3 Br
                                     | |
(f) 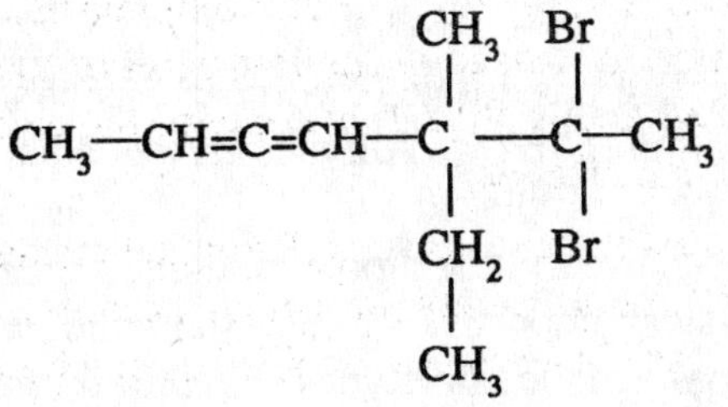 $CH_3-CH=C=CH-C-C-CH_3$
                                     | |
                                    CH_2 Br
                                     |
                                    CH_3

To see how this process works, examine how structure (f) was written. The parent name is heptadiene. The prefix "hepta" indicates that seven carbons are present in the skeleton: C—C—C—C—C—C—C. The 2, 3 indicates the positions of the two double bonds — it is a diene. So, one can write

$$C-C=C=C-C-C-C$$
$$1 \quad 2 \quad 3 \quad 4 \quad 5 \quad 6 \quad 7$$

With this numbering system, the substituents are now added as specified by the 6 for the bromines and 5 for the methyl (CH_3) and ethyl groups.

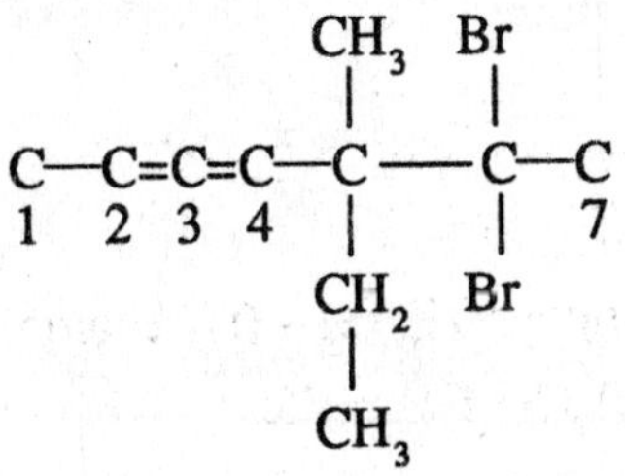

                          CH_3 Br
                           | |
        C—C=C=C—C——C—C
        1 2 3 4 | | 7
                          CH_2 Br
                           |
                          CH_3

And now only the hydrogens need to be added to obtain:

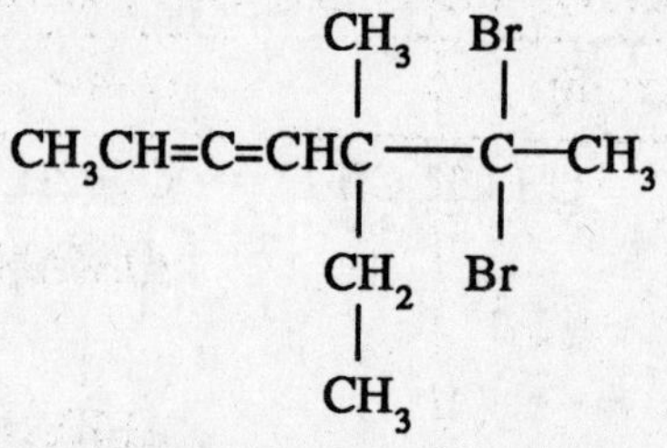

$$CH_3CH=C=CHC \overset{\overset{\displaystyle CH_3}{|}}{\underset{\underset{\displaystyle CH_3}{|}}{\underset{\displaystyle CH_2}{|}}} - \overset{\overset{\displaystyle Br}{|}}{\underset{\underset{\displaystyle Br}{|}}{C}} - CH_3$$

Alkynes are unsaturated hydrocarbons containing triple bonds. They have the general formula C_nH_{2n-2}. Alkynes are named in the same way as alkenes, except the suffix "-ene" is replaced with "-yne." The hydrogens in terminal alkynes are relatively acidic, and the dipole moment is small, but larger than that of an alkene. Other physical properties are essentially the same as alkanes and alkenes. Alkynes undergo many of the same types of addition, cleavage, and dehydration reactions as alkenes.

Organic molecules may also form cyclic structures. Triangular, square planar, and various other regular and irregular polygonal structures may be formed. One of the most important types of structures are aromatic compounds. These are six membered rings with alternating double and single bonds. An example of this is benzene, C_6H_6:

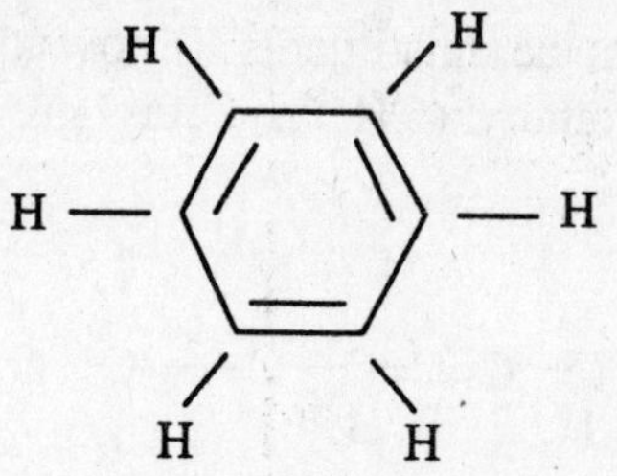

Figure 12. Benzene ring

Other organic compounds include those created by functional groups, additional groups of elements that "add on" to the hydrocarbon skeleton. Functional groups often determine the chemical properties of organic compounds. Some elemental functional groups include oxygen, nitrogen, and sulfuric compounds. Some of the most common are those of oxygen. Alcohols, ethers, aldehydes, ketones, and carboxylic acids are among the most common oxygen containing organic compounds.

Structure/Functional Group	Name
TABLE 8	
NOMENCLATURE OF FUNCTIONAL GROUPS	
—C—C—	ALKANE
C=C	ALKENE
—C≡C—	ALKYNE
(pentagon)	CYCLOALKANE (cyclopentane)
(benzene)	AROMATIC RING (benzene)
—CH_3	METHYL GROUP
—CH_2CH_3	ETHYL GROUP
—$CH_2CH_2CH_3$	PROPYL GROUP
—OH	ALCOHOL
(benzene)—OH	PHENOL
—NH_2, —NA—, —N—	AMINE
—C—O—C—	ETHER
—C—C— O	EPOXIDE
—C(=O)OH	CARBOXYLIC ACID
—C(=O)H	ALDEHYDE
—C(=O)C—	KETONE

REVIEW PROBLEMS

PROBLEM 1

Balance the following by filling in missing species and proper coefficient:

(1) $NaOH +$ _____ $\rightarrow NaHSO_4 + HOH$,

(2) $PCl_3 +$ _____ $HOH \rightarrow$ _____ $+ 3HCl$,

(3) $CH_4 +$ _____ $\rightarrow CCl_4 + 4HCl$.

SOLUTION: To balance chemical equations you must remember that ALL atoms (and charges) must be accounted for. The use of coefficients in front of compounds is a means to this end. Thus,

(1) $$NaOH + \underline{\hspace{2cm}} \rightarrow NaHSO_4 + HOH$$

On the right side of the equation, you have 1 Na, 3 H's, 5 O's, and 1 S. This same number of elements must appear on the left side. However, on the left side, there exists only 1 Na, 1 O, and 1 H. You are missing 2 H's, 1 S, and 4 O's. The missing species is H_2SO_4, sulfuric acid. You could have anticipated this since a strong base (NaOH) reacting with a strong acid yields a salt ($NaHSO_4$) and water. The point is, however, that H_2SO_4 balances the equation by supplying all the missing atoms.

(2) $$PCl_3 + 3HOH \rightarrow \underline{\hspace{2cm}} + 3 HCl$$

Here, the left side has 1 P, 3 Cl's, 6 H's, and 3 O's. The right side has 3 H's and 3 Cl's. You are missing 1 P, 3 O's, and 3 hydrogens. Therefore, $P(OH)_3$ is formed.

(3) $$CH_4 + \underline{\hspace{2cm}} \rightarrow CCl_4 + 4 HCl$$

Here, there are 1 C, 8 Cl's, and 4 H's on the right and 1 C and 4 H's on the left. The missing compound, therefore, contains 8 Cl's and thus it is $4 Cl_2$. One knows that it is $4 Cl_2$ rather than Cl_8 or 8Cl because elemental chlorine gas is a diatomic or 2 atom molecule.

PROBLEM 2

Calculate the mole fractions of ethyl alcohol, C_2H_5OH, and water in a solution made by dissolving 9.2 g of alcohol in 18 g of H_2O. The M.W. of $H_2O = 18$, M.W. of $C_2H_5OH = 46$.

SOLUTION: Mole fraction problems are similar to percent composition problems. A mole fraction of a compound tells us what fraction of 1 mole of solution is due to that particular compound. Hence,

$$\text{mole fraction of solute} = \frac{\text{moles of solute}}{\text{moles of solute } + \text{ moles of solvent}}.$$

The solute is the substance being dissolved into or added to the solution. The solvent is the solution to which the solute is added.

The equation for calculating mole fractions is

$$\frac{\text{moles A}}{\text{moles A } + \text{ moles B}} = \text{mole fraction A}.$$

Moles are defined as grams/molecular weight (M.W.). Therefore, first find the number of moles of each compound present and then use the above equation.

$$\text{moles of } C_2H_5OH \quad = \frac{9.2 \text{ g}}{46.0 \text{ g/mole}} \quad = .2 \text{ mole}$$

$$\text{moles of } H_2O \quad = \frac{18 \text{ g}}{18 \text{ g/mole}} \quad = 1 \text{ mole}$$

$$\text{mole fraction of } C_2H_5OH \quad = \frac{.2}{1 + .2} \quad = .167$$

$$\text{mole fraction of } H_2O \quad = \frac{1}{1 + .2} \quad = .833$$

Note: The sum of the mole fractions is equal to 1.

PROBLEM 3

What is the maximum weight of SO_3 that could be made from 25.0 g of SO_2 and 6.0 g of O_2 by the following reaction?

$$2SO_2 + O_2 \rightarrow 2SO_3$$

SOLUTION: From the reaction, one knows that for every 2 moles of SO_3 formed, 2 moles of SO_2 and 1 mole of O_2 must react. Thus, to find the amount of SO_3 that can be formed, one must first know the number of moles of SO_2 and O_2 present. The number of moles is found by dividing the number of grams present by molecular weight:

$$\text{number of moles} = \frac{\text{number of grams}}{\text{M. W.}}.$$

For O_2: M.W. = 32

$$\text{no. of moles} = \frac{6.0 \text{ g}}{32.0 \text{ g/mole}} = 1.875 \times 10^{-1} \text{ moles}$$

For SO_2: M.W. = 64

$$\text{no. of moles} = \frac{25.0 \text{ g}}{64.0 \text{ g/mole}} = 3.91 \times 10^{-1} \text{ moles}$$

Because 2 moles of SO_2 are needed to react with 1 mole of O_2, 3.75×10^{-1} moles of SO_2 will react with 1.88×10^{-1} moles of O_2. This means that $3.91 \times 10^{-1} - 3.75 \times 10^{-1}$ moles or $.16 \times 10^{-1}$ moles of SO_2 will remain unreacted. In this case, O_2 is called the limiting reagent because it determines the number of moles of SO_3 formed. There will be twice as many moles of SO_3 formed as there are O_2 reacting.

$$\text{no. of moles of } SO_3 \text{ formed} = 2 \times 1.875 \times 10^{-1} \text{ moles}$$

$$= 3.75 \times 10^{-1} \text{ moles}$$

The weight is found by multiplying the number of moles formed by the molecular weight (M.W. of SO_3 = 80).

$$\text{weight of } SO_3 = 3.75 \times 10^{-1} \text{ moles} \times 80 \text{ g/mole} = 30.0 \text{ g}$$

PROBLEM 4

The following reaction

$$2H_2S(g) \rightleftharpoons 2H_2(g) + S_2(g)$$

was allowed to proceed to equilibrium. The contents of the two-liter reaction vessel were then subjected to analysis and found to contain 1.0 mole H_2S, 0.20 mole H_2, and 0.80 mole S_2. What is the equilibrium constant K_{eq} for this reaction?

SOLUTION: This problem involves substitution into the equilibrium constant expression for this reaction:

$$K_{eq} = \frac{[H_2]^2[S_2]}{[H_2S]^2}.$$

The equilibrium concentration of the reactant and products are $[H_2S] = 1.0$ mole/2 liters $= 0.50$ M, $[H_2] = 0.20$ mole/2 liters $= 0.10$ M, and $[S_2] = 0.80$ mole/2 liters $= 0.40$ M, Hence, the value of the equilibrium constant is

$$K_{eq} = \frac{[H_2]^2[S_2]}{[H_2S]^2} = \frac{(0.10)^2(0.40)}{(0.50)^2} = 0.016$$

for this reaction.

PROBLEM 5

The ionization constant for acetic acid is 1.8×10^{-5}.

(1) Calculate the concentration of H^+ ions in a 0.10 molar solution of acetic acid.

(2) Calculate the concentration of H^+ ions in a 0.10 molar solution of acetic acid in which the concentration of acetate ions has been increased to 1.0 molar by addition of sodium acetate.

SOLUTION: 1) The ionization constant (K_a) is defined as the concentration of H^+ ions times the concentration of the conjugate base ions of a given acid divided by the concentration of unionized acid. For an acid, HA:

$$K_a = \frac{[H^+][A^-]}{[HA]}$$

where K_a is the ionization constant, $[H^+]$ is the concentration of H^+ ions, $[A^-]$ is the concentration of the conjugate base ions, and [HA] is the concentration of unionized acid. The K_a for acetic acid is stated as

$$K_a = \frac{[H^+][\text{acetate ion}]}{[\text{acetic acid}]} = 1.8 \times 10^{-5}.$$

The chemical formula for acetic acid is $HC_2H_3O_2$. When it is ionized, one H^+ is formed and one $C_2H_3O^-$ (acetate) is formed, thus the concentration of H^+ equals the concentration of $C_2H_3O^-$:

$$[H^+] = [C_2H_3O^-].$$

The concentration of unionized acid is decreased when ionization occurs. The new concentration is equal to the concentration of H^+ subtracted from the concentration of unionized acid:

$$[HC_2H_3O] = 0.10 - [H^+].$$

Since $[H^+]$ is small relative to 0.10, one may assume that $0.10 - [H^+]$ is approximately equal to 0.10:

$$0.10 - [H^+] \cong 0.10.$$

Using this assumption, and the fact that $[H^+] = [C_2H_3O^-]$, K_a can be rewritten as

$$K_a = \frac{[H^+][H^+]}{0.10} = 1.8 \times 10^{-5}.$$

Solving for the concentration of H^+:

$$[H^+]^2 = (1.0 \times 10^{-1})(1.8 \times 10^{-5}) = 1.8 \times 10^{-6}$$

$$[H^+] = \sqrt{1.8 \times 10^{-6}} = 1.3 \times 10^{-3} M$$

The concentration of H^+ is thus $1.3 \times 10^{-3} M$.

(2) When the acetate concentration is increased, the concentration of H^+ is lowered to maintain the K_a. The K_a for acetic acid is stated as

$$K_a = \frac{[H^+][C_2H_3O^-]}{[HC_2H_3O]} = 1.8 \times 10^{-5}.$$

As previously shown for acetic acid equilibria in a solution of 0.10 molar acid, the concentration of acid after ionization is

$$[HC_2H_3O] = 0.10 - [H^+].$$

Because $[H^+]$ is very small compared to 0.10, $0.10 - [H^+] \cong 0.10$ and:

$$[HC_2H_2O] = 0.10 \ M.$$

In this problem, we are told that the concentration of acetate is held constant at 1.0 molar by addition of sodium acetate. Because we now know the concentrations of the acetate and the acid, the concentration of H^+ can be found:

$$\frac{[H^+][C_2H_3O^-]}{[HC_2H_3O]} = 1.8 \times 10^{-5}$$

$$\frac{[H^+][1.0]}{[0.10]} = 1.8 \times 10^{-5}$$

$$[H^+] = 1.8 \times 10^{-6} M$$

PROBLEM 6

A chemist dissolves $BaSO_4$ in pure water at 25°C. If $K_{sp} = 1 \times 10^{-10}$, what is the solubility of the barium sulfate in the water?

SOLUTION: The solubility of a compound is defined as the limiting concentration of the compound in a solution before precipitation occurs. To find the solubility of the barium sulfate, you need to know the concentration of its ions in solution. $BaSO_4$ will dissociate into ions because it is salt. There will be an equilibrium between these ions and the $BaSO_4$. The equilibrium can be measured in terms of a constant, K, called the solubility constant. The K_{sp} is expressed in terms of the concentrations of the ions. As such, to answer the question, you want to represent this K_{sp}. For this reaction, the equation is

$$BaSO_4 \rightleftharpoons Ba^{++} + SO_4^{=}$$
$$K_{sp} = [Ba^{+2}]\{SO_4^{-2}\} = 1 \times 10^{-10}.$$

Let $x = [Ba^{+2}]$. Thus, $x = [SO_4^{-2}]$, also, since both ions will be formed in equimolar amounts. Therefore, $x \times x = 1 \times 10^{-10}$. Solving:

$$x = 1 \times 10^{-5} M = [Ba^{+2}] = [SO_4^{-2}].$$

PROBLEM 7

"Hard" water contains small amounts of the salts calcium bicarbonate $(Ca(HCO_3)_2)$ and calcium sulfate $(CaSO_4$, molecular weight $= 136$ g/mole). These react with soap before it has a chance to lather, which is responsible for its cleansing ability. $Ca(HCO_3)_2$ is removed by boiling to form insoluble $CaCO_3$. $CaSO_4$ is removed by reaction with washing soda $(Na_2CO_3$, molecular weight $= 106$ g/mole) according to the following equation:

$$CaSO_4 + Na_2CO_3 \rightarrow CaCO_3 + Na_2SO_4.$$

If the rivers surrounding New York City have a $CaSO_4$ concentration of 1.8×10^{-3} g/liter, how much Na_2CO_3 is required to "soften" (remove $CaSO_4$) the water consumed by the city in one day (about 6.8×10^9 liters)?

SOLUTION: We must determine the amount of $CaSO_4$ present in 6.8×10^9 liters and, from this, the amount of Na_2CO_3 required to remove it.

The number of moles per liter, or molarity, of $CaSO_4$ corresponding to 1.8×10^{-3} g/liter is obtained by dividing this concentration by the molecular weight of $CaSO_4$. Multiplying by 6.8×10^9 liters gives the number of moles of $CaSO_4$ that must be removed. Hence:

$$\text{moles } CaSO_4 = \frac{\text{concentration}(\text{g/liter})}{\text{molecular weight of } CaSO_4} \times 6.8 \times 10^9 \text{ liters}$$

$$= \frac{1.8 \times 10^{-3} \text{ g/liter}}{136 \text{ g/mole}} \times 6.8 \times 10^9 \text{ liters}$$

$$= 9.0 \times 10^4 \text{ moles}.$$

From the equation for the reaction between $CaSO_4$ and Na_2CO_3, we see that one mole of $CaSO_4$ reacts with one mole of Na_2CO_3. Hence, 9.0×10^4 moles of Na_2CO_3 are required to remove all the $CaSO_4$. To convert this to mass, we multiply by the molecular weight of Na_2CO_3 and obtain

$$\text{mass } Na_2CO_3 = \text{moles } Na_2CO_3 \times \text{molecular weight } Na_2CO_3$$
$$= 9.0 \times 10^4 \text{ moles} \times 106 \text{ g/mole}$$
$$= 9.5 \times 10^6 \text{ g} = 9.5 \times 10^6 \text{ g} \times 1 \text{ kg/1,000 g}$$
$$= 9.5 \times 10^3 \text{ kg}$$

which is about 10 tons.

PROBLEM 8

For the following voltaic cell, write the half-reactions, designating which is oxidation and which is reduction. Write the cell reaction and calculate the voltage ($E°$) of the cell from the given electrodes. The cell is

$$Cu; Cu^{+2} \| Ag^{+1}; Ag.$$

SOLUTION: In a voltaic cell, the flow of electrons creates a current. Their flow is regulated by two types of reactions that occur concurrently—oxidation and reduction. Oxidation is a process where electrons are lost and reduction where electrons are gained. The equation for these are the half-reactions. From the cell diagram, the direction of the reaction is always left to right.

$$Cu \rightarrow Cu^{2+} + 2e^- \qquad \text{oxidation}$$
$$\leftarrow 2Ag^+ + 2e^- \rightarrow 2Ag \qquad \text{reduction}$$

Therefore, the combined cell reaction is

$$\leftarrow Cu + 2Ag^+ \rightarrow Cu^{2+} + 2Ag.$$

To calculate the total E^0, look up the value for the E^0 of both half-reactions as reductions. To obtain E^0 for oxidation, reverse the sign of the reduction E_0. Then, substitute into $E^0_{cell} = E^0_{red} + E^0_{ox}$. If you do this, you find

$$E^0_{cell} = -(E^0_{red} \, Cu) + E^0_{red} \, Ag^{+1}$$
$$= -.34 + .80$$
$$= .46 \text{ volt}$$

PROBLEM 9

For the following oxidation-reduction reaction, 1) write out the two half-reactions and balance the equation, 2) calculate ΔE^0, and 3) determine whether the reaction will proceed spontaneously as written:

$$Fe^{2+} + MnO_4^- + H^+ \rightarrow Mn^{2+} + Fe^{3+} + H_2O$$

(1) $\quad Fe^{3+} + e^- \rightleftharpoons Fe^{2+}$, $E^0 = 0.77eV$

(2) $\quad MnO_4^- + 8H^+ + 5e \rightleftharpoons Mn^{2+} + 4H_2O$, $E^0 = 1.51eV$

SOLUTION: (1) The two half-reactions of an oxidation = reduction reaction are the equation for the oxidation process (loss of electrons) and the reduction process (gain of electrons). In the overall reaction, you begin with Fe^{2+} and end up with Fe^{3+}. It had to lose an electron to accomplish this. Thus you have oxidation:

$$Fe^{2+} \rightarrow Fe^{3+} + e^-.$$

Note: This is the reverse of the reaction given with $E^0 = .77eV$. As such, the oxidation reaction in this problem has $E^0 = -.77eV$. The reduction must be

$$MnO_4^- + 8H^+ + 5e^- \rightarrow Mn^{2+} + 4H_2O,$$

since in the overall reaction, you see $MnO_4^- + H^+$ go to Mn^{2+}, which suggests a gain of electrons. This is the same reaction as the one given in the problem, $E^0 = 1.51eV$. To balance the overall reaction, add the oxidation reaction to the reduction reaction, such that all electron charges disappear. If you multiply the oxidation reaction by 5, you obtain:

$$5Fe^{2+} \rightarrow 5Fe^{3+} + 5e^-$$

$$MnO_4^- + 8H^+ + 5e^- \rightarrow Mn^{2+} + 4H_2O$$

$$5Fe^{2+} + MnO_4^- + 8H^+ \rightarrow 5Fe^{3+} + Mn^{2+} + 4H_2O$$

Note: Since both equations contained $5e^-$ on different sides, they canceled out. This explains why the oxidation reaction is multiplied by five. Thus, you have written the balanced equation.

(2) The ΔE^0 for the overall reaction is the sum of the E^0 for the half-reactions, i.e.,

$$\Delta E^0 = E_{red} + E_{oxid}.$$

You know E_{red} and E_{oxid}; $\Delta E^0 = 1.51 - .77 = 0.74eV$.

(3) A reaction will only proceed spontaneously when $\Delta E^0 = $ a positive value. You calculated a positive ΔE^0, which means the reaction proceeds spontaneously.

PROBLEM 10

The ketone acid $(CH_2CO_2H)_2CO$ undergoes a first-order decomposition in aqueous solution to yield acetone and carbon dioxide:

$$(CH_2CO_2H)_2CO \rightarrow (CH_3)_2CO + 2CO_2$$

(1) Write the expression for the reaction rate.

(2) The rate constant k has been determined experimentally as 5.48×10^{-2}/sec at 60°C. Calculate $t_{\frac{1}{2}}$ at 60°C.

(3) The rate constant at 0°C has been determined as 2.46×10^{-5}/sec. Calculate $t_{\frac{1}{2}}$ at 0°C.

(4) Are the calculated half-lives in accord with the stated influence of temperature on reaction rate?

SOLUTION: (1) For a chemical decomposition, the rate of the reaction is equal to the product of the rate constant (k) and the concentration of the compound decomposing. Thus:

$$\text{Rate} = k \, [(CH_2CO_2H)_2CO].$$

(2) Because the rate is only proportional to $[(CH_2CO_2H)_2CO]$, the reaction is first-order. For a first-order reaction, the half-life $\left(t_{\frac{1}{2}} \right)$ is related to k by the following equation:

$$t_{\frac{1}{2}} = \frac{0.693}{k}$$

Solving for $t_{\frac{1}{2}}$:

$$t_{\frac{1}{2}} = \frac{0.693}{5.48 \times 10^{-2} / \text{sec}} = 12.65 \text{ sec.}$$

(3) One can solve for $t_{\frac{1}{2}}$ at 0°C using the same equation:

$$t_{\frac{1}{2}} = \frac{0.693}{2.46 \times 10^{-5} / \text{sec}} = 2.82 \times 10^4 \text{ sec.}$$

(4) In general, the speed of a chemical change is approximately doubled for each ten degrees rise in temperature. The temperature rises 60°, from 0°C to 60°C. Therefore, the rate should double six times or the ratio of the $t_{\frac{1}{2}}$ at 0°C to the $t_{\frac{1}{2}}$ at 60°C is $2^6 \times 2^6 = 64$.

$$\frac{t_{\frac{1}{2}} 0°}{t_{\frac{1}{2}} 60°} = \frac{2.82 \times 10^4 \text{ sec}}{12.65 \text{ sec}} = 2.23 \times 10^3$$

This is much greater than the expected ratio of 64.

PROBLEM 11

(1) A reaction proceeds five times as fast at 60°C as it does at 30°C. Estimate its energy of activation. (2) For a gas phase reaction with $E_A = 40,000$ cal/mole, estimate the change in rate constant due to a temperature change from 1,000°C to 2,000°C.

SOLUTION: The actuation energy E_A can be related to the rate constants k_1 (at temperature T_1) and k_2 (at temperature T_2) by the Arrhenius equation:

$$\log\frac{k_2}{k_1} = -\frac{E_a}{2.303R}\left(\frac{1}{T_2} - \frac{1}{T_1}\right)$$

where R = universal gas constant.

(1) You are told a reaction proceeds five times as fast at 60° as it does at 30°C. Therefore, if k_1 = rate constant at 30°C = 303K with T_1 = 303K, then k_2 = 5k_1 at 60°C = 333K with T_2 = 333K. You are given R. Substitute these values into the Arrhenius equation and solve for E_A. Rewriting and substituting:

$$E_a = \frac{-2.303R}{\dfrac{1}{T_2} - \dfrac{1}{T_1}} \log\frac{k_2}{k_1} = \frac{(-2.303)(1.987)\left(\dfrac{1\ kcal}{1,000\ cal}\right)}{\left(\dfrac{1}{333} - \dfrac{1}{303}\right)}\log 5$$

$$= (15.4\ kcal/mole)(.699) = 10.8\ kcal/mole.$$

Note: 1 kcal/1,000 cal is a conversion factor to obtain the correct units.

To answer (2) find $\dfrac{k_2}{k_1}$ from the Arrhenius equation. Rewriting and substituting:

$$\frac{k_2}{k_1} = \text{anti}\log\left(\frac{E_a}{2.303R}\left(\frac{1}{T_2} - \frac{1}{T_1}\right)\right)$$

$$= \text{anti}\log\left(\frac{-40,000}{(2.303)(1.987)}\left(\frac{1}{2,273} - \frac{1}{1,273}\right)\right)$$

$$= \text{anti}\log 3.02 = 1.05 \times 10^3$$

That is, the rate should be about 1,050 times as great at 2,000°C as at 1,000°C.

PROBLEM 12

Four liters of octane gasoline weigh 3.19 kg. Calculate the volume of air required for its complete combustion at STP.

SOLUTION: To answer this problem, you need to write the balanced equation for the combustion of octane gasoline. This means knowing the molecular formula of octane gasoline and what is meant by combustion. Octane is a saturated hydrocarbon, i.e., it is an alkane. A saturated hydrocarbon means a compound that contains only single bonds between the carbon-to-carbon and carbon-to-hydrogen bonds. Alkanes have the general formula C_nH_{2n+2}, where N = number of carbon atoms. Since the prefix "oct" means eight, you know there are 8 carbon atoms, which indicates that 18 hydrogen atoms are present. Thus, gasoline octane has the formula C_8H_{18}. Now, combustion is the reaction of an organic compound with oxygen to produce CO_2 and H_2O. With this in mind, you can write the balanced equation for the reaction:

$$2C_8H_{18} + 25O_2 \rightarrow 16CO_2 + 18H_2O.$$

To determine the volume of air required for combustion, you need the volume of O_2 required, since 21 percent of air is oxygen (O_2). To find the amount of O_2 involved, use the fact that at STP (standard temperature and pressure) 1 mole of any gas occupies 22.4 liters. Thus, if you know how many moles of O_2 were required, you would know its volume. You can find the number of moles by using stoichiometry. You have 3.19 kg or 3,190 g (1,000 g = 1 kg) of octane gasoline. The molecular weight (M.W.) of octane is 114 grams/mole. Thus, since

$$\text{mole} = \frac{\text{grams (weight)}}{\text{M.W.}}, \text{ you have } \frac{3,190}{114} = 27.98 \text{ moles of gasoline.}$$

From the equation's coefficients, you see that for every 2 moles of gasoline, 25 moles of O_2 are required. Thus, for this number of moles of gasoline, you need

$$(27.98)\frac{25}{2} = 349.78 \text{ moles of } O_2.$$

Recalling that 1 mole of gas occupies 22.4 liters at STP, 349.78 moles of O_2 occupies (349.78)(22.4) = 7,835.08 liters. Oxygen is 21% of the air. Thus, the amount of air required is

$$\frac{7,835.08 \text{ liters}}{.21 \text{ liters } O_2} O_2 = 37,309.9 \text{ liters air.}$$

CHAPTER NINE

Thermodynamics

Chapter 9

THERMODYNAMICS

PROPERTIES

The state of a medium is defined by the properties of that medium. Properties are divided into two major categories:

(1) **Intensive properties** which are independent of the mass. Pressure (P), density (ρ), and temperature (T) are examples of intensive properties.

(2) **Extensive properties** which vary directly with the mass of the medium. Volume (V), total enthalpy (H), total internal energy (U), and mass (m) are examples of extensive properties.

An extensive property divided by the mass is called a specific property and can be used in the same manner as the intensive property. Specific volume (v), enthalpy (h), entropy (s), and internal energy (u) are examples of specific properties.

The specific volume is the total volume divided by the mass

$$v = \frac{V}{m}$$

and the density is the inverse of the specific volume

$$\rho = \frac{1}{v}.$$

The enthalpy, a derived property, is equal to the internal energy plus the product of pressure and specific volume. This relationship can be developed by considering the effect of fluid flow on the internal energy. Thus, enthalpy is

$$h = u + Pv.$$

For a pure substance, one that is homogeneous, of constant chemical composition, and with only one work mode (compressibility), two independent intensive or specific properties are required to fix the state of the medium. Thus, the necessity for strict understanding of the difference between extensive and intensive properties.

The utility of properties is their ability to define the state of a medium and to relate to each other to define new, useful relationships. Some properties are defined as a result of physical occurrences, pressure being a prime example. Defined as a force per unit area normal to the force,

$$P = \frac{F}{A}$$

pressure is considered absolute in this case. Most pressure is measured as gauge pressure, the difference between the absolute and the atmospheric. Gauge pressure is either positive or negative (vacuum). Pressure relationships in equation form are

$$P_{abs} = P_{atm} + P_{gauge}$$

where P_{abs} = Absolute pressure

P_{atm} = Atmospheric pressure

P_{gauge} = Gauge pressure (positive or vacuum)

Figure 1 shows the relationship between the gauge, absolute, and atmospheric pressures.

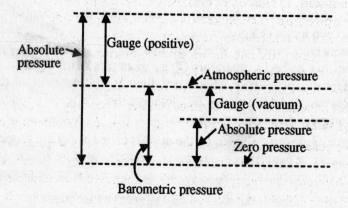

Figure 1. Pressure measurement for both positive gauge and vacuum situations

In every case, the value of the property is defined at a finite state point. Properties are designated as point functions, in contrast to energy transfers (heat and work) which are path functions. Heat and work will be defined later.

Thermodynamic Systems

Throughout the study of thermodynamics, the medium, and the actions associated with the medium, i.e., the flow of matter and energy transfer, must be described in terms of the boundary used to isolate the medium from other media or the surroundings. Generally, the boundary is the means of identifying the type of analysis to be performed. As will be seen later, the type of analysis may determine which properties are logically used. Two types of thermodynamic systems are frequently used:

    (1) An **open system** is one that allows for the flow of matter and the transfer of energy across the system boundary. Problems analyzed using this analysis

are referred to as flow problems. Additionally, the term **control volume** is used to specify open systems. The control volume is any fixed volume in space through which fluid flow takes place. Its surface is called a control surface. Turbines, compressors, and nozzles are analyzed using the control volume.

(2) A **closed system** allows energy to cross the system boundary without the flow of matter. A fixed mass characterizes this type of problem.

The importance of choosing the correct system will be clear after the discussion of the First and Second Laws, processes, and cycles.

PHASE CHANGE

Thermodynamic studies concentrate on the properties of substances and the effect of energy transfers. In most cases the medium under consideration is either a gas or a liquid, or a mixture of both. Seldom is a complete study made of a solid, at least in the introductory levels. Transformation from liquid to vapor is important, as is any phase change, primarily due to the amount of energy transferred (required or liberated) during the phase change. There are also important property definitions and relationships that come from study of phase interactions. The most important medium that experiences phase changes is water. The study of water, especially in the form of steam, lead to the development of the science of thermodynamics.

Since we generally deal with pure substances as defined above, we will consider only those in this discussion. Water is a pure substance because it retains its chemical composition through all three phases (solid, liquid, vapor). On the other hand, air will decompose into individual elements as the temperature is reduced, altering the chemical composition. For this reason mixtures of gases, such as air, can be considered pure at temperatures and pressures that keep them in the gaseous phase.

Figures 2 and 3 are three-dimensional schematics of the *P-V-T* surface for a pure substance. They show that pure substances can exist only in the vapor, liquid, or solid phase in certain regions.

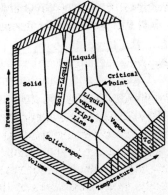

Figure 2. *P-V-T* **surface for a substance that contracts on freezing**

345

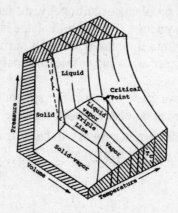

Figure 3. *P-V-T* **surface for a substance that expands on freezing**

The following information applies to these diagrams:

(1) The **critical point** is the point beyond which the substance exists as a gas. At the critical point, the saturated liquid and saturated vapor are identical, and the heat of vaporization, h_{fg}, is zero.

(2) The pressure, temperature, and specific volume at the critical point are called **critical properties:** P_c, T_c, and v_c.

(3) In the **liquid-vapor region**, liquid and vapor exist as a saturated mixture. Any change in the heat transfer (energy) at constant pressure will change the ratio of the liquid to the vapor. One hundred percent liquid describes the saturated liquid line, SLL, while 100 percent vapor describes the saturated vapor line, SVL. Any increase in energy from the SLL causes vapor to form. Any decrease in energy from the SVL causes liquid to form. The temperature is constant during the vaporization process and is referred to as the **saturation temperature**. Temperature and pressure are not independent intensive properties in the liquid-vapor region.

The change in enthalpy associated with a phase change from solid to liquid is the latent heat of fusion. The change in enthalpy associated with a phase change from liquid to vapor is the latent heat of vaporization.

A useful representation of the three-dimensional phase diagram is the two-dimensional equivalent. The *P-T* diagram describes the interaction of the phases experienced by the medium. Figures 4 and 5 are for substances which expand on freezing and for substances which contract on freezing, respectively. For these diagrams, the following terms apply:

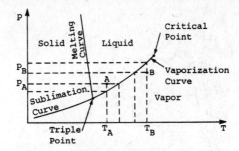

Figure 4. *P-T* diagram for a substance that expands on freezing

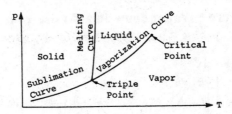

Figure 5. *P-T* diagram for a substance that contracts on freezing

(1) The **triple point** is the point at which all three phases can coexist in equilibrium.
(2) The **sublimation curve** is the curve along which the solid phase may exist in equilibrium with the vapor phase.
(3) The **vaporization curve** is the curve along which the liquid phase may exist in equilibrium with the vapor phase.
(4) The **melting curve** is the curve along which the solid phase may exist in equilibrium with the liquid phase.
(5) In Figure 4, State A is known as a subcooled liquid or a compressed liquid. State B is known as a super heated vapor.

Several other useful diagrams can be obtained as a result of plotting different property comparisons.

These diagrams are used to describe property relationships and to solve problems in which the medium is either water or Freon. In each case a vapor dome is described by the SLL and SVL. Understanding the vapor dome and the property interactions associated with the dome is important in the solution of cycles, especially vapor power cycles. Figures 6 to 9 represent the common combinations used in thermodynamics.

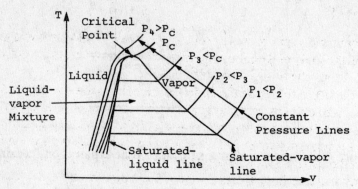

Figure 6. Vapor dome on a *T-v* diagram

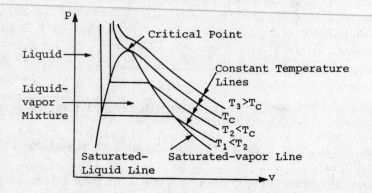

Figure 7. Vapor dome on a *P-v* diagram

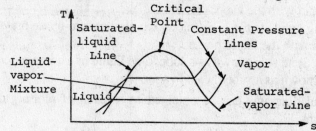

Figure 8. Vapor dome on a *T-s* diagram

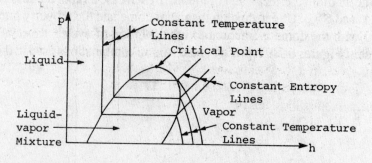

Figure 9. Vapor dome on a *P-h* (Mollier) diagram

Vapor Dome

Since liquid and vapor phases of substances, especially Freon and water, are so important in the study of thermodynamics, a few additional notes are in order. Refer to Figure 6 during the following discussion.

a. In the liquid region on the far left, the medium exists as a saturated liquid. Any reduction of the temperature below the saturation temperature at the existing pressure, or increase of the pressure above the saturation pressure at a given temperature will produce a **subcooled** or **compressed liquid**.

b. On the far right of the vapor dome the medium exists as a saturated vapor. Any increase in temperature above the saturation temperature at a given pressure results in a **superheated vapor**.

c. The line used to construct the left side of the vapor dome is called the saturated liquid line and is where the medium exists as a **saturated liquid**. Any increase in the temperature when the medium is at its saturation temperature at a given pressure will result in vaporization. Similarly, the saturation pressure at a given temperature is the pressure, below which, the medium will vaporize.

d. The line used to construct the right side of the vapor dome is called the saturated vapor line and is where the medium exists as a **saturated vapor**. Any decrease in the temperature when the medium is at its saturation temperature at a given pressure will result in condensation. Similarly, the saturation pressure at a given temperature is the pressure, above which, the medium will condense.

e. Finally, between the saturated liquid and the saturated vapor line lies the **saturated mixture** region or the **liquid-vapor mixture**. Here the temperature and pressure are not independent intensive properties. The quality, as defined below, becomes important when fixing the state of the medium. Saturated mixtures increase in vapor content as more energy is added, or become more liquid as energy is removed.

Common notation used around the vapor dome includes:

f = Saturated liquid

g = Saturated vapor

fg = Difference between values of properties for a liquid and a gas, i.e., $h_{fg} = h_g - h_f$, heat of vaporization.

In the liquid-vapor mixture region, temperature and pressure are not independent properties. To assist in the fixing of states in this region, the ratio of the mass of the vapor to the total mass defines the quality, which is used as an intensive property. The quality is defined only under the dome and ranges from 0 on the SLL to 1 on the SVL. The quality in equation form is

$$x = \frac{m_g}{m_g + m_f} = \frac{m_{vapor}}{m_{total}}$$

where m_g = Mass of vapor

$$m_f = \text{Mass of liquid}$$

Knowledge of the quality is useful in the calculation of other properties such as enthalpy and entropy using the following relationship:

$$P = P_f + x P_{fg} = x P_g + (1-x) P_f$$

where $P = $ Any property (v, u, h, s)

   $x = $ Quality

The moisture content is extremely important in the design and operation of steam turbines. It is defined as

$$y = 1 - x.$$

To illustrate how properties are determined in the various regions, water will be used as an example, since steam tables are readily available:

(1) With T and P known, determine where on the vapor dome the state point is located. This is accomplished by looking up T_{sat} for the given pressure and comparing it to the state point temperature, T. If,

   $T > T_{sat}$, then **superheated vapor**, or

   $T = T_{sat}$, then **saturated mixture**, or

   $T < T_{sat}$, then **compressed liquid**, or

   $T > T_{critical}$, then **superheated vapor**.

(2) If a **compressed liquid** state exists at pressure less than 7.5 MPa, a good approximation is to look up the properties for the saturated liquid (f) at the state point temperature.

(3) If a **superheated state** enters the superheated vapor tables with the known properties and determine the needed values.

(4) If the quality is given, then the state point must be in the **saturated mixture** region. Here, use the procedures outlined above.

(5) With T and v known, look up v_f and v_g at the prescribed T. If

   $v < v_f$, then **compressed liquid**, or

   $v > v_g$, then **superheated vapor**, or

   $v_f < v < v_g$, then **saturated mixture** region,

follow steps 2, 3, or 4 above as appropriate.

(6) Given any two properties which are independent intensive properties, the procedures above can be used if one property is either temperature, T, or pressure, P.

THERMODYNAMIC PROCESSES

When a thermodynamic system changes from one state to another, it is said to execute a process. The process is described in terms of the end states and is influenced by the energy transfers that occur as the medium changes states. When a medium at an initial state experiences changes that cause it to undergo several processes and then returns to its original state, it has experienced a cycle.

Throughout thermodynamics, special processes are used to model actual devices in an attempt to predict the outcome of the actions of these devices. Cycles are combinations of these processes and are the root of the study of heat engines and refrigerators. Processes most commonly experienced are:

(1) **Isothermal process**: one that occurs at constant temperature
(2) **Isobaric process**: one that occurs at constant pressure
(3) **Isometric/ isochoric process**: one that occurs at constant volume
(4) **Adiabatic process**: one that occurs with no heat transfer across the system boundary
(5) **Quasiequilibrium process**: one that occurs as a succession of equilibrium states such that at every instant the system involved departs only infinitesi- mally from the equilibrium state
(6) **Reversible process**: one that occurs such that the initial state of the system can be restored with no observable effect on the system or the surroundings. Also known as an ideal process
(7) **Irreversible process**: one that occurs such that the initial state of the system cannot be restored without observable effects on the system or the surround- ings
(8) **Isentropic process**: one that occurs at constant entropy. Also known as an adiabatic-reversible process
(9) **Polytropic process**: one that obeys the relationship PV^n = constant. Nor- mally a reversible process with an associated heat transfer

IDEAL GASES

It is generally accepted that gases at low density obey what is known as the ideal gas equation of state:

$$Pv = RT$$

where R = specific gas constant.

Rearranging this equation and introducing the compressibility factor, Z, we have

$$Z = \frac{RT}{Pv}$$

When the compressibility factor equals one, then an ideal gas exists. The assumption of an ideal gas can be made even if the compressibility factor differs slightly from one. If the pressure is low, below 10 MPa or so, depending on the gas, and if the temperature is about twice the critical temperature, the ideal gas assumption is considered valid. Other forms of the ideal gas equation of state are

$$PV = mRT$$

$$PV = n\overline{R}T$$

where m = Mass of the gas

n = Number of moles of the gas

$\overline{R}$ = Universal gas constant

In addition to the equation of state, ideal gases have other important relationships used throughout thermodynamic analysis. For an ideal gas, internal energy and enthalpy are functions of temperature only,

$$u = u(T), \qquad h = h(T)$$

From the previous definition of enthalpy, replacing Pv with RT in accordance with the equation of state produces

$$h = u + RT.$$

Two important relationships used to connect internal energy and enthalpy to temperature variations are the constant volume and constant pressure specific heats. The constant volume specific heat is defined as

$$C_v = \left(\frac{\partial u}{\partial T}\right)_v$$

and the constant pressure specific heat is defined as

$$C_p = \left(\frac{\partial h}{\partial T}\right)_p.$$

Several other specific heat relations are frequently used,

$$C_p - C_v = R$$

and

$$k = \frac{C_p}{C_v}.$$

As will be seen later in cycle analysis, changes in internal energy, enthalpy, and entropy are important to the analysis. For ideal gases, the evaluation of these changes are directly related to the properties at the beginning and ending states. Often all that is needed is the temperature variation, i.e., $T \rightarrow T_o$. Using this notation, the following relationships are presented:

(1) Internal energy change

$$u - u_o = \int_{T_o}^{T} C_v dT = C_v (T - T_o)$$

(2) Enthalpy change

$$h - h_o = \int_{T_o}^{T} C_p dT = C_p (T - T_o)$$

(3) Entropy change

$$s - s_o = \int_{T_o}^{T} \frac{C_v dT}{T} + R\ln\frac{v}{v_o} = C_v\ln\frac{T}{T_o} + R\ln\frac{v}{v_o}$$

$$s - s_o = \int_{T_o}^{T} \frac{C_p dT}{T} - R\ln\frac{P}{P_o} = C_p\ln\frac{T}{T_o} - R\ln\frac{P}{P_o}$$

The values obtained using constant specific heats are reasonable approximations. For better results, variable specific heats can be used in concert with tabulated data for the various gases. When using the tables, knowledge of the temperature is sufficient for direct evaluation of u, h, and s at a given state point.

For an ideal gas, variations in properties can be represented on T-s and P-v diagrams as shown in Figures 10 and 11:

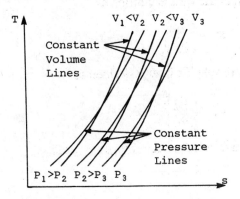

Figure 10. *T-s* diagram for an ideal gas

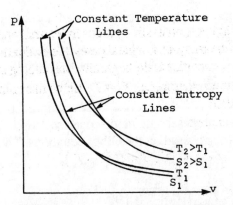

Figure 11. *P-v* diagram for an ideal gas

The reversible polytropic process for an ideal gas is one for which the pressure-volume relation is given by

$$Pv^n = \text{Constant.}$$

The polytropic processes for various values of *n* are shown on the *P-v* and *T-s* diagrams, in Figures 12 and 13:

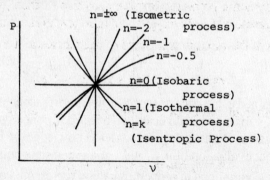

Figure 12. Polytropic processes on a *P-v* diagram

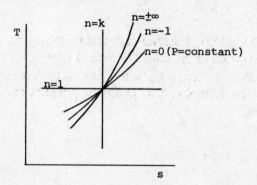

Figure 13. Polytropic processes on a *T-s* diagram

For a polytropic process the properties are related according to the relationship below:

$$\frac{T_2}{T_1} = \left(\frac{P_2}{P_1}\right)^{\frac{(n-1)}{n}} = \left(\frac{v_1}{v_2}\right)^{(n-1)}$$

Many thermodynamic processes are modeled as isentropic. These require additional property relationships based on the Gibbs equations. Thus, for an isentropic process,

$$\frac{T_2}{T_1} = \left(\frac{P_2}{P_1}\right)^{\frac{k-1}{k}} = \left(\frac{v_1}{v_2}\right)^{k-1}$$

Mixture of Gases

Many thermodynamic problems involve mixtures of ideal gases. Air, itself a mixture of ideal gases, is mostly oxygen and nitrogen, at normal temperatures.

The total mass of the mixture is the sum of the masses of the components:

$$m = m_1 + m_2 + m_3 + \ldots + m_n = \sum_i^n m_i$$

The total number of moles of the mixture is the sum of the moles of the components:

$$n = n_1 + n_2 + n_3 + \ldots + n_n = \sum_i^n n_i$$

Define the mole fraction of a component, x_i, as

$$x_i = \frac{n_i}{n} = \frac{P_i}{P} = \frac{v_i}{v}.$$

with the pressure ratio and the volume ratio a result of applying the ideal gas equation of state. Recall that the sum of the mole fractions is equal to one. Using the mole fraction and molar values of h, u, s, C_p, and C_v, along with pressure, provides the following series of relationships used to determine properties of mixtures of ideal gases:

Enthalpy

$$\overline{h} = \sum_i x_i \overline{h}_i$$

Internal energy

$$\overline{u} = \sum_i x_i \overline{u}_i$$

Entropy

$$\overline{s} = \sum_i x_i \overline{s}_i$$

Specific heat

$$\overline{C}_p = \sum_i x_i \overline{C}_{pi}$$

Specific heat

$$\overline{C}_v = \sum_i x_i \overline{C}_{vi}$$

Pressure

$$P_i = x_i P$$
$$P = \sum_i x_i P$$

Finally, the equivalent molecular weight of the mixture, M, is found using

$$M = \sum_i x_i M_i .$$

A similar series of equations can be generated by defining the mass fraction as the mass of a component divided by the total mass. Then using property values based on mass fractions instead of mole fractions, the mixture values are calculated. Both provide necessary information concerning the mixture, and the use of one over the other is merely a convenience of the problem solution.

ENERGY, HEAT, AND WORK

Since for a given closed system the work done is the same in all adiabatic processes between equilibrium states, a fundamental property of the medium in the system can be defined such that the change between equilibrium states is equal to the adiabatic work, as below:

$$E_2 - E_1 = W_{adiabatic}$$

Work will be discussed later. The energy is a fundamental property and is defined in the following word equation:

E = Internal Energy + Kinetic Energy + Potential Energy

where

(1) **Internal energy, U,** is an extensive property since it depends on the mass of the system. It represents the energy modes on the microscopic level, such as the energy associated with nuclear spin, molecular binding, magnetic dipole moment, etc.

(2) **Kinetic energy, KE,** is energy a body possesses due to bulk motion. For example, the kinetic energy of a system of mass, m, with velocity v is given by

$$KE = \frac{1}{2}mv^2 .$$

(3) **Potential energy, PE,** is the energy a body possesses due to its position in a potential field. For example, the potential energy of a system having a mass, m, and an elevation, z, above a defined plane in a gravitational field with a constant gravitational constant, g, is given by

$$PE = mgz.$$

Whereas energy is a property and has a finite value at a fixed point, heat and work are transient phenomena, and are not defined at a point. Systems never possess heat or work. Heat and work cross the boundary of a system undergoing a change of state, and are only observable at the boundary. Both are path functions and are represented by inexact differentials. Heat is represented by δQ, and work by δW. When integrated across a process, in a closed system from state 1 to 2 the amount of heat that crosses the boundary is represented by $_1Q_2$, and similarly for work, $_1W_2$.

Heat, Q, is the form of energy that is transferred across a system boundary as a result of temperature differences. Heat travels from the highest temperature to the lowest temperature. Positive heat transfer is heat addition to a system, and negative heat transfer is heat removed from a system. The details of heat transfer are saved for a complete series of courses that investigate the three modes of heat transfer, conduction, convection, and radiation. These topics are reviewed later.

Work, W, is classically defined as a force, F, applied through a distance, dx. In integral form,

$$W = \int_1^2 F\,dx.$$

In a thermodynamic sense, work is an interaction between a system and its surroundings where the sole effect of the system on the surroundings is the raising of a weight. Work done by a system is considered positive, and work done on a system is considered negative.

There are many work modes used in the analysis of thermodynamic processes. These include compressibility, stretched wire work, surface film work, magnetic work, and electrical work. Since we are concentrating on the pure substance, compressibility is the only work mode being considered. In Figure 14, a gas contained in a closed system is expanded from state 1 to state 2 as a result of a higher pressure inside than outside.

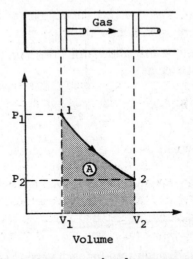

Figure 14. Work done on a simple compressible system

For any small expansion in which the volume of the gas increases by dV, the work done by the gas is

$$W = \int_1^2 P\,dV.$$

The integral value is the area under the curve on the *P-V* diagram. Since we can go from state 1 to state 2 along many different paths, it is evident that the amount of work represented under the curve is a function of both the end states and the path the process follows while going from state 1 to state 2. As previously mentioned, work is a path function, represented mathematically as an inexact differential.

During the analysis of energy transfers associated with the first law, work will be described as:

(1) **System boundary work**: work associated with the movement of a boundary such as that which occurs during a *P-V* expansion or compression.

(2) **Shaft work**: work associated with the rotation of a shaft.

(3) **Flow work**: work associated with the flow of a fluid. This is most often tied to the definition of enthalpy.

First Law

Before any discussion of the first law, it is customary to discuss the conservation of mass. Mass can neither be created nor destroyed; it must be strictly accounted for. For any system, conservation of mass states that:

mass added – mass removed = change in the mass stored

For the closed system the mass is fixed since there is no exchange of mass with the surroundings. Thus, the mass at any state point is constant. Symbolically, this is represented by

$$m_1 = m_2 = m_3 = \ldots$$

For the open system, one that allows for mass transfer across the boundary, conservation of mass is stated as:

$$\dot{m}_{in} - \dot{m}_{out} = \frac{dm}{dt}$$

where $\dot{m}_{out}$ = Mass flow rate out of the control volume

$\dot{m}_{in}$ = Mass flow rate into the control volume

$\frac{dm}{dt}$ = Rate of change in the mass in the control volume

Specific applications will be addressed for each version of the first law discussed. **Note:** The conservation of mass is often referred to as the continuity equation and will be referred to as such from now on.

As with the continuity equation, the first law of thermodynamics can be simply stated via a word equation:

energy input – energy output = change in stored energy

This rather simple equation is the basis for the development of every application of the first law, and many of the applications used in heat transfer. Mastery of the first law is essential in the analysis of work producing machines, and in any device that exchanges heat with the surroundings.

Observations have led to the formulation of the first law for cycles, which in equation form is

$$\oint dQ = \oint dW$$

where $\oint dQ$ = Cyclic integral of the heat transfer

$\oint dW$ = Cyclic integral of work

and the units are System International.

The first law applied to the closed system undergoing a process and changing from state 1 to state 2 is

$$\delta Q - \delta W = dE$$

where δQ = Heat transferred to the system during the process

δW = Work transferred from the system during the process

E = The total energy of the system and a property of the medium.

The net change of the energy of the system is always equal to the net transfer of energy across the system boundary in the form of heat and work.

For analysis purposes it is important to look at continuity for the closed system. Via the continuity equation, the mass is constant. Thus, the mass at any state point can be represented as m. The integrated form of the first law becomes

$$_1Q_2 - {}_1W_2 = U_2 - U_1 + \frac{m\left(v_2^2 - v_1^2\right)}{2} + mg\left(z_2 - z_1\right)$$

where $_1Q_2$ = The heat transferred during the process from 1 to 2

$_1W_2$ = The work done by or in the system during the process from 1 to 2

$U_2 - U_1$ = The change in internal energy

$\dfrac{m\left(v_2^2 - v_1^2\right)}{2}$ = The change in kinetic energy

$mg\left(z_2 - z_1\right)$ = The change in potential energy

Two additional notes concerning this equation:

(1) Only changes in internal energy and kinetic and potential energy can be determined with this equation. Absolute values are not easily obtained.

(2) The first step in applying the first law is to determine the appropriate boundary description—open or closed.

A sign convention used in Thermodynamics, although not universal, and applied equally to open and closed systems is

work in is –   work out is +

heat in is +   heat out is –

The general forms of the equations of continuity and the first law for an open system with multiple inlets and exits are respectively

$$\sum \dot{m}_{in} - \sum \dot{m}_{out} = \frac{dm}{dt}$$

$$\dot{Q}_{cv} - \dot{W}_{cv} = \frac{dE_{cv}}{dt} + \sum \dot{m}_e \left(h_e + \frac{V_e^2}{2} + gz_e \right) - \sum \dot{m}_i \left(h_i + \frac{V_i^2}{2} + gz_i \right)$$

where

$\dot{Q}_{cv}$ = Rate of heat transfer into the control volume

$\dot{W}_{cv}$ = Work rate that crosses or displaces the control volume

$\dfrac{dE_{cv}}{dt}$ = Rate of change of the energy inside the control volume

$\sum \dot{m}_e \left(h_e + \dfrac{v_e^2}{2} + gz_e \right)$ = rate of energy flowing out as a result of mass transfer

$\sum \dot{m}_i \left(h_i + \dfrac{v_i^2}{2} + gz_i \right)$ = rate of energy flowing in as a result of mass transfer

Taking the basic equation and applying it to the steady-state, steady-flow process, a primary form used in thermodynamic analysis requires the following assumptions:
(1) The control volume does not move relative to the coordinate frame.
(2) The state of the mass at each point in the control volume does not change with time.
(3) The mass flux does not vary with time.
(4) The rates at which heat and work cross the control surface remain constant.
(5) And with the requirement of one inlet, one exit produces:

Continuity equation, $\dot{m}_i = \dot{m}_e = \dot{m}$

First law

$$\dot{Q}_{cv} - \dot{W}_{cv} = \sum \dot{m}_e \left(h_e + \frac{v_e^2}{2} + gz_e \right) - \sum \dot{m}_i \left(h_i + \frac{v_i^2}{2} + gz_i \right)$$

Start up operations and time dependent, unsteady situations are analyzed using the uniform-state, uniform-flow equations. Assumptions for this model are:
(1) The control volume remains constant relative to the coordinate frame.
(2) The state of the mass may change with time in the control volume, but at any instant of time the state is uniform throughout the entire control volume.
(3) The state of the mass crossing all the areas of flow is constant with respect to the control surface, but the mass flow rates may vary with time. Thus:

Continuity equation, $\left(m_2 - m_1\right)_{cv} + \sum m_e - \sum m_i = 0$

First law

$$Q_{cv} - W_{cv} = \sum m_e \left(h_e + \frac{v_e^2}{2} + gz_e \right) - \sum m_i \left(h_i + \frac{v_i^2}{2} + gz_i \right)$$

$$+ \left(m_2 \left(u_2 + \frac{v_2^2}{2} + gz_2 \right) - m_1 \left(u_1 + \frac{v_1^2}{2} + gz_1 \right) \right)_{cv}$$

In this equation, the rate expressions for the heat, work, and mass flow terms have not been forgotten: integration over time in the development provides total quantities in lieu of rates.

Second Law

The study of the second law and entropy is predicated on the understanding of heat engines, the Carnot cycle, and reversible and irreversible processes. The latter have been previously discussed. Here, the study will begin with the heat engine and refrigerator, followed by the Carnot cycle. The purpose is to place limitations on real devices not obvious by the first law. For example, it is possible to satisfy the first law and violate the second law. Without this check, heat could flow from cold to hot, a concept that is naturally alien to intuition.

A heat engine is a system that operates in a cycle while only heat and work cross its boundaries. The work is the desired result of a heat engine, having a positive (out) sign. Referring to Figure 15, heat is transferred from the high-temperature reservoir, T_H, to the low-temperature reservoir, T_L.

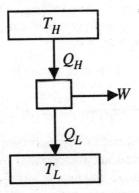

Figure 15. Heat engine

A steam power plant is a heat engine that receives heat from a high-temperature reservoir at the boiler, rejects heat to a low-temperature reservoir at the condenser, and delivers useful work. There are many other heat engines that will be discussed later.

The efficiency of a heat engine is defined as the ratio: the net work delivered to the surroundings divided by the heat received from the high-temperature source. Additionally, the work produced in our simple model is equal to the difference between the heat added and the heat rejected. In equation form,

$$\eta = \frac{W}{Q_H} = \frac{Q_H - Q_L}{Q_H} = 1 - \frac{Q_L}{Q_H}$$

where Q_H = Amount of heat added to the heat engine

$\quad\;\; Q_L$ = Amount of heat rejected by the heat engine

$\quad\;\; W$ = Net work produced by the engine, $Q_H - Q_L$

Refrigerators and heat pumps are heat engines working in reverse. Work is required (input) in order to move heat from a low-temperature reservoir to a high-temperature reservoir. While this appears to fail intuition, remember that heat cannot travel from low to high on its own, only when aided by another energy transfer, work. Figure 16 illustrates the concept of a refrigerator or heat pump,

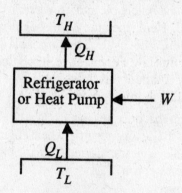

Figure 16. Refrigerator/heat pump

where the values of Q_H, Q_L, and W have particular functions depending on the device. For example,

$\quad Q_H$ = Heating capacity of a heat pump

$\quad Q_L$ = Cooling/refrigeration capacity of the refrigerator or air conditioner

$\quad W$ = Work required to make the cycle operate

The effectiveness of refrigerators and heat pumps is not measured by efficiency, rather by using the coefficient of performance, COP. Values for the coefficient of performance greater than one are not uncommon, and are actually expected. The COP for a refrigerator is

$$\beta_R = \frac{Q_L}{W} = \frac{Q_L}{Q_H - Q_L} = \frac{1}{\dfrac{Q_H}{Q_L} - 1}$$

and the COP for a heat pump is

$$\beta_{HP} = \frac{Q_H}{W} = \frac{Q_H}{Q_H - Q_L} = \frac{1}{1 - \dfrac{Q_L}{Q_H}}$$

where in general terms

Q_L = Amount of heat transferred from the low-temperature reservoir

Q_H = Amount of heat transferred from the high-temperature reservoir

W = Net amount of work required

With the definitions of the heat engine, the refrigerator, and the heat pump in hand, the first of the second law statements can be presented. These are:

(1) **The Kelvin-Plank Statement**: It is impossible to construct a device that will operate in a cycle and produce no effect other than the raising of a weight and the exchange of heat with a single reservoir. In short, there is no such thing as a perfect heat engine—there must be heat rejected.

(2) **The Clausius Statement**: It is impossible to construct a device that operates in a cycle and produces no effect other than the transfer of heat from a cooler body to a hotter body. In short, heat does not flow uphill, cooler to hotter.

There are other limits on the effectiveness of the devices previously discussed. The Carnot cycle is the reversible approximation of the ideal heat engine or refrigerator. It consists of reversible processes that form a reversible cycle. Since the cycle is reversible, a Carnot heat engine can be reversed to operate as a Carnot refrigerator/heat pump. When the cycle operates between two temperature reservoirs, high and low, the cycle consists of the four processes depicted in Figure 17. These four processes are the same for any Carnot cycle.

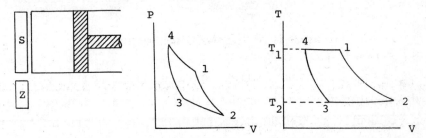

Figure 17. Carnot cycle on *P-v* and *T-v* diagrams

Process 1-2

A reversible adiabatic process in which the temperature of the working fluid decreases from the high temperature to the low temperature.

Process 2-3

A reversible isothermal process in which heat is transferred to or from the low-temperature reservoir.

Process 3-4

A reversible adiabatic process in which the temperature of the working fluid increases from the low temperature to the high temperature.

Process 4-1

A reversible isothermal process in which heat is transferred to or from the high-temperature reservoir.

As a result of the development of the Carnot cycle, two propositions have been formulated regarding the efficiency of the Carnot cycle.

    (1) It is impossible to build an engine that operates between two thermal reservoirs and is more efficient than a reversible engine operating between the same thermal reservoirs.

    (2) All Carnot engines operating between the same thermal reservoirs have the same efficiency.

From the development of a Thermodynamic Temperature Scale, the efficiency of the reversible heat engines and the COP of refrigerators and heat pumps can be expressed as a function of the thermal reservoir temperatures. Thus,

$$\eta_{th} = 1 - \frac{T_L}{T_H}$$

$$\beta_R = \frac{1}{\dfrac{T_H}{T_L} - 1}$$

$$\beta_{HP} = \frac{1}{1 - \dfrac{T_L}{T_H}}$$

where T_L = Low-temperature reservoir

       T_H = High-temperature reservoir, both temperatures are absolute

Finally, the efficiency of a heat engine is always less than one, $\eta_{th} < 1$, and the efficiency of an irreversible engine is less than that of a reversible engine, $\eta_I < \eta_R$.

So far we have discussed cycles made up of processes, most of which have been reversible. Before turning to the evaluation of processes to determine if they are progressing in accordance with the second law, it is important to mention the Inequality of Clausius. Simply stated, for any irreversible cycle, the following cyclic integral must apply:

$$\oint \frac{\delta Q}{T} \leq 0$$

Satisfaction of this requirement indicates that a cycle is obeying the second law; therefore, it is capable of occurring if it also satisfies the first law.

It can be shown that for two reversible processes operating in a cycle, the end points alone define the quantity:

$$dS = \left(\frac{\delta Q}{T}\right)_{rev}$$

where δQ = Heat supplied to the system

T = Absolute temperature of the system

dS = Represents the change in the property called entropy

A few notes are required at this point.

(1) The equation is valid for any reversible process.

(2) Entropy, S, is an extensive property, and it is a function of the end points of the process only, a point function. Thus, since it is independent of the path, the value of the change in entropy is the same for reversible and irreversible processes. The change in the entropy of a closed system can be found by integrating

$$S_2 - S_1 = \int_1^2 \left(\frac{\delta Q}{T}\right)_{rev}$$

In the case of an irreversible process, the entropy change will be exactly the same as for the reversible process. So what is the difference between the two? In the irreversible process a certain amount of the energy transferred is lost, not available for use later on. The entropy change for an irreversible process in a closed system becomes

$$S_2 - S_1 \overset{>}{=} \int_1^2 \frac{\delta Q}{T}$$

The second law, just like the first law, is applied to many different situations: open and closed systems, and steady-state and unsteady problems. Since continuity has been previously reviewed with the first law, only the second law equations for each situation will be presented.

The general form of the second law for the control volume is

$$\frac{dS_{cv}}{dt} + \sum \dot{m}_e s_e - \sum \dot{m}_i s_i = \int_A \left(\frac{\dot{Q}_{cv}/A}{T}\right) dA + \int_V \left(\frac{L\dot{W}_{cv}/V}{T}\right) dVA$$

This expression states that the rate of change of entropy inside the control volume, plus the net rate of entropy flow out, is equal to the sum of two terms: the integrated heat transfer term and the positive, internal irreversibility term.

For the steady-state, steady-flow case, the general equation reduces to

$$\sum \dot{m}_e s_e - \sum \dot{m}_i s_i \overset{>}{=} \int_A \left(\frac{\dot{Q}_{cv}/A}{T}\right) dA$$

where the removal of the irreversibility term requires the inequality. For an adiabatic process, with a single inlet and a single exit, the equation reduces to $s_e \geq s_i$.

For the uniform-state, uniform-flow case, the general equation becomes

$$\left(m_2 s_2 - m_1 s_1\right)_{cv} + \sum m_e s_e - \sum m_i s_i = \int_o^t \left(\frac{\dot{Q}_{cv} + L\dot{W}_{cv}}{T}\right) dt \; .$$

Additional notes concerning the second law are necessary to understand the total importance it has in thermodynamics. Looking first at the reversible steady-state, steady-flow process with one inlet and one exit, there are three important results

1) When the process is both reversible and adiabatic,

$$w = -\int_i^e v\,dP + \frac{\left(v_i^2 - v_e^2\right)}{2} + g\left(z_i - z_e\right).$$

2) Taking this equation one step further by specifying that the work is zero and the fluid is incompressible, after integration we obtain Bernoulli's equation:

$$v\left(P_e - P_i\right) + \frac{\left(v_e^2 - v_i^2\right)}{2} + g\left(z_e - z_i\right) = 0$$

3) If the process is reversible and isothermal,

$$T\left(s_e - s_i\right) = \frac{\dot{Q}_{cv}}{\dot{m}} = q.$$

Entropy is a way to determine the direction of time. As with time, entropy will always be positive in the total universe. This is referred to as the principle of the increase in entropy. For any isolated system,

$$ds_{isol} \geq 0$$

and for a control volume interacting with the surroundings

$$\frac{dS_{cv}}{dt} + \frac{dS_{surr}}{dt} \geq 0.$$

Finally, there are two important property relationships that are applicable for reversible or irreversible processes since they provide a means for evaluation of the change in entropy needed above.

$$TdS = dU + PdV$$
$$TdS = dH - VdP$$

AVAILABILITY-IRREVERSIBILITY

The maximum work that can be done by a system is called the **availability**. This maximum work is achieved when the work is reversible, with the system undergoing a reversible process until it achieves equilibrium with the surroundings. Calculation of the availability depends, in part, on the type of analysis being performed, that is, closed or open system. For the closed system, boundary variations and the work associated with them must be considered since this reduces the total available work. Availability, per unit mass, neglecting kinetic and potential energy effects, is given by

$$\phi = \left(w_{rev}\right)_{max} - w_{surr}$$

where $\left(w_{rev}\right)_{max} = \left(u - T_o s\right) - \left(u_o - T_o s_o\right)$

$$w_{surr} = -P_o\left(v - v_o\right)$$

thus $\phi = \left(u - u_o\right) + P_o\left(v - v_o\right) - T_o\left(s - s_o\right)$

where u, v, and s are the internal energy, specific volume, and entropy of the system and u_o, v_o, and s_o are the internal energy, specific volume, and entropy of the surroundings.

For the open system there is no boundary work, and thus no work to the surroundings. The availability is the reversible work, and is in the general form:

$$\dot{W}_{rev} = \sum \dot{m}_i \psi_i - \sum \dot{m}_e \psi_e$$

where $\psi = \left(h - T_o s + \dfrac{V^2}{2} + gz\right) - \left(h_o - T_o s_o + gz_o\right),$

and those symbols without subscript are either inlet or exit values as per the previous equation.

Irreversibility is the difference between the reversible work and the actual work accomplished. It is defined as

$$I = W_{rev} - W_{cv}$$

where W_{rev} = The reversible work

W_{cv} = The work crossing the control volume

The irreversibility is expressed for a control volume experiencing a uniform-state, uniform-flow process as

$$I = \sum m_e T_o s_e - \sum m_i T_o s_i + m_2 T_o s_2 - m_1 T_o s_1 - Q_{cv}$$

where the subscripts e = Exit

i = Inlet

o = Surroundings

1 = State 1

2 = State 2

This is the most general form of the irreversibility relationship from which the others are developed. For a steady-state, steady-flow process the equation becomes

$$I = \sum m_e T_o s_e - \sum m_i T_o s_i - Q_{cv}$$

and for the system of fixed mass the equation reduces to

$$_1 I_2 = m T_o\left(s_2 - s_1\right) - {}_1 Q_2.$$

Finally, if the process taking place was a reversible one, the irreversibility would be equal to zero. Such is the case in the ideal processes that describe ideal cycles.

Components

Cycles used to produce power are designed based on specific needs and are improved by addition of components which increase efficiency or otherwise influence operating conditions. This section will introduce these components and briefly discuss their operation and application.

Pumps and **compressors** are used to increase the pressure of the medium flowing through them. Pumps are used when the medium is a liquid and compressors are used when gases are flowing. Both devices require work and while increasing the pressure they also add energy to the medium.

Boilers, superheaters, and **evaporators** all take energy from some source and increase the energy of the flowing medium. In the case of boilers and superheaters the energy required comes from a large source such as a nuclear reactor or the flame from an oil or gas burner. In the boiler the phase of the medium usually changes from liquid to vapor whereas in the superheater additional energy is added to a vapor to move farther into the superheated region. The evaporator also uses phase changes but the energy source is often much smaller and at lower temperatures such as air at ambient conditions. As the medium in the evaporator "boils" the energy source cools.

Condensers, as the name implies, change the phase of the medium from a vapor to a liquid. Condensers are heat exchangers which normally operate by having two distinct flow channels, one with the vapor that is condensed and one with a cooling fluid, usually water. The cooling fluid absorbs enough energy to cause condensation.

Turbines and **throttling valves** are both used to reduce the pressure and the associated energy level of the vapor. Turbines make this energy change as the fluid flows over a series of blades which change the thermal energy to a mechanical form, usually shaft work. The shaft is connected to a generator or the blades of a helicopter or to a compressor in jet engines. The throttling valve reduces the pressure and causes some of the liquid to flash to vapor prior to evaporation.

Economizers and **regenerators** use waste heat in the exhaust to precondition incoming air or liquids from the condenser prior to pumping. In both cases the preconditoning reduces the need for energy in the form of fuel. A method for warming the incoming water is the use of **feedwater heaters**. These come in two types, open, where the water is mixed directly with bleed steam, and closed which is more like a conventional heat exchanger in that the hot and cold streams are kept separated.

Nozzles and **diffusers** change the velocity of the medium, increasing or decreasing the velocity depending on the conditions of the flow. Diffusers are often used to slow the flow while increasing the pressure, an important aspect in the design of supersonic aircraft engines. Nozzles are used to accelerate the flow to produce thrust for the operation of jet aircraft.

For any of these components, there is an efficiency known as the **component efficiency** or the **isentropic efficiency** which compares the actual operation to the operation if it took place reversibly. In all cases this efficiency must be less than one. For components receiving work, the actual work required will be more than the reversible work provided. For components producing work the actual work will be less than the reversible work which could be generated. Component efficiencies will be seen in example problems.

CYCLES

Cycles are divided into two categories: power and refrigeration. This section will concentrate on power cycles; the next section will concentrate on air conditioning and refrigeration.

Power cycles are divided into two major categories: vapor power cycles and air-standard cycles. Vapor power cycles use external heat to produce steam, the working fluid used to power the cycle. Air-standard cycles use combustion of a fuel within the engine as the source of the energy to drive the cycle. Since vapor power cycles have been around the longest, we will start with them.

The Rankine cycle is the idealization of the steam (vapor) power cycle. Figure 18 illustrates the simple steam power plant.

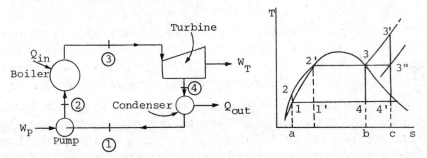

Figure 18. Simple steam power plant that operates on the Rankine cycle

The Ranking cycle consists of the following idealized processes:

Process 1-2

Reversible adiabatic (isentropic) pumping

Process 2-3

Constant-pressure heat addition in the boiler

Process 3-4

Reversible adiabatic (isentropic) expansion

Process 4-1

Constant pressure heat rejection from the condenser.

Assuming steady-state, steady-flow processes throughout, the continuity equation becomes

$$\dot{m}_1 = \dot{m}_2 = \dot{m}_3 = \dot{m}_4.$$

Neglecting kinetic and potential energy in each component, the first law for each component reduces to

Boiler $\qquad \dot{Q}_{in} = \dot{m}(h_3 - h_2)$

Turbine $\qquad \dot{W}_T = \dot{m}(h_3 - h_4)$

Condenser $\qquad \dot{Q}_{out} = \dot{m}(h_4 - h_1)$

Pump $\qquad \dot{W}_p = \dot{m}(h_2 - h_1) = \dot{m}v_1(P_2 - P_1)$

The last equation is possible since the fluid is incompressible and the process reversible. The thermal efficiency is the net work output divided by the energy added. In the efficiency equation, the mass flow rate has been divided out of the equation.

$$\eta_{th} = \frac{w_{net}}{q_{in}} = \frac{w_T - w_P}{q_{in}} = \frac{(h_3 - h_4) - (h_2 - h_1)}{h_3 - h_2}$$

and

$$w_{net} = q_{in} - q_{out} = (h_3 - h_2) - (h_4 - h_1)$$

The Rankine cycle efficiency can be increased by lowering the exhaust pressure from the turbine, increasing the pressure during heat addition, or superheating the steam.

The Rankine cycle with superheater seen in Figure 19 is used to increase the efficiency by increasing the mean temperature of heat addition with no increase in the maximum cycle pressure.

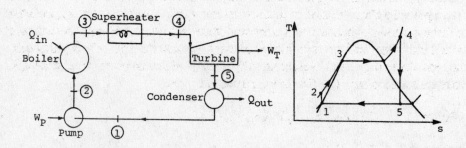

Figure 19. Rankine cycle with superheater

The continuity equation has not changed; however, the first law has a few subtle differences.

Boiler/Superheater $\quad \dot{Q}_{in} = \dot{m}(h_4 - h_2)$

Turbine $\quad\quad\quad\quad \dot{W}_T = \dot{m}(h_4 - h_5)$

Condenser $\quad\quad\quad \dot{Q}_{out} = \dot{m}(h_5 - h_1)$

Pump $\quad\quad\quad\quad\quad \dot{W}_P = \dot{m}(h_2 - h_1) = \dot{m}v_1(P_2 - P_1)$

and the thermal efficiency becomes

$$\eta_{th} = \frac{w_{net}}{q_{in}} = \frac{w_T - w_p}{q_{in}} = \frac{(h_4 - h_5) - (h_2 - h_1)}{h_4 - h_2}$$

and

$$w_{net} = q_{in} - q_{out} = (h_4 - h_2) - (h_5 - h_1)$$

A second improvement over the original Rankine cycle is the Reheat cycle that was developed to take advantage of the increased efficiency associated with higher pressures. Figure 20 depicts an ideal Reheat cycle.

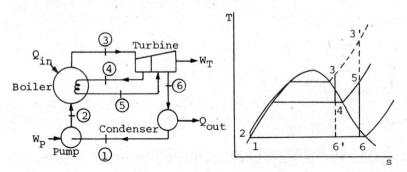

Figure 20. The ideal Reheat cycle

Notice that with the higher pressure comes the penalty of increased moisture content at state point 6'. The reheating of the flow after it leaves the high-pressure portion of the turbine allows expansion to a more reasonable moisture content. The turbine is divided into high- and low-pressure portions; however, these may be nothing more than taps. As in the other cases, continuity remains essentially the same, except for two extra state points. The first law analysis becomes

Boiler $\quad\quad\quad \dot{Q}_{in} = \dot{m}[(h_3 - h_2) + (h_5 - h_4)]$

Turbine $\quad\quad\quad \dot{W}_T = \dot{m}[(h_3 - h_4) + (h_5 - h_6)]$

Condenser $\quad \dot{Q}_{out} = \dot{m}\left(h_6 - h_1\right)$

Pump $\quad\quad \dot{W}_p = \dot{m}\left(h_2 - h_1\right) = \dot{m}v_1\left(P_2 - P_1\right)$

and thermal efficiency becomes

$$\eta_{th} = \frac{w_{net}}{q_{in}} = \frac{w_T - w_p}{q_{in}} = \frac{\left(h_3 - h_4\right) + \left(h_5 - h_6\right) - \left(h_2 - h_1\right)}{h_4 - h_2}$$

and

$$w_{net} = q_{in} - q_{out} = \left(h_3 - h_2\right) + \left(h_5 - h_4\right) - \left(h_6 - h_1\right)$$

The last of the vapor power cycles is the regenerative cycle. This cycle increases the average temperature at which heat is added in the boiler by taking some of the flow out of the turbine early and mixing it with the remaining flow, thus increasing the average temperature of the water entering the pump. The two flows are mixed in a feedwater heater, the number of heaters is determined by economic considerations, and are fed back into the boiler via the pump. Figure 21 illustrates the regenerative cycle.

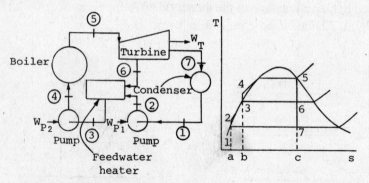

Figure 21. Regenerative cycle with open feedwater heater

Continuity for this cycle is somewhat more complex. Based on a one kilogram flow at points 3, 4, 5, the remaining flows are some fraction of the initial kilogram. The flow at points 7, 1, 2 is $(1 - m_1)$ and the flow at 6 is m_1. Based on the one kilogram flow, the following energy transfers apply:

$$q_{in} = h_5 - h_4$$
$$w_T = \left(h_5 - h_6\right) + \left(1 - m_1\right)\left(h_6 - h_7\right)$$
$$q_{out} = h_1 - h_7$$
$$w_{p2} = h_4 - h_3 = v_3\left(P_4 - P_3\right)$$
$$w_{p1} = h_2 - h_1 = v_1\left(P_2 - P_1\right)$$

where the energy balance around the feedwater heater is

$$m_1 h_6 + (1 - m_1) h_2 = h_3$$

producing the thermal efficiency

$$\eta_{th} = \frac{w_T - (1 - m_1) w_{p1} - w_{p2}}{h_5 - h_4}$$

TABLE 1
RANKINE CYCLE COMPARISON CHART

Device	Ideal Rankine	With Superheater	Reheat	Regenerative*
Boiler $\dot{Q}_{in} =$	$\dot{m}(h_3 - h_2)$	$\dot{m}(h_4 - h_2)$	$\dot{m}\left[(h_3 - h_2) + (h_5 - h_4)\right]$	$q_{in} = h_5 - h_4$
Turbine $\dot{W}_T =$	$\dot{m}(h_3 - h_4)$	$\dot{m}(h_4 - h_5)$	$\dot{m}\left[(h_3 - h_4) + (h_5 - h_6)\right]$	$w_T = (h_5 - h_6) + (1 - m_1)(h_6 - h_7)$
Condenser $\dot{Q}_{out} =$	$\dot{m}(h_4 - h_1)$	$\dot{m}(h_5 - h_1)$	$\dot{m}(h_6 - h_1)$	$q_{out} = h_1 - h_7$
Pump(s) $\dot{W}_p =$	$\dot{m}(h_2 - h_1) = \dot{m}v_1(P_2 - P_1)$	$\dot{m}(h_2 - h_1) = \dot{m}v_1(P_2 - P_1)$	$\dot{m}(h_2 - h_1) = \dot{m}v_1(P_2 - P_1)$	$w_{P_2} = h_4 - h_3 = v_3(P_4 - P_3)$ $w_{P_1} = h_2 - h_1 = v_1(P_2 - P_1)$
Efficiency $\eta_{th} =$	$\frac{(h_3 - h_4) - (h_2 - h_1)}{h_3 - h_2}$	$\frac{(h_4 - h_5) - (h_2 - h_1)}{h_4 - h_2}$	$\frac{(h_3 - h_4) + (h_5 - h_6) - (h_2 - h_1)}{h_4 - h_2}$	$\frac{w_T - (1 - m_1)w_{P_1} - w_{P_2}}{h_5 - h_4}$

* The regenerative Rankine cycle is best analyzed using a per mass basis equation

The next series of power cycles are the air-standard power cycles. While the working fluid is not all air and the cycles are actually open, power cycles can be effectively modeled using air as the primary working fluid, and a closed cycle can approximate the actual engine operation. The following assumptions apply to air-standard engines:

(1) A fixed mass of air is the working fluid and the air is always an ideal gas.
(2) The combustion process is replaced by a heat transfer from an external source.
(3) The cycle is completed by heat transfer to the surroundings.
(4) All processes are internally reversible.
(5) Air has a constant specific.

The Air-Standard Carnot cycle is the standard against which all other air-standard heat engines are compared. The Carnot cycle has been previously discussed, as have the isentropic relations that assist in specifying state point data. The isentropic relations can also be used to fix the thermal efficiency in terms of the isentropic pressure ratio and isentropic compression ratio as follows:

Isentropic pressure ratio: $\quad r_{ps} = \dfrac{P_1}{P_4} = \dfrac{P_2}{P_3} = \left(\dfrac{T_3}{T_2}\right)^{\frac{k}{(1-k)}}$

Isentropic compression ratio: $r_{vs} = \dfrac{V_4}{V_1} = \dfrac{V_3}{V_2} = \left(\dfrac{T_3}{T_2}\right)^{\frac{1}{(1-k)}}$

Thus, the efficiency of the Carnot cycle is

$$\eta_{th} = 1 - r_{ps}^{\frac{(1-k)}{k}} = 1 - r_{vs}^{1-k}.$$

Engines such as the spark ignition engine, the compression ignition engine, and the gas turbine engine have been modeled using ideal cycles. These cycles are then compared to Carnot to see how they measure up. Remember, no cycle can be *more* efficient than the Carnot cycle.

The spark ignition engine has been modeled by the Air-Standard Otto cycle using a closed system. Figure 22 illustrates the processes used to model the spark ignition engine.

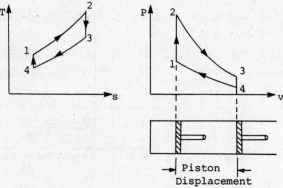

Figure 22. Air-Standard Otto cycle

The processes associated with the Otto cycle are:

Process 1-2

Constant volume heat addition

Process 2–3

Isentropic expansion

Process 3–4

Constant volume heat rejection

Process 4–1

Isentropic compression

For the closed system, the mass is fixed and is given by m. Application of the first law for the closed system results in

$$Q_{in} = {}_1Q_2 = U_2 - U_1 = mC_v(T_2 - T_1)$$

$$Q_{out} = {}_3Q_4 = U_3 - U_4 = mC_v(T_3 - T_4)$$

and a thermal efficiency of

$$\eta_{th} = \frac{W_{net}}{Q_{in}} = \frac{Q_{in} - Q_{out}}{Q_{in}} = 1 - \frac{(T_3 - T_4)}{(T_2 - T_1)} = 1 - \frac{1}{r_v^{(k-1)}}$$

where $r_v = \dfrac{V_3}{V_2} = \dfrac{V_4}{V_1}$ is known as the compression ratio. It is interesting to note that the Otto cycle efficiency increases with increased compression ratio. Also, the network is simple heat in, minus heat out.

The compression ignition engine has been modeled by the Air Standard Diesel cycle using a closed system. Figure 23 illustrates the process used to model the compression ignition engine.

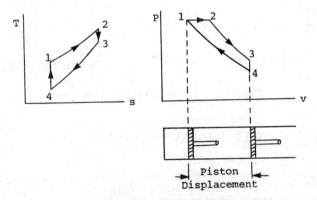

Figure 23. The Diesel cycle

The processes associated with the Diesel cycle are:

Process 1-2

Constant pressure heat addition

Process 2-3

Isentropic expansion

Process 3-4

Constant volume heat rejection

Process 4-1

Isentropic compression

For the closed system, the mass is fixed and is given by m. Application of the first law for the closed system results in

$$Q_{in} = {}_1Q_2 = H_2 - H_1 = mC_p(T_2 - T_1)$$

$$Q_{out} = {}_3Q_4 = U_3 - U_4 = mC_v(T_3 - T_4)$$

with the C_p in process 1–2 a result of the combination of properties. The thermal efficiency is

$$\eta_{th} = \frac{W_{net}}{Q_{in}} = \frac{Q_{in} - Q_{out}}{Q_{in}} = 1 - \frac{(T_3 - T_4)}{k(T_2 - T_1)}$$

$$\eta_{th} = 1 - \frac{1}{r_v^{k-1}}\left[\frac{r_c^k - 1}{k(r_c - 1)}\right]$$

where $r_v = \dfrac{V_4}{V_1}$ = the compression ratio

$r_c = \dfrac{V_2}{V_1}$ = the cutoff ratio

Recall that the cutoff ratio is a measure of amount of time that fuel is injected, and is expressed as volume ratio changes as the piston moves.

TABLE 2

OTTO, DIESEL COMPARISONS

Process	Otto	Diesel
Heat Addition Q_{in} =	$U_2 - U_1 = mC_v(T_2 - T_1)$	$H_2 - H_1 = mC_p(T_2 - T_1)$
Heat Rejection Q_{out} =	$U_3 - U_4 = mC_v(T_3 - T_4)$	$U_3 - U_4 = mC_v(T_3 - T_4)$
Compression Ratio r_v =	$\dfrac{V_3}{V_2} = \dfrac{V_4}{V_1}$	$\dfrac{V_4}{V_1}$
Efficiency ηth	$1 - \dfrac{(T_3 - T_4)}{(T_2 - T_1)} = 1 - \dfrac{1}{r_v^{(k-1)}}$	$1 - \dfrac{(T_3 - T_4)}{k(T_2 - T_1)} = 1 - \dfrac{1}{r_v^{(k-1)}}\left[\dfrac{r_c^k - 1}{k(r_{c-1})}\right]$

For both Otto and Diesel cycles there is one power stroke for every four strokes represented on the *T-s* diagram. The strokes are: intake, compression, power, and exhaust. Thus, if an engine operates at 4,000 revolutions per minute (RPM), there are 2,000 power strokes in that period. Recall that a revolution will include two strokes. Each piston will have a power stroke every three one-hundredths of a second at 4,000 RPM.

The gas turbine engine has been modeled by the Air-Standard Brayton cycle using both a closed and an open system. In both cases there is flow through each component, thus necessitating a control volume analysis of the components. Figure 24 illustrates the processes used to model the gas turbine engine.

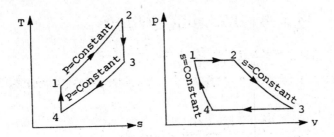

Figure 24. Brayton cycle

The processes associated with the Brayton cycle are:

Process 1-2

Constant pressure heat addition

Process 2-3

Isentropic expansion

Process 3-4

Constant pressure heat rejection

Process 4-1

Isentropic compression

As stated above, a control volume analysis is required on each component of the Brayton cycle. Since each device has a single inlet/exit, the mass flow rate will be constant. Application of the first law produces

$$\dot{Q}_{in} = \dot{m} C_p (T_2 - T_1)$$

$$\dot{Q}_{out} = \dot{m} C_p (T_4 - T_3)$$

and a thermal efficiency of

$$\eta_{th} = \frac{\dot{W}_{net}}{\dot{Q}_{in}} = \frac{\dot{Q}_{in} - \dot{Q}_{out}}{\dot{Q}_{in}} = 1 - \frac{(T_3 - T_4)}{(T_2 - T_1)}$$

The turbine produces the power; however, some of that power is used to turn the compressor. Thus, the net work can be calculated using either the heat difference or

$$\dot{W}_{net} = \dot{W}_T - \dot{W}_c = \dot{m} C_p (T_2 - T_3) - \dot{m} C_p (T_1 - T_4)$$

$$\eta_{th} = 1 - \frac{1}{r_p^{\frac{k-1}{k}}}$$

where $r_p = \dfrac{P_1}{P_4} = \dfrac{P_2}{P_3} =$ the pressure ratio.

Figure 25 depicts the open and closed Brayton cycles. In the open cycle the heat is rejected to the air, much the same as is done in most turbine applications. In the closed cycle there is a heat exchanger present to remove heat.

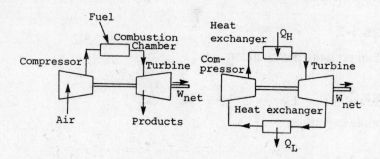

Figure 25. A gas turbine operating on the Brayton cycle

In an effort to recover some of the heat that escapes as exhaust, a regenerator is added to the Brayton cycle. This addition increases the cycle efficiency of the turbine. Assuming a perfect regenerator, one where the amount of heat carried to the regenerator as exhaust is exactly equal to the amount of heat picked up by the incoming air, we have the cycle seen in Figure 26.

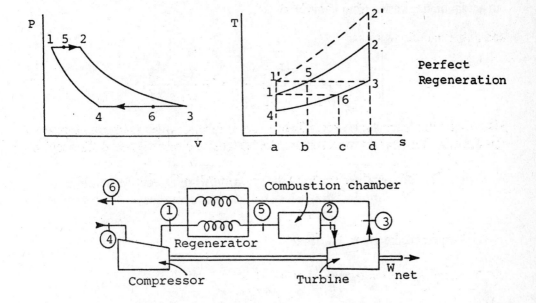

Figure 26. Brayton cycle with regenerator

Continuity remains the same, and the first law produces

$$\dot{W}_{net} = \dot{m}C_p(T_2 - T_3) - \dot{m}C_p(T_2 - T_4)$$

$$\dot{Q}_{in} = \dot{m}C_p(T_2 - T_5) = \dot{m}C_p(T_2 - T_3)$$

and a thermal efficiency of

$$\eta_{th} = \frac{(T_2 - T_3) - (T_1 - T_4)}{(T_2 - T_3)} = 1 - \frac{T_4}{T_2}\left(\frac{P_1}{P_4}\right)^{\frac{(k-1)}{k}}$$

The ideal cycles are just that, ideal. In real cycles there are deviations caused by many reasons. If there is piping, as there is in the steam power plant, then there will be piping losses. These are due to frictional effects and heat transfer that cause a loss in pressure. These losses cause a decrease in the entropy and the availability. Condensers have losses associated with cooling the liquid below the saturation temperature, thus causing it to be heated more than necessary.

Pumps and turbines have losses associated with nonisentropic (irreversible) behavior. For this case the turbine/pump efficiencies are defined as a comparison of the actual work to the adiabatic-reversible (isentropic) work. Figure 27 shows how an actual device varies from isentropic.

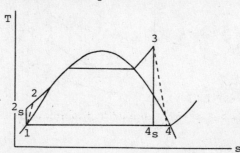

Figure 27. Temperature-entropy diagram showing effect of turbine and pump inefficiencies on cycle performance

The turbine efficiency is defined as

$$\eta_t = \frac{w_t}{h_3 - h_{4s}} = \frac{h_3 - h_4}{h_3 - h_{4s}}$$

where   4 is the actual state leaving the turbine, and

4s is the state after the isentropic expansion.

A similar equation can be developed for the pump. Here the actual work required will be more than the isentropic, thus keeping the efficiency less than 100 percent; the actual work must be in the denominator.

$$\eta_p = \frac{h_{2s} - h_1}{w_p} = \frac{h_{2s} - h_1}{h_2 - h_1}$$

Every mechanical device has a similar efficiency.

AIR CONDITIONING AND REFRIGERATION

Vapor compression refrigeration is the most common form of refrigeration and "air conditioning." Conditioning is accomplished by changing the makeup of the air by removing dust or water vapor, or adding water vapor if necessary. The description of the vapor compression refrigeration cycle will be given first; followed by a problem that addresses "conditioning" the air.

The vapor compression refrigeration cycle is essentially the same as the Rankine cycle but in reverse. The only difference is that the pump is replaced by an expansion valve. The following processes make up the cycle:

Process 1-2

Isentropic compression

Process 2-3

Constant pressure heat rejection

Process 3-4

Adiabatic throttling process

Process 4-1

Constant pressure heat addition

Figure 28 shows the Vapor Compression Refrigeration cycle.

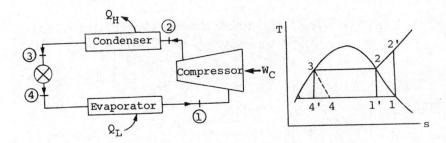

Figure 28. The ideal Vapor Compression Refrigeration cycle

For a refrigerator or air conditioner, $\dot{Q}_L$ is the quantity of interest. Used as a heat pump, $\dot{Q}_H$ is the quantity of interest. In both cases, the capacity of the device may be specified in BTUs or in tons. Cooling capacity is a term frequently used when referring to $\dot{Q}_L$.

The continuity equation is based on steady-state, steady-flow, with one inlet/exit. Thus , the mass flow rate is constant. Applying the first law produces the following:

Compressor $\dot{W}_c = \dot{m}(h_2 - h_1) = \dot{m}C_p(T_2 - T_1)$

Condenser $\dot{Q}_H = \dot{m}(h_3 - h_2) = \dot{m}C_p(T_3 - T_2)$

Valve $h_3 = h_4$

Evaporator $\dot{Q}_L = \dot{m}(h_1 - h_4) = \dot{m}C_p(T_1 - T_4)$

The coefficient of performance for the refrigerator or air conditioner is

$$\beta_R = \frac{\dot{Q}_L}{\dot{W}_c} = \frac{h_1 - h_4}{h_2 - h_1}$$

and for the heat pump

$$\beta_{HP} = \frac{\dot{Q}_H}{\dot{W}_c} = \frac{h_2 - h_3}{h_2 - h_1}.$$

The easiest way to solve vapor compression refrigeration problems is to plot them on a Freon chart (see Figure 29). The availability of constant entropy lines makes this chart extremely easy to use.

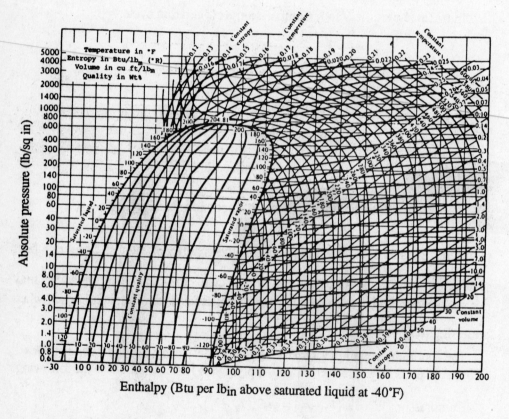

Figure 29. Pressure-enthalpy diagram for Freon-22 refrigerant

The problem below is an excellent review of the use of the psychrometric chart when dealing with conditioning of the air.

EXAMPLE: Psychrometric charts are indispensable when calculating the mass of water removed and the refrigeration required in certain systems. Using a psychrometric chart, determine these two quantities for the following system:

Initial state:

$t_i = 80° F$

humidity = 40%

Final state:

$t_f = 50°$ F

humidity = 100%

Amount of airflow

2,000 cfm incoming air processed

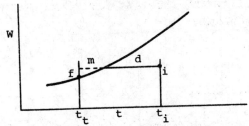

Figure 30. Schematic of refrigeration system

SOLUTION: By using the psychrometric chart, the initial (and final) enthalpy, weight of moisture (per lb dry air), and volume are obtained. By using mass and energy balances, the mass of H_2O removed and the refrigeration required can be calculated.

Initial state properties

$t_i = 80°$ F

humidity = 40%

enthalpy = h_i = 29 BTU/lb dry air

mass of moisture = m_i = 62 grains/lb dry air

volume = V_i = 13.78 ft³/lb dry air

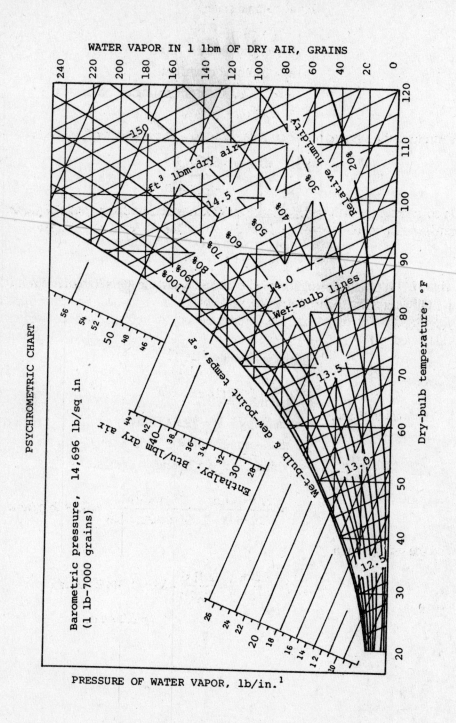

Figure 31. Psychrometric chart for air

Final state properties

$t_f = 50°$ F

humidity = 100%

$h_f = 20.4$ BTU/lb dry air

$m_f = 53$ grains/lb dry air

$V_f = 13.0$ ft³/lb dry air

Therefore, the amount of water to be separated out per pound of dry air is

$$m_{sep.} = m_i - m_f = 62 - 53$$

$$= 9 \text{ grains}$$

In conjunction with the mass balance on air and water, the energy balance yields, the refrigeration requirements (per lb of dry air) as the decrease in enthalpy:

$$-Q = h_i - h_f - \left(m_i - m_f\right)h_w$$

where h_w is the enthalpy of saturated H_2O liquid at 50° F, $h_w = 18.1$ Btu. Since 1 lbm = 7,000 grains, m_i and m_f are divided by 7,000.

Therefore,

$$-Q = 29 - 20.4 - \frac{9(18.1)}{7,000}$$

$$= 8.6 \text{ BTU/lbm dry air}$$

The total moisture removed is given by

$$M_T = \frac{2,000 \text{ cfm}}{V_i} \frac{\left(m_i - m_f\right)}{7,000}$$

$$= \frac{2,000 \text{ cfm}}{13.78 \text{ cf}} \frac{(9 \text{ grains})}{7,000 \text{ grains} / \text{lbm}} = 0.19 \text{ lbm/min}$$

and the refrigeration

$$(-Q) \bullet \left(\frac{\text{cfm}}{\text{cf}}\right) = \frac{2,000}{13.78} \times 8.6 = 1,248 \text{ BTU/min}$$

$$= 6.24 \text{ tons}$$

COMBUSTION AND CHEMICAL REACTIONS

Combustion is a process involving the reaction of a fuel and an oxidizer in which stored chemical energy is released. The combustion process is often studied separately or as a special subset of chemical reactions because it moves quickly toward completion and is so important to the production of power from fossil fuels. A complete combustion reaction, one in which all carbon atoms result in the formation of carbon dioxide, can be represented as

$$C_xH_y + aO_2 + 3.76aN_2 \rightarrow bCO_2 + cH_2O + dO_2 + eN_2$$

$$\underbrace{}_{\text{air}}$$

where x and y determine fuel type, and $a, b, c, d,$ and e are moles of other participants when one mole of fuel is burned.

Definitions frequently experienced during combustion calculations and discussions include:

(1) **Theoretical Air (TA)** is the minimum amount of air that supplies enough oxygen to ensure complete combustion of all elements in the fuel. Realize that the nitrogen is a nonparticipant in the oxidation of the fuel, although it can form pollutants if excess oxygen is present.

(2) **Excess Air (EA)** is the amount of air supplied over and above the theoretical air. This will normally ensure complete combustion, but it can lead to the formation of pollutants, especially compounds involving nitrogen.

(3) **Air-Fuel Ratio (AF)** is the ratio of the mass of theoretical air to the mass of the fuel. The inverse is the Fuel-Air Ratio (FA).

(4) The combustion efficiency η_{comb} is

$$\eta_{comb} = \frac{FA \text{ ideal}}{FA \text{ actual}}$$

The useful result of the combustion process is the liberation of heat used to power the heat engines previously discussed. The combustion process occurring in a steady-state, steady-flow device at constant pressure and with no work is depicted in Figure 32.

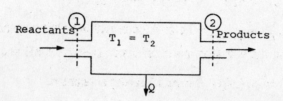

Figure 32. Combustion process

Applying the first law and neglecting potential and kinetic energy leads to

$$Q = H_2^P - H_1^R = \Delta H$$

where Q = Heat flow in

H_1^R = Enthalpy of the reactants at state 1

H_2^P = Enthalpy of the products at state 2

If the reactants and products are at the same temperature, the quantity ΔH is the enthalpy of reaction.

Before discussing the first and second laws as they apply to the combustion process, several explanations concerning enthalpy of reacting components are necessary:

(1) **The standard enthalpy of formation** (h_f^o) of a compound is the enthalpy of reaction for the formation of the compound from its elements at 25°C and 1 atm. The enthalpy of formation of an element is zero.

(2) **Exothermic reactions** $(\Delta H < 0)$ are those that liberate heat.

(3) **Endothermic reactions** $(\Delta H > 0)$ are those that absorb heat.

(4) **The heating value of a fuel** is numerically equal to its **enthalpy of reaction** but with opposite sign:

$$H_1^R - H_2^P = -\Delta H$$

(5) **Heating values** are normally presented as "higher" and "lower." The higher heating value occurs when the water in the products is still a liquid, while the lower heating value occurs when the water is a vapor. The difference in the heating values is the enthalpy of evaporation of the water. In equation form:

$$HHV = LHV + m_{H_2O} h_{fg}$$

where HHV = Higher heating value of fuel

LHV = Lower heating value of fuel

m_{H_2O} = Amount of water formed

h_{fg} = Enthalpy of evaporation of water

(6) The **total molal enthalpy** at any temperature and pressure is:

$$\overline{h}_{T,P} = \overline{h}_f^o + \left(\overline{h}_{T,P} - \overline{h}_{298(atm)}\right)$$

where $\overline{h}_{T,P} - \overline{h}_{298(atm)}$ = Difference in enthalpy between any given state and the enthalpy at the reference state of 298K and 1 atm.

$\overline{h}_f^o$ = enthalpy of formation of a substance.

(7) **Standard enthalpy** of reaction is $\Delta H° = H_2^\circ - H_1^\circ$.

(8) **Adiabatic flame temperature** is the temperature of the products when the combustion occurs adiabatically and with no work or changes in kinetic or potential energy.

With the definitions in hand and an understanding of the basic combustion process, application of the first law to a steady-state, steady-flow chemically reacting process while neglecting kinetic and potential energy produces

$$Q_{cv} - W_{cv} = \sum_P n_e \left(\overline{h}_f + \overline{h} \right)_e - \sum_R n_i \left(\overline{h}_f + \overline{h} \right)_i$$

where R,P = The reactants and products, respectively

n_i = Moles of reactants and n_e = moles of products

$\overline{h}_f$ = Molal enthalpy of formation

$\overline{h}$ = $\overline{h}_f^o - \overline{h}_{298}^o$ of a substance, which can be found directly from the tables.

Similarly for the second law, for any reactive process we have

$$\Delta S = S_P - S_R - \sum \frac{Q_{cv}}{T} \geq 0$$

where S_P = Entropy of products

S_R = Entropy of reactants

$\sum \dfrac{Q_{cv}}{T}$ = Entropy change due to heat transfer

For the steady-state, steady-flow process under consideration, the following apply:

(1) Reversible work

$$W_{rev} = \sum_R n_i \left(h_f^o + \Delta \overline{h} - T_o \overline{s} \right)_i - \sum_P n_e \left(h_f^o + \Delta \overline{h} - T_o \overline{s} \right)_e$$

(2) Irreversibility

$$I = \sum_P n_e T_o \overline{s}_e - \sum_R n_i T_o \overline{s}_i - Q_{cv}$$

(3) Availability

$$\psi = \left(h - T_o s \right) - \left(h_o - T_o s_o \right)$$

Determination of the absolute base entropy value is accomplished via the third law of thermodynamics. The third law states that the entropy of a pure crystalline substance is zero at the absolute zero of temperature. The value can be determined using

$$\overline{S}_{T,P} = \overline{S}_T^o - \overline{R} \ln \left(\frac{P}{0.1} \right)$$

where

$\overline{S}_{T,P}$ = Absolute entropy at 0.1 MPa and temperature T

P = Pressure in MPa

$\overline{R}$ = Universal gas constant

388

For any chemical reaction to proceed as discussed, it must satisfy the Gibbs function criteria, that is, the chemical reaction is only possible if the Gibbs function for the products is less than the Gibbs function of the reactants. The Gibbs function itself is defined as

$$G = H - TS.$$

For a chemical reaction carried out at constant temperature and pressure,

$$\Delta G = \Delta H - T\Delta S \le 0.$$

The last topic dealing with reactions is chemical equilibrium. Applying the Gibbs function to a reactive system, we find that a chemical reaction carried out at constant pressure and temperature can proceed only if the Gibbs function of the system will continually decrease. The reaction will stop when the Gibbs function of the system has reached a minimum. Thus, we can say that the equilibrium composition of any reactive system of known temperature and pressure is governed by $dG_{T,P} = 0$. To find the equilibrium composition, an expression for dG in terms of the moles of reactants and products is needed:

$$dG = VdP - SdT + \sum_i u_i dN_i$$

where

$$u_i = \left(\frac{\partial G}{\partial N_i}\right)_{P,T,N_i}$$

N_i = number of moles of each chemical species within the system at some time.

When $dG_{T,P} = 0$, a minor amount of mathematical manipulation produces

$$\ln K = -\frac{\Delta G^o}{RT}$$

where K = the equilibrium constant

$$\Delta G^o = c_c \overline{g}_C^o + d_d \overline{g}_D^o - a_a \overline{g}_A^o - b_b \overline{g}_B^o$$

$\overline{g}_C^o, \overline{g}_D^o, \overline{g}_B^o, \overline{g}_A^o$ = Standard Gibbs functions

a, b, c, d = The stoichiometric coefficients

The value of K takes on many forms depending on available information. If partial pressures are known, then

$$K = \frac{(P_C)^c (P_D)^d}{(P_A)^a (P_B)^b}$$

where P_A, P_B, P_C, P_D = Partial pressures of the chemical constituents

A similar expression for K can be obtained by replacing the partial pressures with the activity coefficients. If mole fractions are used in place of the partial pressures, then the fraction must be multiplied by

$$(P)^{c+d-a-b}$$

HEAT TRANSFER

While thermodynamics concerns itself with the macroscopic exchange of energy, heat transfer deals with the specifics. Three modes of heat transfer—conduction, convection, and radiation—are commonly studied separately and then combined to model real problems. The following is a brief discussion of these modes.

(1) **Conduction** is the transfer of heat in a material due to molecular motion in the material. A temperature gradient must exist to act as the potential for the flow of heat. The heat will flow from the high temperature to the cooler temperature. Fourier's law of conduction expresses the rate of heat transfer within the medium,

$$q = -kA\frac{dT}{dx}$$

where q = Heat transfer rate [W]

k = Thermal conductivity [W/m^3K]

A = Area normal to the direction of the flow [m^2]

T = Temperature [K]

x = Direction [m]

and, the negative sign is necessary since the flow of heat is in the direction opposite to the thermal gradient.

There are similar equations for each coordinate direction. Often the heat transfer is equated to a simple electrical circuit with a flow, a potential, and a resistance. The heat transfer is the flow, the temperature gradient is the potential, and the resistance is determined by the physical dimensions and properties of the material concerned. The following equations apply in this case:

$$q = \frac{kA\Delta T}{L} = \frac{\Delta T}{R_{th}}$$

where L = the thickness of the material in the direction of the heat flow

R_{th} = the conductive resistance

$$R_{th} = \frac{L}{kA}$$

Conduction through several materials of the same area normal to the flow can be treated as a series circuit with the resistance summed. Parallel circuits are needed when heat flows through different materials with varying normal areas. It is important to note that the rate of heat transfer is constant through a composite material. This fact allows for calculation of intermediate surface temperatures.

(2) **Convection** is the transfer of heat due to motion of a fluid near the surface of an object. Forced convection occurs when the fluid is placed into motion by a fan, pump, moving object, or due to the wind. Free convection occurs when the flow of the fluid is induced by buoyancy—a situation noticeable on roads during the summer.

Newton's law of cooling applies to convection problems, regardless of the type: free or forced.

$$q = hA\left(T_s - T_\infty\right)$$

where h = Convective heat transfer coefficient

   T_s = Surface temperature

   T_∞ = Fluid temperature

   A = Area normal to the heat transfer

The convective heat transfer coefficient can be found through an energy balance at the surface or through evaluation of the Nusselt number:

$$Nu = \frac{hx}{k}$$

where x can be the length of a flat plate, diameter of a cylinder or sphere, or other characteristic dimension specified. Solution for h is simple once the Nusselt number is known. For forced convection, the Nusselt number is a function of the Reynolds and Prandtl numbers,

$$Nu = f(Re, Pr)$$

where

$$Re = \frac{\rho U_\infty x}{\mu}$$

$$Pr = \frac{\upsilon}{\alpha}$$

and U_∞ = Fluid velocity

   ρ = Density

   μ = Dynamic viscosity

   υ = Kinematic viscosity

   x = Characteristic dimension

   α = Thermal diffusivity

For free convection, the Nusselt number is a function of the Grashof number and the Prandtl number:

$$Nu = f_2(Gr, Pr)$$

where

$$Gr = \frac{g\beta(T_s - T_\infty)L^3}{\upsilon^2}$$

and g = Gravitational acceleration

β = Volumetric thermal expansion coefficient

L = Surface length

with other variables being previously defined. As with conduction, convection can be expressed as a circuit with the convection resistance defined as

$$R_{th} = \frac{1}{hA}$$

and the convection and conduction resistances are summed when a combined mode problem is solved. Again, the heat transfer rate is constant through the circuit.

(3) **Radiation heat transfer** is due to thermal radiation. Radiation heat transfer is temperature dependent, to the fourth power, and can occur without a medium. Black bodies are perfect emitters and absorbers. A perfect emitter obeys the Stefan-Boltzmann Law,

$$q = \sigma A T^4$$

where A = Surface area

σ = Stefan-Boltzmann constant

T = Absolute temperature

There are few perfect emitters, such as black bodies; thus, it is normal to deal with gray bodies that emit some fraction of the energy of the black body. The emissivity of such bodies is defined as the ratio of the energy emitted compared to the energy emitted by a black body at the same temperature.

$$\varepsilon = \frac{q_{gray}}{q_{black}}$$

Thus, $q_{gray} = \sigma \varepsilon A T^4$

The net exchange of energy due to radiation heat transfer involves shape, view, or configuration factors. These factors geometrically predict the amount of energy departing one body and arriving at another. We all realize that if all the sun's energy arrived on earth it would be very hot. It is the shape factor that predicts the actual fraction that arrives. Thus, the net radiation exchange between two bodies is

$$q_{1 \to 2} = \sigma A_1 F_{1-2}(T_1^4 - T_2^4)$$

where F_{1-2} represents the net effect of shape view and configuration factors between the two bodies.

GAS DYNAMICS:
FLOW THROUGH NOZZLES AND BLADE PASSAGES

The differential from of the conservation equations is most often abandoned at this point in favor of the integral equivalents. Thus the conservation of mass for a control volume (open system) becomes:

$$\frac{\partial}{\partial t} \int_V \rho \, dV + \int_A \rho \vec{v} \bullet d\vec{A} = 0$$

where $\frac{\partial}{\partial t} \int_V \rho \, dV$ = Rate of change of mass within the control volume

$\int_A \rho v \, dA$ = Net rate of mass efflux through the control surface

For an incompressible flow (ρ = Constant),

$$\int_A \rho \vec{v} \bullet d\vec{A} = 0$$

Similarly, the conservation of momentum expressed as an integral becomes:

$$\sum F_j = \frac{1}{g_c} \left[\frac{d}{dt} \int_V V_j \rho \, dV + \int_A v_j \rho v_{rn} \, dA \right]$$

For the steady-state, steady-flow process we have:

$$\sum F_j = \frac{1}{g_c} \left[\sum \dot{m}_e (v_e)_j - \sum \dot{m}_i (v_i)_j \right]$$

where $\dot{m}_i, \dot{m}_e$ = Rate of mass entering and leaving the C.V.

v_i, v_e = Velocity of the mass entering and leaving the C.V.

i = x, y, z (directions)

At this juncture it is important to introduce the concepts of the speed of sound and the Mach number. The Mach number is defined by:

$$M = \frac{v}{c}$$

where v = Local gas speed

c = Local speed of sound

For an ideal gas $c = \sqrt{kRT}$, where $k = \dfrac{C_p}{C_v}$ as previously defined.

The Mach number is used to define the relative speed of the flow as follows:

if $M > 1$,   the flow is supersonic

if $M = 1$,   the flow is sonic

if $M < 1$,   the flow is subsonic

Within the flowfield it is necessary to calculate properties. Of particular importance are the local isentropic properties that would be obtained at any point in a flowfield if the fluid at that point were decelerated from local conditions to zero velocity following a frictionless adiabatic (isentropic) process.

For an ideal gas the isentropic stagnations (denoted by the subscript 0) are:

$$\frac{P_0}{P} = \left[1 + \frac{k-1}{2}M^2\right]^{\frac{k}{(k-1)}}$$

$$\frac{T_0}{T} = 1 + \frac{k-1}{2}M^2$$

$$\frac{\rho_0}{\rho} = \left[1 + \frac{k-1}{2}M^2\right]^{\frac{1}{(k-1)}}$$

Another important property set represents the conditions at the throat of a nozzle. These conditions can be found by noting that $M = 1$ at the throat. These properties (denoted by an asterisk *) are referred to as critical pressure, critical temperature, and critical density and are given below as ratios to the stagnation properties.

$$\frac{P*}{P_0} = \left(\frac{2}{k+1}\right)^{\frac{k}{(k-1)}}$$

$$\frac{T*}{T_0} = \left(\frac{2}{k+1}\right)$$

$$\frac{\rho*}{\rho_0} = \left(\frac{2}{k+1}\right)^{\frac{1}{(k-1)}}$$

Area variations directly effect the flow properties in an isentropic flow. The following equation is applicable:

$$\frac{dA}{A} = \frac{dP}{\rho V^2}\left(1 - M^2\right)$$

where M = Mach number

dA = Change in the area

dP = Change in the pressure

Since we have considered normal shocks, it is now possible to complete the discussion of flow in a converging-diverging nozzle. The pressure distribution through the nozzle for different back pressures is shown in Figure 33.

In Regime 1 the flow is subsonic throughout (i and ii). At condition (iii), the flow at the throat is sonic, that is $M_t = 1$.

In Regime 2 the exit flow is subsonic, a consequence of $P_e = P_b$.

In Regime 3 the back pressure is higher than the exit pressure but not sufficiently high to sustain a normal shock in the exit plane.

In Regime 4 the flow adjusts to the lower back pressure through a series of oblique expansion waves.

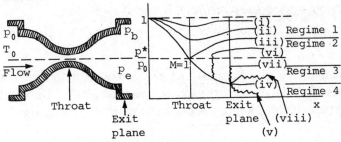

Figure 33. Pressure distribution vs. nozzle position

Operating conditions for nozzles and diffusers are optimized to ensure peak performance. Efficiencies and the coefficients of discharge and velocity make comparison between like devices a simple task. Comparison is followed by optimization in the design process.

(1) Nozzle efficiency is defined:

$$n_N = \frac{\text{Actual kinetic energy at nozzle exit}}{\text{Kinetic energy at nozzle exit with isentropic}} \cdot$$
$$\text{flow to same exit pressure}$$

(2) The coefficient of discharge C_p is defined by the relation:

$$C_p = \frac{\text{Actual mass rate of flow}}{\text{Mass rate of flow with isentropic flow}} \cdot$$

(3) The efficiency of a diffuser is defined:

$$n_p = \frac{\left(1 + \frac{k-1}{2} M_1^2\right)\left(\frac{P_{02}}{P_{01}}\right)^{\frac{(k-1)}{K}} - 1}{\frac{k-1}{2} M_1^2}$$

where
(a) states 1 and 01 are the actual and stagnation states of the fluid entering the diffuser.
(b) states 2 and 02 are the actual and stagnation states of the fluid leaving the diffuser
(4) The velocity coefficient C_v is defined:

$$C_v = \frac{\text{Actual velocity at nozzle exit}}{\text{Velocity at nozzle exit with isentropic flow and same exit pressure}}$$

REVIEW PROBLEMS

PROBLEM 1

Determine the final equilibrium state when 2 lbm of saturated liquid mercury at 1 psia is mixed with 4 lbm of mercury vapor at 1 psia and 1400°F. During the process the pressure in the cylinder is kept constant and no energy is lost between the cylinder and mercury.

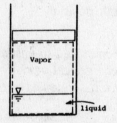

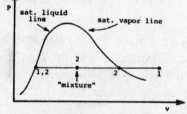

Figure 34. a) The control mass   **b) The process representation**

SOLUTION: Since the amount of liquid might change during the process, the liquid or only the vapor cannot be taken as the control mass. Instead take the entire 6 lbm of mercury. By assumption, no energy transfer as heat occurs, but the volume is expected to change, resulting in an energy transfer as work. The only energy stored within the control mass is the internal energy of the mercury; the energy balance, made over the time for the process to take place, is therefore (Figures 34 and 35)

$$\begin{array}{ccc} W & = & \Delta U \\ \text{energy} & & \text{increase in} \\ \text{input} & & \text{energy storage} \end{array}$$

where $\Delta U = U_2 - U_1$

The work calculation is made easy by the fact that the pressure is constant. When the piston moves an amount dx, the energy transfer as work from the environment to the control mass is

$$dW = PAdx = -P\,dV.$$

Integrating,

$$W = \int_1^2 -P\,dV = P(V_1 - V_2).$$

Combining with the energy balance obtain

$$U_2 + PV_2 = U_1 + PV_1. \tag{1}$$

TABLE 3 PROPERTIES OF SATURATED MERCURY				
		Enthalpy, Btu/lbm		
P, psia	T,°F	Sat. liq.	Evap.	Sat. vap.
0.010	233.57	6.668	127.732	134.400
0.020	259.88	7.532	127.614	135.146
0.030	276.22	8.068	127.540	135.608
0.050	297.97	8.778	127.442	136.220
0.100	329.73	9.814	127.300	137.114
0.200	364.25	10.936	127.144	138.080
0.300	385.92	11.639	127.047	138.086
0.400	401.98	12.159	126.975	139.134
0.500	415.00	12.568	126.916	139.484
0.600	425.82	12.929	126.868	139.797
0.800	443.50	13.500	126.788	140.288
1.00	457.72	13.959	126.724	140.683
2.00	504.93	15.476	126.512	141.988
3.00	535.25	16.439	126.377	142.816
5.00	575.70	17.741	126.193	143.934

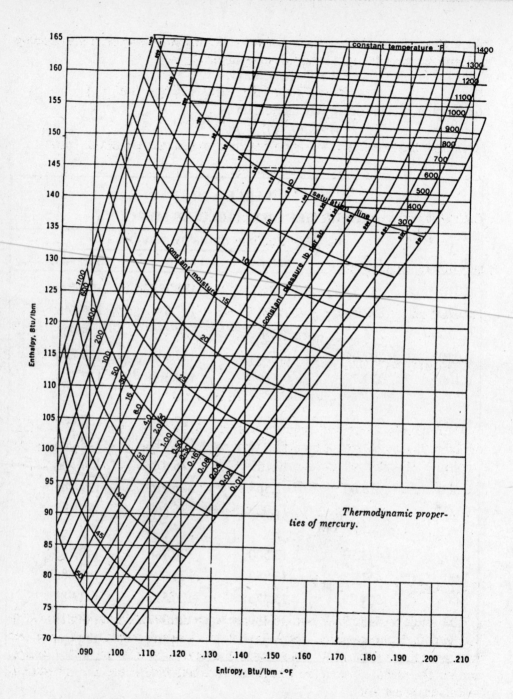

Figure 35. Thermodynamic properties of mercury

To evaluate the initial terms assume that the liquid is in an equilibrium state and the vapor is in an equilibrium state, even though they are not in equilibrium with one another. The graphical and tabular equations of state, Figure 35 and the Table for the thermodynamic properties of saturated mercury, may then be employed for each phase. Since the available equation-of-state information is in terms of the enthalpy property, express the right-hand side of equation (1) as

$$U_1 + PV_1 = M_{l_1} u_{l_1} + M_{v_1} u_{v_1} + P\left(M_{l_1} v_{l_1} + M_{v_1} v_{v_1}\right)$$
$$= M_{l_1} h_{l_1} + M_{v_1} h_{v_1}.$$

Now, from the tables, the initial liquid enthalpy is (saturated liquid at 1 psia)

$$h_{l_1} = 13.96 \text{ Btu/lbm}$$
$$T_1 = 457.7°F$$

The initial vapor enthalpy is found from Figure 35 as

$$h_{v_1} = 164 \text{ Btu/lbm}.$$

Substituting the numbers,

$$U_1 + PV_1 = 2 \times 13.96 + 4 \times 164 = 684 \text{ Btu}.$$

The final state is a state of equilibrium, for which

$$U_2 + PV_2 = M(u + Pv)_2 = Mh_2.$$

The enthalpy in the final state is therefore

$$h_2 = \frac{684 \text{ Btu}}{6 \text{ lbm}} = 114 \text{ Btu / lbm}.$$

The final pressure and enthalpy may be used to fix the final state. Upon inspection of Figure 35 the final state is a mixture of saturated liquid and vapor at 1 psia and the "moisture" $(1 - x)$ is about 21 percent (0.79 quality). Alternatively, the information in Table 1, could have been used.

$$114 = (1 - x_2) \times 13.96 + x_2 \times 140.7$$
$$x_2 = 0.79$$

PROBLEM 2

The gauge pressure in an automobile tire when measured during winter at 32°F was 30 pounds per square inch (psi). The same tire was used during the summer, and its temperature rose to 122°F. If we assume that the volume of the tire did not change, and no air leaked out between winter and summer, what is the new pressure as measured on the gauge?

SOLUTION: From one season to another, the only properties of the gas that will change are pressure and temperature. The mass (hence the number of moles) and the volume will remain the same. If it is assumed that this gas is ideal, then

$$PV = n\bar{R}T \tag{1}$$

where P = Pressure of the gas

         V = Volume of the gas

n = Number of moles

$\overline{R}$ = Gas constant

T = Temperature of the gas

Rearranging equation (1) to solve for P gives

$$P = \left(\frac{n}{V}\right)\overline{R}T. \tag{2}$$

Since n and V are constant, equation (2) shows that pressure is directly proportional to temperature. That is, $\dfrac{P}{T} = \dfrac{n\overline{R}}{V}$ = constant. Therefore

$$\frac{P_1}{T_1} = \frac{P_2}{T_2} = \frac{n_1\overline{R}}{V_1} = \frac{n_2\overline{R}}{V_2} \tag{3}$$

where P_1 = Initial pressure

T_1 = Initial temperature

P_2 = Final pressure

T_2 = Final temperature

n_1 and n_2 are initial and final moles, respectively. V_1 and V_2 are initial and final volume, respectively.

The moles and volume are not changing; therefore, $n_1 = n_2$ and $V_1 = V_2$. Consequently, equation (3) can be written as

$$\frac{P_1}{T_1} = \frac{P_2}{T_2}. \tag{4}$$

Before equation (4) can be used, the pressure and temperature must be in absolute scales.

$$\frac{T_C}{5} = \frac{T_F - 32}{9} \tag{5}$$

and

$$P = 14.7 \text{ psia} + \text{psig} \tag{6}$$

where T_C = Temperature in degrees centigrade

T_F = Temperature in degrees fahrenheit

psia = Absolute psi

psig = Gauge psi

Using equations (5) and (6),

$$122°F = 50°C = (50 + 273)K = 323K$$

and

$$P = 14.7 + 30 = 44.7 \text{ psia}.$$

These can now be inserted into equation (4) to give

$$\frac{44.7}{273} = \frac{P_2}{323}.$$

Therefore,

$$P_2 = \left[\frac{(44.7)(323)}{273}\right] \text{psia}$$
$$= 52.9 \text{psia}$$

or from equation (6),

$$52.9 \text{ psia} = 14.7 \text{ psia} + x \text{ psig}$$
$$P_2 = (52.9 - 14.7) \text{ psig}$$
$$= 38.2 \text{ psig}$$

PROBLEM 3

A container which has a volume of 0.1m^3 is fitted with a plunger enclosing 0.5 kg of steam at 0.4 MPa. Calculate the amount of heat transferred and the work done when the steam is heated to 300°C at constant pressure.

SOLUTION: For this system changes in kinetic and potential energy are not significant. Therefore

$$Q = m(u_2 - u_1) + W$$
$$W = \int_1^2 P dV = P \int_1^2 dV = P(V_2 - V_1) = m(P_2 v_2 - P_1 v_1)$$

Therefore

$$Q = m(u_2 - u_1) + m(P_2 v_2 - P_1 v_1) = m(h_2 - h_1)$$
$$v_1 = \frac{V_1}{m} = \frac{0.1}{0.5} = 0.2 = 0.001084 + x_1(0.4614)$$
$$x_1 = \frac{0.1989}{0.4614} = 0.4311$$

Then

$$h_1 = h_f + x_1 h_{fg}$$
$$= 604.74 + 0.4311 \times 2133.8 = 1524.6$$
$$h_2 = 3066.8$$
$$Q = 0.5(3066.8 - 1524.6) = 771.1 \text{ kJ}$$
$$W = mP(v_2 - v_1) = 0.5 \times 400(0.6548 - 0.2)$$
$$= 91.0 \text{ kJ}$$

Therefore

$$U_2 - U_1 = Q - W = 771.1 - 91.0 = 680.1 \text{ kJ}.$$

The heat transfer can be calculated from u_1 and u_2 by using

$$Q = m(u_2 - u_1) + W$$

$$u_1 = u_f + x_1 u_{fg}$$

$$= 604.31 + 0.4311 \times 1949.3 = 1444.6$$

$$u_2 = 2804.8$$

and

$$Q = 0.5(2804.8 - 1444.6) + 91.0 = 771.1 \text{ kJ}$$

PROBLEM 4

Steam at 3 MPa, 300°C leaves the boiler and enters the high-pressure turbine (in a reheat cycle) and is expanded to 300 kPa. The steam is then reheated to 300°C and expanded in the second stage turbine to 10 kPA. What is the efficiency of the cycle if it is assumed to be internally reversible?

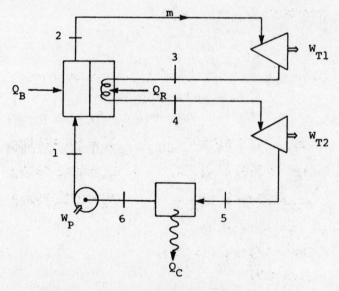

Figure 36. Schematic of heating cycle

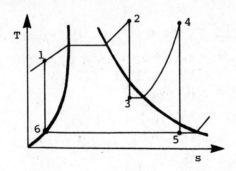

Figure 37. *T-s* **diagram for heating cycle**

SOLUTION: The efficiency η can be obtained from the following equation:

$$\eta = \frac{\dot{W}_{t_1} + \dot{W}_{t_2} - \dot{W}_p}{\dot{Q}_b - \dot{Q}_r} \qquad (1)$$

To calculate $\dot{W}_{t_1}$ assume that the turbine is adiabatic and neglect kinetic and potential energy changes. Applying the first law to the turbine,

$$\dot{W}_{t_1} = \dot{m}(h_2 - h_3).$$

From the steam tables

$$h_2 = 2993.5 \text{ kJ/kg}$$
$$s_2 = 6.5390 \text{ kJ/kg} - \text{K}$$

To find h_3, for the internally reversible adiabatic process $2 \rightarrow 3$:

$$s_2 = s_3 = 6.5390 \text{ kJ/kg} - \text{K}$$

At state 3,

$$s_{f_3} = 1.6718 \text{ kJ/kg} - \text{K} \qquad h_{f_3} = 561.47 \text{ kJ/kg}$$

$$s_{fg_3} = 5.3201 \text{ kJ/kg} - \text{K} \qquad h_{fg_3} = 2163.8 \text{ kJ/kg}$$

$$s_{g_3} = 6.9919 \text{ kJ/kg} - \text{K} \qquad h_{g_3} = 2725.3 \text{ kJ/kg}$$

$$s_2 = s_3 = s_{f_3} + x_3 s_{fg_3}$$
$$6.5390 = 1.6718 + x_3(5.3201)$$
$$x_3 = 0.915$$
$$h_3 = h_{f_3} + x_3 h_{fg_3}$$
$$h_3 = 561.47 + 0.915(2163.8)$$
$$h_3 = 2542 \text{ kJ/kg}$$

$$\frac{\dot{W}_{t_1}}{\dot{m}} = h_2 - h_3$$

$$= 2993.5 - 2542$$

$$= 452 \text{ kJ/kg}$$

Similarly, to find $\dot{W}_{t_2}$

$$\dot{W}_{t_2} = \dot{m}\left(h_4 - h_5\right)$$

From the steam tables,

$$h_4 = 3069.3 \text{ kJ/kg}$$

$$s_4 = 7.7022 \text{ kJ/kg} - \text{°K}$$

To find h_5, note that

$$s_4 = s_5$$

At state 5

$$s_{f_5} = 0.6493 \text{ kJ/kg} - \text{°K}$$

$$h_{f_5} = 191.83 \text{ kJ/kg}$$

$$s_{fg_5} = 7.5009 \text{ kJ/kg} - \text{°K}$$

$$h_{fg_5} = 2392.8 \text{ kJ/kg}$$

$$s_{g_5} = 8.1502 \text{ kJ/kg} - \text{°K}$$

$$h_{g_5} = 2584.7 \text{ kJ/kg}$$

$$s_4 = s_5 = s_{f_5} + x_5 s_{fg_5}$$

$$x_5 = 0.949$$

$$h_5 = h_{f_5} + x_5 h_{fg_5}$$

$$h_5 = 191.83 + 0.949(2392.8)$$

$$h_5 = 2463 \text{ kJ/kg}$$

$$\therefore \frac{\dot{W}_{t_2}}{\dot{m}} = h_4 - h_5$$

$$= 3069.3 - 2463$$

$$= 606 \text{ kJ/kg}$$

To obtain $\dot{W}_p$, assume that $\dot{W}_p = \dot{m} v_6 (p_1 - p_6)$.

From the steam tables,

$$v_6 = v_{f_6}$$
$$= 1.0102 \times 10^{-3} \, \text{m}^3/\text{kg}$$

Thus,

$$\frac{\dot{W}_p}{\dot{m}} = 1.0102(30 - 0.1)10^5 \times 10^{-6}$$
$$= 3.0 \, \text{kJ/kg}$$

To obtain $\dot{Q}_b$, use

$$\dot{Q}_b = \dot{m}(h_2 - h_1)$$

$$h_1 = h_6 + \frac{\dot{W}_p}{\dot{m}}$$
$$= 191.8 + 3.0$$
$$= 194.8 \, \text{kJ/kg}$$

$$\frac{\dot{Q}_b}{\dot{m}} = 2993.5 - 194.8$$
$$= 2799 \, \text{kJ/kg}$$

To find $\dot{Q}_r$

$$\dot{Q}_r = \dot{m}(h_4 - h_3)$$

$$\frac{\dot{Q}_r}{\dot{m}} = 3069.3 - 2542$$
$$= 527 \, \text{kJ/kg}$$

From equation (1) then

$$\eta = \frac{452 + 606 - 3}{2799 + 527}$$
$$= 0.317$$

PROBLEM 5

Steam leaves the boiler in a steam turbine plant at 2 MPa, 300°C and is expanded to 3.5 kPa before entering the condenser. Compare the following four cycles:

1. A superheated Rankine cycle.
2. A reheat cycle, with steam reheated to 300°C at the pressure where it becomes saturated vapor.
3. A regenerative cycle, with an open feedwater heater operating at the pressure where steam becomes saturated vapor.
4. A regenerative cycle, with a closed feedwater heater operating at the pressure where steam becomes saturated vapor.

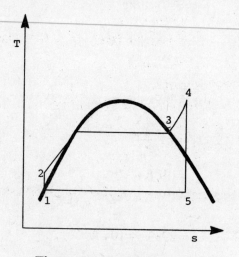

Figure 38. Rankine cycle

SOLUTION: 1 Referring to Figure 38, the steam tables show that

$$h_4 = 3025 \text{ kJ/kg}$$
$$s_4 = 6.768 \text{ kJ/kg} - °K$$

At $P = 3.5$ kPa,

$$s_g = 8.521 \text{ kJ/kg} - °K$$
$$s_f = 0.391 \text{ kJ/kg} - °K$$

Since $s_5 = s_4$, steam at 5 is a mixture of liquid and vapor. The quality is found as

$$x_5 = \frac{s_5 - s_f}{s_{fg}}$$
$$= \frac{6.768 - 0.391}{8.130}$$
$$= 0.785$$

Therefore

$$h_5 = h_f + x_5 h_{fg}$$
$$= 112 + 0.785(2438)$$
$$= 2023 \text{ kJ/kg}$$

hence

$$w_{45} = h_4 - h_5$$
$$= 3025 - 2023$$
$$= 1002 \text{ kJ/kg}$$

Now

$$w_{12} = h_1 - h_2$$
$$= v_f(p_1 - p_2)$$
$$= 0.0010(0.0035 - 2) \times 10^3 \text{ kJ/kg}$$
$$= -2 \text{ kJ/kg}$$

Therefore the net work output is

$$w = w_{45} + w_{12} = 1000 \text{ kJ/kg}$$

Heat input is

$$q_{42} = h_4 - h_2$$

But

$$h_2 = h_1 - w_{12} = 112 + 2 = 114 \text{ kJ/kg}$$

therefore

$$q_{42} = 3025 - 114 = 2911 \text{ kJ/kg}$$

Thus,

$$\eta = \frac{w}{q_{42}} = \frac{1000}{2911} = 0.344$$

Also

$$\text{Specific Steam Consumption} = \frac{3600}{w} = \frac{3600}{1000} = 3.6 \text{ kg/kWh}$$

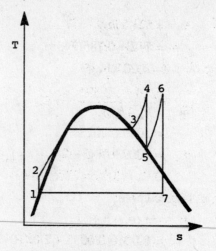

Figure 39. Reheat cycle

2. Refer to Figure 39, and note that since

$$s_5 = s_{sat} = s_4 = 6.768 \text{ kJ/kg} - {}^{\circ}\text{K}$$

the pressure at reheat point 5 can be found using the steam tables. Interpolating between 0.55 MPa and 0.6 MPa gives

$$P_5 = 0.588 \text{ MPa.}$$

Then

$$h_5 = 2753 + \frac{0.588 - 0.55}{0.60 - 0.55}(2757 - 2753)$$

$$= 2753 + \frac{0.038}{0.05} \times 4$$

$$= 2756 \text{ kJ/kg}$$

As 6 and 5 are on the same isobar, by interpolation

$$h_6 = 3065 + \frac{0.588 - 0.5}{0.60 - 0.5}(3062 - 3065)$$

$$= 3065 + \frac{0.088}{0.1}(-3)$$

$$= 3062.4 \text{ kJ/kg}$$

$$s_6 = 7.460 + 0.88(7.373 - 7.460)$$

$$= 7.460 + 0.88(-0.087)$$

$$= 7.384 \text{ kJ/kg} - {}^{\circ}\text{K}$$

At $P = 3.5$ kPa,

$$s_g = 8.521 \text{ kJ/kg} - °\text{K}$$
$$s_f = 0.391 \text{ kJ/kg} - °\text{K}$$

Since $s_7 = s_6$, the quality at 7 is found as

$$x_7 = \frac{7.384 - 0.391}{8.130} = 0.86.$$

Then

$$h_7 = 112 + 0.86(2438)$$
$$= 112 + 2095 = 2207 \text{ kJ/kg}$$

The net work output is given by

$$w = w_{45} + w_{67} + w_{12}$$
$$= (3025 - 2765) + (3062.4 - 2207) - 2$$
$$= 1122.4$$

The heat input is

$$q = q_{42} + q_{65}$$
$$= 2911 + (h_6 - h_5)$$
$$= 2911 + (3062.4 - 2756)$$
$$= 3217.4$$

Therefore

$$\eta = \frac{1122.4}{3217.4} = 0.349$$

and

$$\text{s.s.c.} = \frac{3600}{w} = \frac{3600}{1122.4} = 3.2 \text{ kg/kWh}.$$

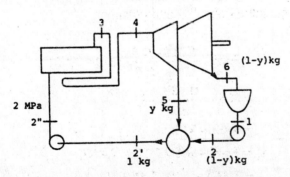

Figure 40. a) Equipment schematic for regenerative cycle

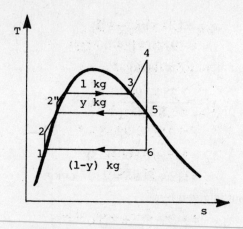

Figure 40. b) Regenerative cycle

3. Refer to Figures 40 a) and 40 b). The work is as in (b)

$$w_{45} = 269 \text{ kJ/kg}$$

Next determine the amount of steam bled off at 5. Consider an energy balance for the open feedwater heater with

$$h_{2'} = yh_s - (1-y)h_2$$

which gives

$$y = \frac{h_{2'} - h_2}{h_5 - h_2}$$

To find the value for $h_{2'}$, enter the steam tables. At 5 the pressure is known ($P = 0.588$ MPa) and the state of the steam is given as saturated vapor. Therefore, by interpolating between the values of 0.5 MPa and 0.6 MPa, obtain

$$h_{2'} = 656 + \frac{0.588 - 0.55}{0.60 - 0.55}(670 - 656)$$

$$= 656 + \frac{0.038}{0.05} \times 14$$

$$= 666.6 \text{ kJ/kg}$$

Then

$$y = \frac{666.6 - 114}{2756 - 114}$$

$$= \frac{552.6}{2642}$$

$$= 0.209$$

Hence

$$w_{56} = (1 - y)(h_5 - h_6)$$
$$= 0.791(2756 - 2023)$$
$$= 580 \text{ kJ/kg}$$

also

$$w_{2'2''} = v_f(P_{2'} - P_{2''})$$
$$= 0.0011(0.588 - 2) \times 10^3$$
$$= -1.1 \times 1.412$$
$$= -1.55 \text{ kJ/kg}$$

Therefore

$$w = w_{45} + w_{56} + w_{12} + w_{2'2''}$$
$$= 269 + 580 - 0.791 \times 2 - 1.55$$
$$= 845.87 \text{ kJ/kg}$$

The heat input is

$$q_{42''} = 3025 - (666.6 + 1.55)$$
$$= 2356.8 \text{ kJ/kg}$$

The efficiency of this cycle is

$$\eta = \frac{w}{q_{42''}} = \frac{845.87}{2356.8} = 0.3595$$

and

$$\text{s.s.c.} = \frac{3600}{w} = \frac{3600}{845.9} = 4.25 \text{ kg/kWh}.$$

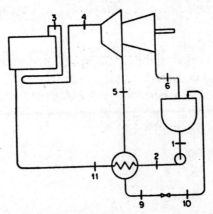

Figure 41. a) Equipment diagram including closed heater

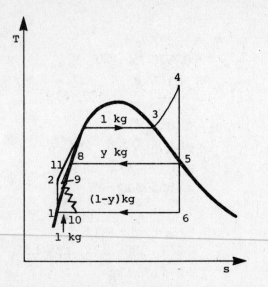

Figure 41. b) A regenerative cycle with closed heater

4. Refer to Figures 41 (a) and 41 (b). The work is as in part (b).

$$w_{45} = 269 \text{ kJ/kg}$$

Heat balance for the heater as a closed system gives

$$h_{21} = yh_5 - (1-y)h_2$$

giving

$$y = \frac{h_{11} - h_2}{h_5 - h_9}$$

Now in finding the enthalpies in the feed line, it is usual to make the following assumptions:

i.   Neglect the feed pump term.

ii. Assume the enthalpy of the compressed liquid to be the same as that of the saturated liquid at the same temperature.

iii. Assume the states of the condensate extracted from the turbine, before and after throttling, to be the same as that of the saturated liquid at the lower pressure of the throttled liquid.

Using these assumptions

$$h_2 = h_1$$
$$h_{11} = h_8$$
$$h_9 = h_{10} = h_1$$

whence

$$y = \frac{h_8 - h_1}{h_5 - h_1}$$

$$= \frac{666.6 - 112}{2756 - 112} = 0.209 \text{ kJ/kg}$$

Also

$$w_{56} = 580 \text{ kJ/kg.}$$

Therefore

$$w = w_{45} + w_{56} + w_{12}$$
$$= 269 + 580 - 2 = 847 \text{ kJ/kg}$$

Heat input $q_{411} = 2358.4$ kJ/kg.

Then

$$\eta = \frac{w}{q_{411}} = \frac{847}{2358.4} = 0.360$$

and

$$\text{s.s.c.} = \frac{3600}{w} = \frac{3600}{847} = 4.25 \text{ kg/kWh.}$$

PROBLEM 6

1. One kilogram of air at 101.35 kPa, 21°C is compressed in an Otto cycle with a compression ratio of 7 to 1. During the combustion process, 953.66 kJ of heat is added to the air. Compute (a) the specific volume, pressure, and temperature at the four points in the cycle, (b) the air standard efficiency, and (c) the mep (mean effective pressure) and hp of the engine, if it uses 1 kg/min of air.

2. Calculate the efficiency for a Carnot cycle operating between the maximum and minimum temperatures of the Otto cycle.

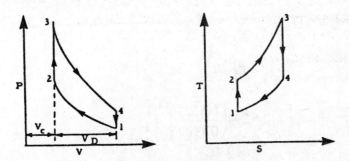

Figure 42. Otto cycle

SOLUTION: 1 a) At state 1,

$$P_1 = 101.35 \text{ kPa}$$
$$T_1 = 294\text{K}$$

The specific volume, v_1, is determined by using the perfect gas equation of state.

$$v_1 = \frac{RT_1}{P_1}$$

$$= \frac{0.287(294)}{101.35}$$

$$= 0.8325 \text{ m}^3/\text{kg}$$

At 2, the specific volume can be obtained by using the compression ratio.

$$\frac{v_2}{v_1} = \frac{1}{7}$$

or

$$v_2 = \frac{v_1}{7}$$

$$= \frac{0.8325}{7}$$

$$= 0.1189 \text{m}^3/\text{kg}$$

The pressure (P_2) is obtained from the isentropic relation

$$P_2 = P_1\left(\frac{v_1}{v_2}\right)^k$$

$$= 101.35\left(\frac{0.8325}{0.1189}\right)^{1.4}$$

$$= 1,545.6 \text{ kPa}$$

The temperature (T_2) is

$$T_2 = T_1\left(\frac{v_1}{v_2}\right)^{k-1}$$

$$= 294\left(\frac{0.8325}{0.1189}\right)^{1.4-1}$$

$$= 640.4\text{K}$$

At state 3, $v_3 = v_2 = 0.1189$ m³/kg. The temperature here can be calculated from the quantity of heat supplied since

$$Q_{in} = mc_v(T_3 - T_2)$$

or solving for T_3,

$$T_3 = \frac{Q_{in}}{mc_v} + T_2$$

$$= \frac{953.66}{1(0.7243)} + 640.4$$

$$= 1957.1 \text{K}$$

The pressure (P_3) is

$$P_3 = \frac{RT_3}{v_3}$$

$$= \frac{0.287(1957.1)}{0.1189}$$

$$= 4,724 \text{ kPa}$$

At 4,

$$v_4 = v_1 = 0.8325 \text{ m}^3/\text{kg},$$

and the pressure is

$$P_4 = P_3\left(\frac{v_3}{v_4}\right)^k$$

$$= 4724\left(\frac{0.1189}{0.8325}\right)^{1.4}$$

$$= 309.7 \text{ kPa}$$

The temperature T_4 is

$$T_4 = T_3\left(\frac{v_3}{v_4}\right)^{k-1}$$

$$= 1957.1\left(\frac{0.1189}{0.8325}\right)^{1.4-1}$$

$$= 898.5 \text{K}$$

(b) The efficiency of the Otto cycle is defined as

$$\eta = \frac{Q_{in} - Q_{out}}{Q_{in}} \times 100 \qquad (1)$$

where

$$Q_{in} = 953.66 \text{ kJ}$$

and

$$\begin{aligned} Q_{out} &= mc_v(T_4 - T_1) \\ &= 1(0.7243)(898.5 - 274) \\ &= 452.3 \text{ kJ} \end{aligned}$$

Therefore,

$$\eta = \frac{953.66 - 452.3}{953.66} \times 100$$
$$= 53\%$$

(c) The mep is

$$\text{mep} = \frac{W_{net}}{v_1 - v_2}$$

where

$$\begin{aligned} W_{net} &= q_{in} - q_{out} \\ &= 953.66 - 452.3 \\ &= 501.36 \text{ kJ} \end{aligned}$$

Thus,

$$\text{mep} = \frac{501.36}{0.8325 - 0.1189} = 702.6 \text{ kPa}$$

The horsepower is

$$\text{hp} = \dot{m} W_{net} = 501.36 \text{ kW}$$
$$= 5.2 \text{ hp}$$

2) The maximum and minimum temperatures of the Otto cycle are

$$T_{max} = 1957.1 \text{K}$$
$$T_{min} = 294 \text{K}$$

The Carnot cycle efficiency is

$$\begin{aligned} \eta &= \frac{T_{max} - T_{min}}{T_{max}} \times 100 \\ &= \frac{1957.1 - 294}{1957.1} \times 100 \\ &= 85\% \end{aligned}$$

which is comparatively higher than the Otto cycle efficiency.

PROBLEM 7

Consider an air standard Diesel cycle. At the beginning of compression the temperature is 300K and the pressure is 101.35 kPa. If the compression ratio is 15 and during the process 1860 kJ/kg of air are added as heat, calculate: (a) the maximum cycle pressure and temperature, (b) the thermal efficiency of the cycle, and (c) the mep.

SOLUTION: a) Referring to the figure for the states, and using the ideal gas equation of state

$$Pv = RT \tag{1}$$

the specific volume at state d is

$$v_d = \frac{RT_d}{P_d}$$
$$= \frac{0.287(300)}{101.35}$$
$$= 0.8495 \text{ m}^3/\text{kg}$$

Process $c \rightarrow d$ is an isochoric (constant volume) process. Hence,

$$v_c = v_d = 0.8495 \text{ m}^3/\text{kg}$$

The compression ratio is

$$r_v = \frac{v_d}{v_a} = 15$$

or

$$= \frac{v_d}{15}$$
$$= \frac{0.8495}{15}$$
$$= 0.0566 \text{ m}^3/\text{kg}$$

Process $d \rightarrow a$ is an isentropic process. Therefore,

$$\frac{T_a}{T_d} = \left(\frac{v_d}{v_a}\right)^{k-1}$$

or

$$T_a = 300\left(\frac{0.8495}{0.0566}\right)^{1.4-1}$$
$$= 886.5 \text{K}$$

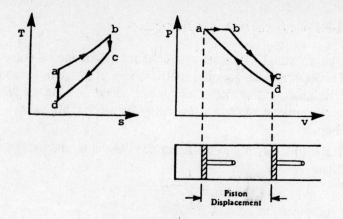

Figure 43. Diesel cycle

Also,

$$\frac{P_a}{P_d} = \left(\frac{v_d}{v_a}\right)^k$$

or

$$P_a = 101.35\left(\frac{0.8495}{0.0566}\right)^{1.4}$$

$$= 4,495 \text{ kPa}$$

$$\therefore P_{\max} = P_a = 4,495 \text{ kPa}$$

The maximum temperature can be obtained as follows. From the first law, assuming constant specific heats, the heat supplied is

$$Q_{\text{in}} = Q_{ab} = C_p(T_b - T_a)$$

or

$$T_b = \frac{Q_{ab}}{C_p} + T_a$$

$$= \frac{1860}{1.0035} + 886.5$$

$$= 2,740 \text{K}$$

$$\therefore T_{\max} = T_b = 2,740 \text{K}$$

(b) The thermal efficiency of the Diesel cycle is

$$\eta_{th} = 1 - \frac{(T_c - T_d)}{k(T_b - T_a)} \qquad (2)$$

T_c can be obtained from the isentropic relation

$$\frac{T_b}{T_c} = \left(\frac{v_c}{v_b}\right)^{k-1} \qquad (3)$$

where

$$v_b = \frac{RT_b}{P_b} \text{ from equation (1)}$$

$$= \frac{0.287(2,740)}{4,495}$$

$$= 0.1749 \text{ m}^3/\text{kg}$$

Substituting into (3),

$$\frac{T_b}{T_c} = \left(\frac{0.8495}{0.1749}\right)^{1.4-1}$$

or

$$\frac{T_b}{T_c} = 1.88$$

Solving for T_c,

$$T_c = \frac{2,740}{1.88}$$

$$= 1,457.5 \text{K}$$

From equation (2), then,

$$\eta_{th} = 1 - \frac{(1,457.5 - 300)}{1.4(2,740 - 886.5)}$$

$$= 1 - \frac{1,157.5}{2,594.9}$$

$$= 0.554$$

(c) The mean effective pressure (mep) is defined as

$$\text{mep} = \frac{W_{net}}{v_d - v_a} \qquad (4)$$

where

$$W_{net} = \eta_{th} \, Q_{in}$$

$$= 0.554(1860)$$

$$= 1030.44 \text{ kJ/kg}$$

Substituting into (4),

$$mep = \frac{1030.44}{(0.8495 - 0.0566)}$$
$$= 1299.6 \text{ kPa}$$

PROBLEM 8

The adiabatic efficiencies of the compressor and turbine used in an air-standard Brayton cycle are 85% and 90%, respectively. If the cycle operates between 14.7 and 55 psia and if the maximum and minimum temperatures are 1500°F and 80°F, respectively, compute the thermal efficiency of the cycle. Assume constant specific heats.

SOLUTION: Use the accompanying figure to refer to the different states. The thermal efficiency of the cycle is calculated using the formula

$$\eta_{th} = \frac{W_{act}}{Q_H} \tag{1}$$

where

$$W_{act} = W_{turb.} + W_{comp.} \tag{2}$$

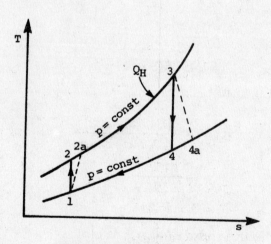

Figure 44. Brayton cycle

For this problem, the processes in the turbine and compressor are not reversible, and so the work done will be less than the work if the processes were reversible. Using the given efficiencies, it can be written

$$\left. W_{act} \right|_{turb.} = \eta_{turb.} \times \left. W_{theo.} \right|_{turb.} \tag{3}$$

and

$$\left. W_{act} \right|_{comp.} = \frac{\left. W_{theo} \right|_{comp.}}{\eta_{comp.}} \tag{4}$$

where

$$\left. W_{theo.} \right|_{turb.} = h_3 - h_4 = c_p(T_3 - T_4) \tag{5}$$

and

$$\left. -W_{theo.} \right|_{comp.} = h_2 - h_1 = c_p(T_2 - T_1) \tag{6}$$

To find the temperatures at states 2 and 4, use the isentropic relation

$$\frac{T_a}{T_b} = \left(\frac{P_a}{P_b}\right)^{\frac{k-1}{k}}.$$

At state 2,

$$T_2 = T_1 \left(\frac{P_2}{P_1}\right)^{\frac{k-1}{k}}$$

$$= 540 \left(\frac{55}{14.7}\right)^{0.286}$$

$$= 787.6°R$$

At state 4,

$$T_4 = T_3 \left(\frac{P_4}{P_3}\right)^{\frac{k-1}{k}}$$

$$= 1960 \left(\frac{14.7}{55}\right)^{0.286}$$

$$= 1343.9°R$$

With these values, and $c_p = 0.24$ Btu/lbm–°R, from equations (5) and (6),

$$\left. W_{theo.} \right|_{turb.} = 0.24(1960 - 1343.9)$$

$$= 147.9 \text{ Btu/lbm}$$

and

$$-w_{theo.}\bigg|_{comp.} = 0.24(787.6 - 540)$$

$$= 59.42 \text{ Btu/lbm}$$

Substituting into equations (3) and (4),

$$w_{act}\bigg|_{turb.} = 0.90(147.9) = 133.1 \text{ Btu/lbm}$$

$$-w_{act}\bigg|_{comp.} = \frac{59.42}{0.85} = 69.9 \text{ Btu/lbm}$$

From equation (2), then,

$$w_{act} = 133.1 - 69.9 = 63.2 \text{ Btu/lbm}$$

The only term unknown in equation (1) is the heat added to the system during process 2-3 (Q_H). However,

$$Q_H = h_3 - h_{2a} = c_p(T_3 - T_{2a}) \tag{7}$$

where T_{2a} is the actual temperature at state 2, and can be found using the efficiency of the compressor. Hence,

$$\eta_{comp} = \frac{T_2 - T_1}{T_{2a} - T_1} = 0.85$$

or

$$T_{2a} = \left(\frac{T_2 - T}{0.85}\right) + T_1$$

$$= \left(\frac{787.6 - 540}{0.85}\right) + 540$$

$$= 831.3°\text{R}$$

Equation (7) then gives

$$Q_H = 0.24(1960 - 831.3)$$

$$= 270.89$$

Finally, using equation (1), the efficiency of the cycle is calculated as

$$\eta_{th} = \frac{63.2}{270.89} = 0.233$$

or

$$\eta_{th} = 23.3\%$$

PROBLEM 9

A standard vapor compression refrigeration cycle uses Freon-22 as the working fluid to provide 3 tons of cooling capacity. If the condenser operates at 140°F and the evaporator operates at 50°F, compute (a) the mass flow rate of the Freon-22, (b) the horsepower required for the compressor, and (c) the heat transferred in the condenser. Assume the compression process to be reversible and adiabatic.

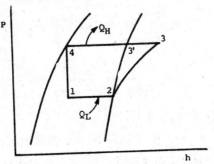

Figure 45. Refrigeration cycle

SOLUTION: Since the process has been assumed to be reversible and adiabatic, it will also be isentropic. Assuming the throttling process to be adiabatic, then it is also isenthalpic (constant enthalpy). Furthermore, assume that the condenser and evaporator operate at constant pressure. Then this cycle can be plotted on a P-h diagram as shown in Figure 45.

The values of the various properties at the different states shown in Figure 45 are taken from the P-h diagram for Freon-22, as shown in Figure 46. The procedure is as follows.

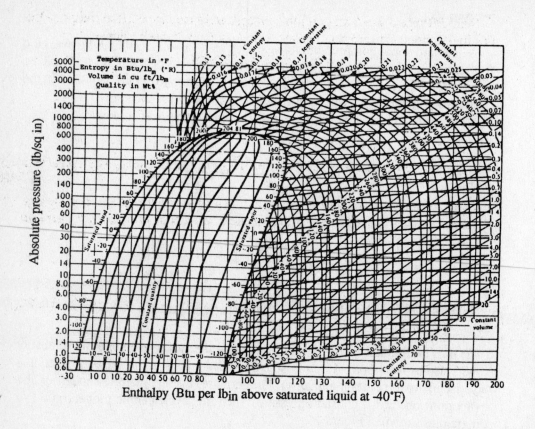

Figure 46. Pressure-enthalpy diagram for Freon-22 refrigerant

At state 3' the temperature is known to be 140°F, and the state is saturated vapor. Therefore, from Figure 46

$$P_{3'} = P_3 = 350 \text{ psia}.$$

At state 4

$$P_4 = P_{3'} = 350 \text{ psia}$$
$$T_4 = T_{3'} = 140°F$$

Hence,

$$h_4 = 52 \text{ Btu/lbm}.$$

At state 1, since process 1-4 is isenthalpic

$$h_1 = h_4 = 52 \text{ Btu/lbm}.$$

At state 2 the temperature is known to be 50°C, and the state is saturated vapor. Hence, from Figure 46,

$$P_2 = 100 \text{ psia}$$
$$h_2 = 109 \text{ Btu/lbm}$$
$$s_2 = 0.218 \text{ Btu/lbm–°R}$$

State 3: Process 2-3 is an isentropic process and state 3, due to the compression, lies in the superheated region. Futhermore, $P_3 = P_{3'} = 350$ psia. Hence,

$$s_3 = s_2 = 0.218 \text{ Btu/lbm-°R}$$
$$h_2 = 109 \text{ Btu/lbm}$$

From Figure 46,

$$h_3 = 123 \text{ Btu/lbm}$$
$$T_2 = 180°F$$

Now that the values of the various properties have been obtained, we can solve the problem.

(a) Consider the evaporator and write an energy balance around it to find the heat absorbed by the Freon-22. Neglecting potential and kinetic energies, we can write

$$q_L = h_2 - h_1$$
$$= 109 - 52$$
$$= 57 \text{ Btu/lbm}$$

It is known, however, that the evaporator is to absorb 3 tons or 36,000 Btu/hr. Hence, the rquired mass flow rate is

$$\dot{m} = \frac{\dot{Q}_L}{q_L}$$
$$= \frac{36,000}{57}$$
$$= 631.58 \text{ lbm/hr}$$

(b) Consider the compressor and write an energy balance around it, neglecting potential and kinetic energies.

$$-w = h_3 - h_2$$
$$= 123 - 109$$
$$= 14 \text{ Btu/lbm}$$

or

$$w = -14 \text{ Btu/lbm}$$

The total work production is

$$\dot{W} = \dot{m}w$$
$$= 631.58(-14)$$
$$= -8842.12 \text{ Btu/hr}$$

However, 1 hp = 2545 Btu/hr, and hence the compressor will need

$$p = \frac{8842.12}{2545} = 3.48 \text{ hp}$$

(c) The heat load of the condenser can be computed in two ways: either by writing an energy balance around it, or by writing the overall energy balance around the refrigerator. Here the second way is used. Hence,

$$\dot{W} = \dot{Q}_H + \dot{Q}_L$$

Solving for $\dot{Q}_H$ gives

$$\begin{aligned}\dot{Q}_H &= \dot{W} - \dot{Q}_L \\ &= -8842.12 - 36,000 \\ &= -44842.12 \text{ Btu/hr}\end{aligned}$$

The heat transferred per unit mass is

$$\begin{aligned}q_H &= \frac{\dot{Q}_H}{\dot{m}} \\ &= \frac{-44842.12}{631.58} \\ &= 71 \text{ Btu/lbm}\end{aligned}$$

PROBLEM 10

A converging nozzle has air flowing through it. Calculate the stagnation temperature T_o and pressure P_o if at point A within the nozzle, $P_A = 40$ psia, $T_A = 2000°$R, $V_A = 500$ ft/sec, and A_A (cross-sectional area) $= 0.2$ ft^2. Also calculate the sonic velocity, Mach number at this section and the exit area A_B, exit pressure P_B, temperature T_B, and velocity V_B if the exit Mach number is one. Assume air to be an ideal gas with $k = 1.40$.

SOLUTION: The stagnation temperature is

$$T_o = T_A + \frac{k-1}{2kR}V_A^2 .$$

Therefore,

$$T_o = 2000°\text{R} + \frac{(1.4-1)(500)^2 \text{ ft}^2 / \text{sec}^2}{2 \times 1.4 \times 53.35 \dfrac{\text{ft} \bullet \text{lb}}{\text{lbm}°\text{R}} \times \dfrac{32 \text{ lbm ft}}{\text{lb sec}^2}}$$

$$= 2021°\text{R}$$

The stagnation pressure is

$$P_o = P_A \left(\frac{T_o}{T_A} \right)^{\frac{k}{(k-1)}}$$

Therefore,

$$P_o = 40 \frac{\text{lb}}{\text{in}^2} \times \left(\frac{2021}{2000} \right)^{\frac{1.4}{(1.4-1)}}$$

$$= 41 \text{ lb/in}^2$$

The sonic velocity is

$$C_A = \sqrt{kRT_A}$$

$$C_A = \left(1.4 \times 53.35 \frac{\text{ft} \cdot \text{lb}}{\text{lbm}^\circ \text{R}} \times 2000^\circ \text{R} \times 32.2 \frac{\text{lbm} \cdot \text{ft}}{\text{lb sec}^2} \right)^{\frac{1}{2}}$$

$$C_A = 2193 \text{ ft/sec}$$

The Mach number is,

$$M_A = \frac{V_A}{C_A} = \frac{500}{2193} = 0.228.$$

The exit pressure is

$$\gamma_c = \frac{P_B}{P_o} = 0.528.$$

Therefore,

$$P_B = 0.528 P_o = 0.528(41 \text{ psia}) = 21.65 \text{ lb/in}^2$$

The exit velocity is

$$V_B = C_B = \left(\frac{2k}{k+1} RT_o \right)^{\frac{1}{2}}$$

$$= \left(\frac{2 \times 1.4}{1.4+1} \times 53.35 \frac{\text{ft} \cdot \text{lb}}{\text{lbm}^\circ \text{R}} \times 32.2 \frac{\text{lbm ft}}{\text{lb sec}^2} \right)^{\frac{1}{2}}$$

$$= 2013 \text{ ft/sec}$$

The exit temperature is

$$T_B = \frac{2}{k+1} T_o = \left(\frac{2}{1.4+1} \right)(2021^\circ \text{R})$$

$$T_B = 1684^\circ \text{R}$$

To calculate the exit area, the mass rate of flow is required. Therefore,

$$\rho_A = \frac{P_A}{RT_A} = \frac{40\frac{\text{lb}}{\text{in}^2} \times 144 \text{ in}^2/\text{ft}^2}{53.35\frac{\text{ft} \bullet \text{lb}}{\text{lbm}^\circ\text{R}} 2000^\circ\text{R}}$$

$$\rho_A = 0.054 \text{ lbm}/\text{ft}^3$$

$$\rho_B = \frac{P_B}{RT_B} = \frac{21.65 \times 144}{53.35 \times 1684}$$

$$\rho_B = 0.035 \text{ lbm}/\text{ft}^3$$

Using the equation of continuity,

$$A_B = \frac{\rho_A A_A V_A}{\rho_B V_B} = \frac{0.054 \times 0.2 \times 500}{0.035 \times 2013} = 0.077 \text{ft}^2.$$

CHAPTER TEN

Fluids

Chapter 10

FLUIDS

FLUID PROPERTIES

A fluid is defined as a substance that cannot resist a shear stress by static deformation. Both liquids and gases are fluids and are distinguished from solids by the above definition. There are many properties of fluids to which numerical values can be given. **Density,** ρ, is defined as the mass of a small fluid element divided by its volume. Often, **specific weight,**

$$\gamma = \rho g,$$

is more useful since density and gravitational acceleration usually occur together. Both density and specific weight are dimensional quantities; common units for density are kg/m^3 and $slug/ft^3$, and lbf/ft^3 is the typical unit for specific weight. **Specific gravity,** on the other hand, is dimensionless and is defined as the ratio of a fluid's density to the density of some reference fluid. For liquids, water is the reference fluid; for gases, air is used (at a standard temperature and pressure).

Viscosity, another important property of fluids, is the ratio of the local shearing stress to the rate of shearing strain of a fluid element in a moving fluid. For simple shear flows where velocity component μ in the x-direction is a function of only the normal coordinate y, the shear stress, τ, :

$$\tau = \mu \frac{du}{dy},$$

where μ is called the coefficient of viscosity. This linear relation applies only to **Newtonian** fluids. Fortunately, most common fluids, such as air, water, oil, etc., are Newtonian. The shear stress on a solid surface is equal but opposite to that applied to a fluid wetting the surface. Thus, frictional forces on surfaces can be found if the velocity gradient $\frac{du}{dy}$ (also called the rate of shearing strain) and coefficient of viscosity μ are known. Units for μ are the pascal-second in the SI system, and lbf-sec/ft^2 in the English system. **Kinematic viscosity,** υ, is defined as:

$$\upsilon = \frac{\mu}{\rho},$$

431

commonly given in units of ft²/sec or cm²/sec.

EXAMPLE: A block weighing 100 lb and having an area of 2 ft² slides down an inclined plane as shown in Figure 2, with a constant velocity. An oil gap between the block and the plane is 0.01 in. thick, the inclination of the plane is 30° to the horizontal, and the velocity of the block is 6 fps. Find the viscosity of the lubricating film.

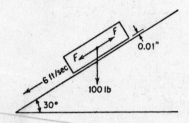

Figure 1. Forces on block

SOLUTION: The coefficient of viscosity μ (mu) is defined as the ratio:

$$\frac{\text{Shearing stress } \tau}{\text{Rate of shearing strain} \left(\dfrac{du}{dy} \right)}$$

and may be compared with the modulus of rigidity of a solid.

Rate of shearing strain is given by $\dfrac{du}{dy}$, and hence:

$$\mu = \frac{\tau}{\dfrac{du}{dy}}$$

or

$$\tau = \mu \frac{du}{dy}.$$

μ is also called the absolute or dynamic viscosity and has units of lbf -ft/s² or slugs/ft-s.

In this problem, the component of the weight acting down the plane is opposed by a viscous force exactly equal and opposite to it. Therefore,

$$F = 100 \sin 30° = 50 \text{ lbf}.$$

Hence:

$$\tau = \frac{F}{A} = \frac{50}{2} = 25 \text{ psf}$$

but

$$\tau = \mu \frac{du}{dy}.$$

Assuming a linear inverse in fluid speed from zero at the surface of the inclined plane to 6 fps at the lower block surface:

$$\frac{du}{dy} = \frac{6 \ \text{ft/s}}{0.01/12 \ \text{ft}} = 7200 \ \text{s}^{-1}$$

Therefore:

$$\mu = \frac{\tau}{\frac{du}{dy}} = \frac{25 \ \text{lbf/ft}^2}{7200 \ \text{s}^{-1}} = 0.00347 \ \text{lbf-ft/s}^2$$

Vapor pressure, p_v, is defined as the pressure at which a liquid will boil at a given temperature; p_v depends greatly on temperature, and its value can be obtained from charts. In flows of liquids, local fluid pressures typically decrease as velocity increases. If the local pressure falls below p_v, local boiling or *cavitation* may occur.

With an increase in pressure, all fluids compress, and an increase in density results. The **coefficient of compressibility**, α, is given by:

$$\alpha = \frac{1}{V_o} \left(\frac{\partial V}{\partial p} \right)_T$$

where p denotes pressure and V, volume, for a constant temperature.

The **bulk modulus**, E, is the reciprocal of compressibility:

$$E = \frac{1}{\alpha} = V_o \left(\frac{\partial p}{\partial V} \right)_T,$$

with typical units of psi, atm, or kPa.

When a liquid forms an interface with a second liquid or gas, a tensional force exists at the interface. The **coefficient of surface tension**, σ, is a measure of this tensional force per unit length of the surface:

$$\sigma = \frac{F}{2L}.$$

The height of capillary rise in a tube can be found using the surface tension and the specific weight of the fluid.

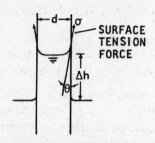

Figure 2. Capillary action in a tube

Taking the summation of forces in the vertical direction of the water in the tube that has risen above the reservoir level yields:

$$\sigma\pi d - \gamma(\Delta h)\left(\frac{\pi d^2}{4}\right) = 0$$

yielding
$$\Delta h = \frac{4\sigma}{\gamma d}$$

assuming the angle, θ, is equal to zero.

FLUID STATICS

The fundamental equation that describes the pressure field in a fluid at rest is

$$\nabla p = \rho g$$

where g is the acceleration vector due to gravity. If we adopt a coordinate system where z is "up" vertically, g acts downward (opposite to the direction of increasing z), the equation above reduces to:

$$\frac{dp}{dz} = -\rho g$$

when applied to gases with large height differences, such as the atmosphere density in a variable. For liquid applications ρ can be assumed to be constant with negligible error. For constant ρ and g:

$$p_2 - p_1 = -\rho g(z_2 - z_1),$$

where 1 and 2 represent any two positions in the same fluid. At a liquid surface, the pressure must equal the pressure of the air (or other fluid) immediately above the surface. For liquids exposed to atmospheric pressure, p_{atm}, the local pressure at some depth h (measured from the surface) is:

$$p = \rho g h + p_{atm},$$

where ρ is the density of the liquid. Often, it is more convenient to use gage (sometimes spelled "gauge") pressure, defined as the absolute pressure minus p_{atm}:

$$P_{gage} = P_{abs} - P_{atm},$$

where absolute pressure is measured with respect to a true zero pressure reference. In the liquid discussed above, the gage pressure would equal ρgh.

The equation: $p_2 - p_1 = -\rho g(z_2 - z_1)$ can be applied in piecewise fashion to determine the pressure difference across a **simple manometer**. A manometer consisting of n different substances with different specific gravities is shown in the figure. The following equation is used to determine the pressure difference $(P_A - P_n)$:

$$\left(P_A - P_n\right) = \left(P_A - P_{A_1}\right) + \left(P_{A_1} - P_{A_2}\right) + \left(P_{A_2} - P_{A_3}\right) + \left(P_{A_3} - P_{A_4}\right) + \ldots + \left(P_{An_{-1}} - P_n\right)$$

$$= -\gamma_1 z_1 - \gamma_2 z_2 + \gamma_3 z_3 + \gamma_4 z_4 \ldots - \gamma_n z_n$$

where

$\gamma_1, \gamma_2, \gamma_3, \gamma_4, \ldots, \gamma_n$ = the specific weights of the substances,

$z_1, z_2, z_3, z_4 \ldots, z_n$ = the distances between two successive points in the columns of the manometer.

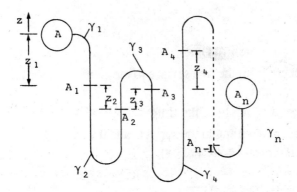

Figure 3. Simple manometer

Note: The distances $z_1, z_2, z_3, z_4 \ldots, z_n$ are considered as positive if the end point, A_3, is at a higher position than the start point, A_2 (Distance z_3), or as negative if the end point, A_1, is at a lower position than the start point, A (Distance z_1).

In order to determine the force acting on a submerged surface, we must specify the magnitude, the direction, and the line of action of the resultant force F_R.

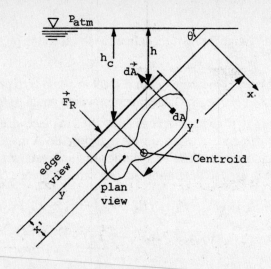

Figure 4. Plane submerged surface

The magnitude of the resultant force acting on the submerged surface is:

$$F_R = \gamma h_c A$$

where

A is the surface area

h_c is the depth from the free surface to the centroid of the area on which the force acts, and

γ is the specific weight of the fluid.

The direction of F_R is normal to the surface, and the line of action passes through the points x', y', which can be located by:

$$y' = y_c + \frac{I_c}{A_{yc}} \qquad\qquad \frac{I_c}{A_{yc}} > 0$$

$$x' = x_c + \frac{(I_{xy})_c}{A_{yc}}$$

where

I_c, $(I_{xy})_c$ are the moments of inertia about its center gravity axes, and

x_c, y_c are the center of gravity coordinates.

For a force acting on a curved surface, the vertical projection of the surface is used to determine the force in the horizontal direction F_{Ry} and z'. The force in the vertical direction F_{Rz} is equal to the volume of fluid displaced multiplied by the specific weight of the fluid. It acts through the center of gravity of the volume of the displaced fluid.

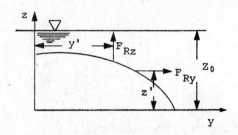

Figure 5. Two-dimensional curved submerged surface

The table below is a review of the geometry needed to solve fluid statics problems.

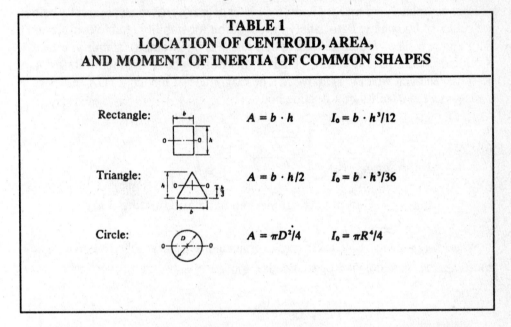

The resultant vertical force exerted on a body by a static fluid in which it is submerged or floating is called the **buoyant force**. The magnitude of this force is

$$F_z = \int \rho g (z_2 - z_1) dA = \rho g V$$

where

ρ = Density of the fluid

g = Gravitational constant

V = Volume of the body

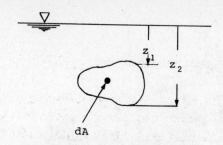

Figure 6. Buoyant force

The buoyant force for an incompressible fluid goes through the centroid of the volume displaced by the body.

For a completely submerged body, the center of gravity, G, must be directly below the center of buoyancy, B, to satisfy the condition for stability (stable equilibrium) as in Figure 6. The vertical alignment of B and G is important for stability. If G and B coincide, neutral equilibrium is obtained. In a floating body stable equilibrium can be achieved even when G is above B. The magnitude of the length GA serves as a measure of the stability of a floating body.

FIG HERE

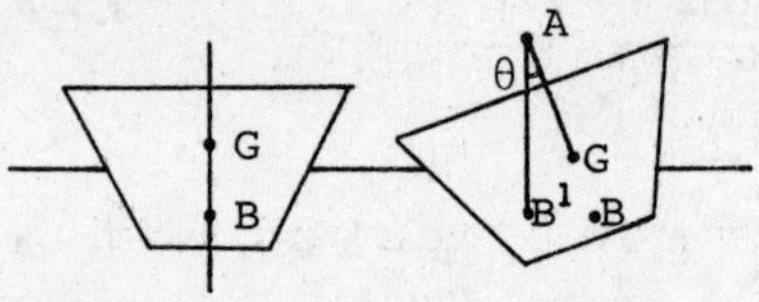

Figure 7. Completely submerged body and floating body

When a container of fluid undergoes constant uniform linear acceleration **a** (in any direction), the equation $\nabla p = \rho g$ may still be applied, by substituting $g - a$ for the vector g, i.e.:

$$\nabla p = \rho \, (g - a).$$

In other words, all the hydrostatic equations above remain valid, but with a different constant of gravity ($g - a$ instead of g). The surface of an accelerating container will align itself perpendicularly to the vector $g - a$. The pressure increases linearly with a coordinate along the direction of $g - a$, rather than simply along the direction of g in hydrostatics.

When a container of liquid rotates at a constant angular velocity about the vertical axis, the equation above is still valid, with the centripetal acceleration equal to:

$$a = -r\omega^2 i_r,$$

where

        r = The radial distance from the axis of rotation

      ω = The magnitude of the angular velocity

      i_r = The unit coordinate in the radial direction,

and pressure is:

$$p = \text{Constant} - \rho g z + \frac{1}{2}\rho r^2 \omega^2.$$

EXAMPLE: A gate 5 ft wide is hinged at point B and rests against a smooth wall at point A. Compute (a) the force on the gate due to seawater pressure; (b) the horizontal force P exerted by the wall at point A; and (c) the reactions at the hinge B (Figure 8).

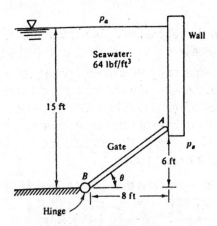

Figure 8. Submerged gate

SOLUTION: (a) By geometry the gate is 10 ft long from A to B, and its centroid is halfway between, or at elevation 3 ft above point B. The depth h_{CG} is thus $15 - 3 = 12$ ft. The gate area is $5 \times 10 = 50$ ft^2. Neglect p_A as acting on both sides of the gate. The hydrostatic force on the gate is:

$$F = p_{CG}A = \rho g h_{CG}A = (64 \text{ lbf/ft}^3)(12 \text{ ft})(50 \text{ ft}^2) = 38{,}400 \text{ lbf}$$

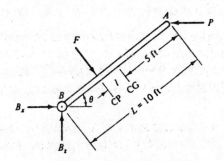

Figure 9. Free body diagram

(b) First we must find the center of pressure of F. A free-body diagram of the gate is shown in Figure 9. The gate is a rectangle and hence $I_{xy} = 0$ and $I_{xx} = \dfrac{bL^3}{12} =$

$$\frac{\left[(5\text{ft})\times(10\text{ft})^3\right]}{12} = 417\text{ ft}^4.$$ The ambient pressure p_a is neglected if it acts on both sides of the plane; e.g., the other side of the plane is inside a ship or on the dry side of a gate or dam. In this case $p_{CG} = \rho g h_{CG}$, and the center of pressure becomes independent of specific weight:

$$F = \rho g h_{CG} A \left| y_{CP} = -\frac{I_{xx}\sin\theta}{h_{CG}A} \right| x_{CP} = -\frac{I_{xy}\sin\theta}{h_{CG}A}$$

The distance I from the CG to the CP is given by the equation above, since p_a is neglected:

$$I = -y_{CP} = +\frac{I_{xx}\sin\theta}{h_{CG}A} = \frac{\left(417\text{ ft}^4\right)\left(\frac{6}{10}\right)}{(12\text{ ft})\left(50\text{ ft}^2\right)} = 0.417\text{ ft}.$$

The distance from point B to force F is thus $10 - I - 5 = 4.583$ ft. Summing moments counterclockwise about B gives:

$$PL \sin\theta - F(5 - I) = P(6\text{ ft}) - (38,400\text{ lbf})(4.583\text{ ft}) = 0$$

or

$$P = 29,300\text{ lbf.}$$

(c) With F and P known, the reactions B_x and B_z in Figure 9 are found by summing forces on the gate

$$\sum F_x = 0 = B_x + F\sin\theta - P = B_x + 38,400(0.6) - 29,300$$

and

$$B_x = 6,300\text{ lbf}$$

$$\sum F_z = 0 = B_z - F\cos\theta = B_z - 38,400(0.8)$$

and

$$B_z = 30,700\text{ lbf.}$$

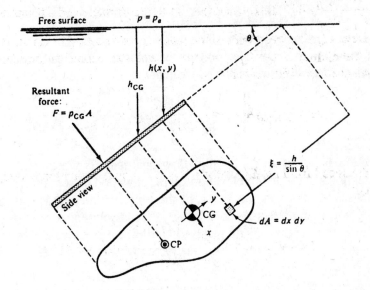

Figure 10. Hydrostatic force and center of pressure on an arbitrary plane surface of area A inclined at an angle θ below the free surface.

HYDRAULICS AND FLUID MACHINES

The first step in the solution of control volume problems is to choose a control volume. This basic step is most critical since the degree of difficulty of the problem can often be greatly reduced by a wise choice of control volume. Next, evaluate the net outflux of mass (the control surface integral). Finally, equate this to the negative of the unsteady control volume integral. In many cases, if the flow is steady, this volume integral will vanish.

Often, a control volume will cut through a duct or pipe where the flow is predominantly in one direction. In such cases, an **average velocity**:

$$v = \frac{Q}{A}$$

is defined where Q is the **volume flow rate** across the pipe's cross section and A is the cross-sectional area. In the control volume equation, then, the **mass flux** across this surface is simply:

$$\dot{m} = \rho v A.$$

The conservation of mass law is:

$$\rho_1 A_1 v_1 = \rho_2 A_2 v_2,$$

or for an incompressible fluid where $\rho_1 = \rho_2$:

$$A_1 v_1 = A_2 v_2.$$

In Cartesian coordinates, incompressible continuity is expressed as:

$$\frac{\partial u}{\partial x}+\frac{\partial v}{\partial y}+\frac{\partial w}{\partial z}=0,$$

where u, v, w are the x, y, z compnents of velocity. For steady incompressible frictionless flow along a streamline between points 1 and 2, the following Bernoulli energy equation applies:

$$\frac{p_1}{\rho}+\frac{v_1^2}{2}+gz_1=\frac{p_2}{\rho}+\frac{v_2^2}{2}+gz_2.$$

This equation is useful for duct or pipe flows, particularly when there are changes in cross-sectional areas and/or elevations. If the elevation increases ($z_2 > z_1$) or if the area decreases ($A_2 < A_1$ and thus $v_2 > v_1$ by the conservation of mass), it can be seen that the pressure p_2 must decrease accordingly to satisfy the equation. Often it is more convenient to rewrite the equation above in terms of the equivalent column height of fluid, or **head**. Dividing by g, the equation above can be written:

$$H=\frac{p_1}{\gamma}+\frac{v_1^2}{2g}+z_1=\frac{p_2}{\gamma}+\frac{v_2^2}{2g}+z_2$$

where $\gamma = \rho g$, and H is called the total head or total Bernoulli head. H is also equivalent to the height of the Energy Grade Line (EGL).

In most practical problems, friction cannot be neglected, and there may be work added to the flow (by a pump), work extracted from the flow (by a turbine), or heat transferred to or from the fluid. In this case, a much more general Bernoulli equation must be developed. Specifically, the total head H in the equation above does not remain constant, but rather changes whenever friction, work, or heat transfers are present. From the Bernoulli energy equation:

$$\frac{p_1}{\rho g}+\frac{v_1^2}{2g}+z_1=\frac{p_2}{\rho g}+\frac{v_2^2}{2g}+z_2+h_s+h_{\text{total}}.$$

where h_s = shaft work head, h_{total} = the total amount of head loss due to friction, heat transfer, etc. Here 1 and 2 are locations at an inlet and outlet of a control volume. The shaft work head, h_s:

$$h_s=\frac{\dot{W}_s}{\left(\dot{m}g\right)},$$

is the head associated with work done by the fluid,

where $\dot{W}_s$ = the shaft power (work per unit time) done by the fluid.

For a pump, let:

$$h_s=-E_p,$$

where E_p = The energy added per unit weight of the fluid (dimensions of head, i.e., height of fluid).

For a turbine, let E_T be the energy extracted from the fluid per unit weight. In this case:

$$h_s = E_T.$$

The power required by a pump is given by:

$$\dot{W}_p = \frac{Q \; E_p}{\eta_p},$$

where

Q = The volumetric flow rate

γ = The specific weight of the fluid

η_p = The pump efficiency.

The power produced by a turbine is given by:

$$\dot{W}_t = Q\,\gamma E_t \eta_t$$

where η_t = The turbine efficiency.

EXAMPLE: (a) Three kilonewtons of water per second flow through this pipeline reducer. Calculate the flow rate in cubic meters per second and the mean velocities in the 300 mm and 200 mm pipes (Figure 11).

(b) Thirty newtons of air per second flow through the reducer of the preceding problem, the air in the 300 mm pipe having a weight density of 9.8 N/m³. In flowing through the reducer, the pressure and temperature will fall, causing the air to expand and producing a reduction of density. Assuming that the weight density of the air in the 200 mm pipe is 7.85 N/m³, calculate the mass and volume flow rates and the velocities in the two pipes.

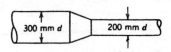

Figure 11. Dimensions of pipeline

SOLUTION: (a) Q: volumetric flow rate is:

$$Q = A_{cs}v$$

if density is essentially constant.

$\dot{G}$ is the weight flow rate:

$$Q = \frac{\dot{G}}{\gamma} = \frac{3 \times 10^3 \, N/s}{9.8 \times 10^3 \, N/m^3} = 0.306 \ \ m^3/s$$

$$v_{300} = \frac{Q}{A} = \frac{0.306}{\frac{\pi}{4}(0.3)^2} = 4.33 \ \ m/s$$

$$v_{200} = \frac{0.306}{\frac{\pi}{4}(0.2)^2} = 9.74 \ \ m/s$$

or,

$$Q_{300} \text{ mm pipe} = Q_{200} \text{ mm pipe:}$$

$$v_{200} = v_{300}\left(\frac{r_{300}}{r_{200}}\right)^2 = 4.33\left(\frac{.3}{.2}\right)^2 = 9.74 \ \ m/s$$

(b)

$$Q_{300} = \frac{\dot{G}}{\gamma_1} = \frac{30 \, N/s}{9.8 \, N/m^3} = 3.06 \ m^3/s$$

$$Q_{200} = \frac{\dot{G}}{\gamma_2} = \frac{30 \, N/s}{7.85 \, N/m^3} = 3.82 \ m^3/s$$

$$v_{300} = \frac{Q_{300}}{A} = \frac{3.06}{\frac{\pi}{4}(0.3)^2} = 43.3 \ \ m/s, \ v_{200} = \frac{3.82}{\frac{\pi}{4}(0.2)^2} = 121.6 \ \ m/s$$

$$\dot{m} = A\rho v = \text{Constant} = \frac{\dot{G}}{g_n} = \frac{30 \, N/s}{9.81 \, m/s^2} = 3.06 \ \ kg/s$$

$\dot{m}$ = Mass flow rate

$\dot{G}$ = Weight flow rate

To check:

$$\dot{m} = \frac{\gamma_{300} Q_{300}}{g_n} = 9.8 \times \frac{3.06}{9.81} = 3.06 \ \ kg/s$$

$$\dot{m} = \frac{\gamma_{200} Q_{200}}{g_n} = 7.85 \times \frac{3.82}{9.81} = 3.06 \ \ kg/s$$

The theoretical velocity of a jet issuing from an orifice can be derived from the Bernoulli energy equation, as shown in the following example.

EXAMPLE: (a) Determine the velocity of efflux from the nozzle in the wall of the reservoir of the figure. (b) Find the discharge through the nozzle.

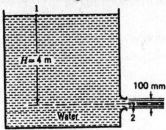

Figure 12. Flow through nozzle from reservoir

SOLUTION: (a) The jet issues as a cylinder with atmospheric pressure around its periphery. The pressure along its centerline is at atmospheric pressure for all practical purposes. Bernoulli's equation is applied between a point on the water surface and a point downstream from the nozzle,

$$\frac{v_1^2}{2g}+\frac{p_1}{\gamma}+z_1=\frac{v_2^2}{2g}+\frac{p_2}{\gamma}+z_2.$$

With the pressure datum as local atmospheric pressure, $p_1=p_2=0$; with the elevation datum through point 2, $z_2=0$, $z_1=H$. The velocity on the surface of the reservoir is zero (practically); hence:

$$0+0+H=\frac{v_2^2}{2g}+0+0$$

and

$$v_2=\sqrt{2gH}=\sqrt{2\times9.806\times4}=8.86 \text{ m/s}$$

which states that the velocity of efflux is equal to the velocity of free fall from the surface of the reservoir. This is known as Torricelli's theorem.

(b) The discharge Q is the product of velocity of efflux and area of stream:

$$Q=A_2v_2=\pi\,(0.05\text{ m})^2\,(8.86\text{ m/s})=0.07\text{ m}^3/\text{s}=70\text{ l/s}.$$

MOMENTUM

In many fluid flow problems, the total force on a solid object or wall is desired, for example, in determining the required bolt strength on flanges in a piping system, or the total thrust produced by a jet engine. The control volume or integral technique may be applied here by utilizing the law of conservation of linear momentum.

The result is the integral conservation of momentum law for a control volume,

$$\frac{d}{dt}\iiint_{cv} \rho v dv + \iint_{cs} \rho v(v_r \bullet \mathbf{n})dA = \sum \mathbf{F}$$

where v_r is the velocity of the fluid relative to the control surface, $\mathbf{n}$ is the unit outward normal vector, and $\sum \mathbf{F}$ is the total force acting on the control volume when the control volume is considered as a free body. Note that this equation is a *vector* equation, and thus represents in general three components that may have to be evaluated separately.

In most cases, the control volume is fixed, $v_r = v$ and the equation above reduces to

$$\frac{\partial}{\partial t}\iiint_{cv} \rho v dv + \iint_{cs} \rho v(v_r \bullet \mathbf{n})dA = \sum \mathbf{F}$$

The technique for solving problems with the integral conservation of momentum equation involves first choosing an appropriate control volume, then determining the flux terms and (if nonzero) the unsteady control volume term. Typically, the unknown — some force on the right-hand side of the equation above — can then be obtained.

In general, the force term consists of surface forces due to pressure and friction, body forces such as gravity (and possibly electromagnetic forces), and other forces acting on the control surface such as the tension force in a bolt through which the control surface is sliced. It is important to keep two facts in mind:

1) $\sum \mathbf{F}$ is the *vector* sum of all forces acting on the control volume, and

2) $\sum \mathbf{F}$ include(s) *all* forces acting on both solid and fluid material in the control volume.

It is usually best to draw a free-body diagram of the control volume, showing all forces acting on it. If the desired result is the force acting on a solid wall *by* the fluid flow, remember to change the sign since any force in the equation above must be applied on the control volume or control surface.

For flowfields involving sections of pipe flow as inlets or exits, the surface integral in the equation above reduces to:

$$\iint_{cs} \rho v(v \bullet \mathbf{n})dA = \sum_{\text{outlets}} \dot{m} v - \sum_{\text{inlets}} \dot{m} v$$

where $\dot{m} = \rho Q$.

Also, for such sections, the pressure is typically constant over the cross-sectional area A; thus, the pressure force on the area is simply pA acting in the direction opposite to the unit outward normal vector $\mathbf{n}$. Although absolute pressure is implied in the above, gage pressure is often more convenient since atmospheric pressure may be subtracted uniformly over the entire control surface without changing the problem.

EXAMPLE: Consider a jet that is deflected by a stationary vane, such as is given in Figure 13. If the jet speed and diameter are 100 ft/s and 1 in., respectively, and the jet is deflected 60°, what force is exerted on the vane by the jet?

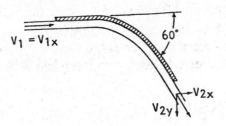

Figure 13. Flow of jet in stationary vane

SOLUTION:
$$F_x = \sum_{cs} \mu \rho v \bullet A + \frac{d}{dt} \int_{cv} \mu \rho dv \qquad (1)$$

If the flow is steady, the second term on the right of Equation (1) will be zero, which leaves the following:

$$F_x = \sum_{cs} \mu \rho v \bullet A. \qquad (2)$$

At section 1, the velocity is constant over the section and the area vector A_1 is in the reverse direction of the velocity vector v. Therefore, for this part of the control surface, we have:

$$\sum_{cs} \mu \rho v \bullet A = v_{1x}\rho(-v_1 A_1)$$

By a similar analysis for section 2 (in this case, the velocity v_2 and area A_2 have the same sense), we get:

$$\sum_{cs} \mu \rho v \bullet A = v_{2x}\rho v_2 A_2$$

However, $v_1 A_1 = v_2 A_2 = Q$, so when these substitutions are made, Equation (2) becomes:

$$F_x = \rho Q(v_{2x} - v_{1x}). \qquad (3)$$

In a similar manner the force in the y direction, F_y, will be:

$$F_y = \rho Q(v_{2y} - v_{1y}). \qquad (4)$$

Since the forces given by Equations (3) and (4) are the forces exerted by the vane on the jet, we obtain the forces of the jet on the vane by simply reversing the sign on F_x and F_y. First solve for F_x, the x component of force of the vane on the jet, by using Equation (3):

$$F_x = \rho Q(v_{2x} - v_{1x}) \qquad (3)$$

Here, the final velocity component in the x direction is:

$$v_{2x} = 100 \cos 60° \text{ ft/s}$$

Hence,

$$v_{2x} = 100 \times 0.500 = 50 \text{ ft/s}$$

also,

$$v_{1x} = 100 \text{ ft/s}$$

and

$$Q = v_1 A_1 = 100 \,\frac{0.785}{144} = 0.545 \text{ ft}^3/\text{s}.$$

Therefore,

$$F_x = 1.94 \text{ lbf-s}^2/\text{ft}^4 \times 0.545 \text{ ft}^3/\text{s} \times (50 - 100) \text{ ft/s} = -52.9 \text{ lbf}.$$

Similarly determined, the y component of force on the jet is:

$$F_y = 1.94 \text{ lbf-s}^2/\text{ft}^4 \times 0.545 \text{ ft}^3/\text{s} \times (-86.6 \text{ ft/s} - 0) = -91.6 \text{ lbf}$$

Then the force on the vane will be the reactions to the forces of the vane on the jet, or:

$$F_x = +52.9 \text{ lbf}$$

$$F_y = +91.6 \text{ lbf}.$$

DIMENSIONAL ANALYSIS AND SIMILITUDE

It is generally desirable to express a given set of dimensional variables in a flowfield in terms of dimensionless parameters (or "Pi's"). These parameters provide universal measures of flow regimes and effectively reduce the number of independent variables. They are also necessary in order to obtain scaling laws to predict prototype performance based on measurements on a (typically smaller) model. Examples of dimensionless parameters are given in the following table.

TABLE 2

FORCE RATIOS

Name of Ratio	Definition	Physical Meaning
Reynolds number, Re	$\dfrac{v\ell\rho}{\mu}$	$\dfrac{\text{Inertia force}}{\text{Viscous force}}$
Froude number, Fr	$\dfrac{v^2}{(\ell g)}$	$\dfrac{\text{Inertia force}}{\text{Gravity force}}$
Weber number, We	$\dfrac{v^2\ell\rho}{\sigma}$	$\dfrac{\text{Inertia force}}{\text{Surface tension force}}$
Mach number, M	$\dfrac{v}{a}$	$\dfrac{\text{Inertia force}}{\text{Elastic force}}$
Euler number, Eu	$\dfrac{\Delta\rho}{\rho v^2}$	$\dfrac{\text{Pressure force}}{\text{Inertia force}}$

The Buckingham Pi technique provides a systematic way to determine dimensionless parameters from a given set of dimensional variables. The following procedure is invoked in the Buckingham Pi technique:

1) Count the total number of variables, n.

2) List the dimensions of each variable. This is typically done in terms of the four primary dimensions—mass, length, time, and temperature. Alternately, force can replace mass as a primary dimension.

3) Count the total number of primary dimensions (typically three or four), and let j equal this number. **Note:** If the dimensional analysis fails in the steps below, return here, decrease j by one, and then repeat the analysis.

4) You now expect to find $k = n - j$ dimensionless parameters $\Pi_1, \Pi_2,..., \Pi_k$. To find these, you first need to select j variables as "repeating variables." Choose variables that do not by themselves form a dimensionless group, but that represent all of the primary dimensions involved in the problem. There are "preferred" choices, such as velocity and density, which will generate recognizable Πs, such as the Reynolds number, etc.

5) Form a power product consisting of the j repeating variables and each of the remaining k variables. By forcing the exponent of each dimension to be zero, $\Pi_1, \Pi_2,..., \Pi_k$ are found.

Note that although your Πs may be dimensionless, they may not be of a form suitable for your particular application. It is perfectly valid to multiply or divide two or more Πs together to form a new Π set. In fact, since each Π by itself is dimensionless, any Π multiplied by another Π raised to any exponent must also be dimensionless. The new set of Πs formed in this manner is no more or less valid than the first set; the "correct" set is that which is most suitable to the problem at hand. Note also that such a rearrangement of parameters may be used sometimes to obtain "standard" $\prod$ s such as the Reynolds number, the Froude number, etc., even if these do not result directly from your dimensional analysis.

The best way to learn this technique is to practice on many problems. True dynamic similarity between a model and a prototype can only exist if each $\prod$ for the model exactly matches the corresponding $\prod$ of the prototype. When such is the case, it is possible to scale up from the model to the prototype to predict its performance. These concepts of dimensional analysis and similitude are thus of paramount significance to designers who test small models before building a full-scale prototype.

TABLE 3
DIMENSIONS OF FLUID-MECHANICS QUANTITIES

Quantity	Symbol	Dimensions	
		$(MLT\Theta)$	$(FLT\Theta)$
Angle	θ	None	None
Angular velocity	ω	T^{-1}	T^{-1}
Area	A	L^2	L^2
Density	ρ	ML^{-3}	FT^2L^{-4}
Force	F	MLT^{-2}	F
Kinematic viscosity	ν	L^2T^{-1}	L^2T^{-1}
Length	L	L	L
Mass flux	$\dot{m}$	MT^{-1}	FTL^{-1}
Moment, torque	M,T	ML^2T^{-2}	FL
Power	P	ML^2T^{-3}	FLT^{-1}
Pressure, stress	p, τ	$ML^{-1}T^{-2}$	FL^{-2}
Speed of sound	a	LT^{-1}	LT^{-1}
Specific heat	c_p, c_v	$L^2T^{-2}\Theta^{-1}$	$L^2T^{-2}\Theta^{-1}$
Strain rate	i	T^{-1}	T^{-1}
Surface tension	σ	MT^{-2}	FL^{-1}
Temperature	T	Θ	Θ
Thermal conductivity	k	$MLT^{-3}\Theta^{-1}$	$FT^{-1}\Theta^{-1}$
Velocity	v	LT^{-1}	LT^{-1}
Viscosity	μ	$ML^{-1}T^{-1}$	FTL^{-2}
Volume	v	L^3	L^3
Volume flux	Q	L^3T^{-1}	L^3T^{-1}

EXAMPLE: The thrust F of a screw propeller is known to depend upon the diameter d, speed of advance v, fluid density ρ, revolutions per second N, and the coefficient of viscosity μ of the fluid. Find an expression for F in terms of these quantities.

SOLUTION: The general relationship must be $F = \phi(d, v, \rho, N, \mu)$, (where ϕ represents "a function of"), which can be expanded as the sum of an infinite series of terms giving:

$$F = A\left(d^m v^p \rho^q N^r \mu^s\right) + B\left(d^{m'} v^{p'} \rho^{q'} N^{r'} \mu^{s'}\right) + \dots,$$

where A, B, etc. are numerical constants and m, p, q, r, s are unknown powers. Since, for dimensional homogeneity, all terms must be dimensionally the same, this can be reduced to:

$$F = K d^m v^p \rho^q N^r \mu^s \qquad (1)$$

where K is a numerical constant.

The dimensions of the dependent variable F and the independent variables d, v, ρ, N, and μ are

$$[F] = [\text{Force}] = [MLT^{-2}]$$
$$[d] = [\text{Diameter}] = [L]$$
$$[v] = [\text{Velocity}] = [LT^{-1}]$$
$$[\rho] = [\text{Mass density}] = [ML^{-3}]$$
$$[N] = [\text{Rotational speed}] = [T^{-1}]$$
$$[\mu] = [\text{Dynamic viscosity}] = [ML^{-1}T^{-1}]$$

Substituting the dimensions for the variables in (1),
$$[MLT^{-2}] = [L]^m [LT^{-1}]^p [ML^{-3}]^q [T^{-1}]^r [ML^{-1}T^{-1}]^s.$$

Equating powers of [M], [L], and [T]:

$$[M], \; 1 = q + s; \qquad (2)$$
$$[L], \; 1 = m + p - 3q - s; \qquad (3)$$
$$[T], \; -2 = -p - r - s. \qquad (4)$$

Since there are five unknown powers and only three equations, it is impossible to obtain a complete solution, but three unknowns can be determined in terms of the remaining two. If we solve for p and q, we get

$$q \; = 1 - s \text{ from (2)}$$
$$p \; = 2 - r - s \text{ from (4)}$$
$$m \; = 1 - p + 3q + s = 2 + r - s \text{ substituting (2) and (4)}$$
$$\quad \text{into (3).}$$

Substituting these values in (1),

$$F = K d^{2+r-s} v^{2-r-s} \rho^{1-s} N^r \mu^s,$$

and regrouping the powers,

$$F = K\rho v^2 d^2 \left(\frac{\rho vd}{\mu}\right)^{-s} \left(\frac{dN}{v}\right)^r$$

Since s and r are unknown, this can be written

$$F = K\rho v^2 d^2 \phi\left\{\frac{\rho vd}{\mu}, \frac{dN}{v}\right\}. \tag{5}$$

At first sight, this appears to be a rather unsatisfactory solution, and (5) indicates that:

$$F = KC\rho v^2 d^2$$

where C is a constant to be determined experimentally and is dependent on the values of $\frac{\rho vd}{\mu}$ and $\frac{dN}{V}$.

EXAMPLE: A marine research facility uses a towing basin to test models of proposed ship hull configurations. A new hull shape utilizing a bulbous underwater bow is proposed for a nuclear-powered aircraft carrier that is to be 300 m long. A 3-m model has been tested in the towing tank and found to have a maximum practical hull speed of 1.4 m/s. What is the anticipated hull speed for the prototype?

SOLUTION: In the study of ship hulls, surface tension and compressibility effects are not significant. Therefore, for geometrically similar bodies, dynamic similarity occurs when:

$$\left(\frac{v^2}{\ell g}\right)_m = \left(\frac{v^2}{\ell g}\right)_p \text{ and } \left(\frac{\rho v\ell}{\mu}\right)_m = \left(\frac{\rho v\ell}{\mu}\right)_p$$

Experience has shown that the Froude number is of greater significance than the Reynolds number in this particular application. Thus, the fluid used in the towing tank is generally water. The Froude number alone is maintained between model and prototype, and empirical corrections are made to compensate for the differences that exist between the Reynolds numbers.

Hence, we ignore the viscous effects, as measured by the Reynolds number, and concentrate on the hull's wave-making characteristics, as measured by the Froude number:

$$\left(\frac{v^2}{\ell g}\right)_m = \left(\frac{v^2}{\ell g}\right)_p$$

Since the gravitational acceleration is the same for model and prototype, the anticipated prototype velocity becomes:

$$v_p = v_m \left(\frac{\ell_p}{\ell_m}\right)^{\frac{1}{2}}$$

or

$$v_p = 1.4 \ \text{m/s} \left(\frac{300\,\text{m}}{3\,\text{m}}\right)^{\frac{1}{2}} = 14 \ \text{m/s}$$

PIPE FLOW AND CHANNEL FLOW

Fluid flow can be categorized into two fundamental types: internal (or bounded) flow and external (or unbounded) flow. This chapter considers the former, where the flow is surrounded by walls, as in the flow of water through a pipe. An important dimensionless parameter in pipe flows is the **Reynolds number**, defined as

$$\text{Re} = \frac{\rho v d}{\mu} = \frac{v d}{\nu},$$

where

v = The average velocity through a cross section of the pipe

d = The pipe diameter

μ = The viscosity

ρ = the density

ν = The kinematic viscosity $\left(\nu = \frac{\mu}{\rho}\right)$

Average velocity is simply the volume flow rate Q divided by cross-sectional area $\frac{\pi d^2}{4}$. For pipe flows where Re is less than about 2,300, the flow is **laminar**, i.e., smooth and steady. For Re $\geq$ 2,300, the pipe flow becomes **turbulent**, i.e., unsteady, three-dimensional, and irregular fluctuating eddies or vortices. Laminar pipe flows can be predicted analytically, while experiments must be used to guide any attempts at analyzing turbulent flow. Most practical pipe flow problems are turbulent.

For flow along a pipe, the steady one-dimensional energy equation may be applied:

$$\frac{p_1}{\rho g} + \alpha_1 \frac{v_1^2}{2g} + z_1 = \frac{p_2}{\rho g} + \alpha_2 \frac{v_2^2}{2g} + z_2 + \frac{\dot{W}_s}{\dot{m} g} + h_{\text{total}}$$

where,

$\dot{W}_s$ = The shaft power (work per unit time) done by the fluid,

z = The vertical height,

$\dot{m}$ = The mass flow rate through the pipe,

α = The kinetic energy correction factor, and

h_{total} = The total head loss (dimensions of length) from inlet 1 to outlet 2.

$\dot{W}_s$ is positive for a turbine, which draws power *from* the fluid, but negative for a pump, which supplies power *to* the fluid. The total head loss is typically split into two parts:

$$h_{total} = \sum h_f + \sum h_m$$

The first term on the right is the sum of all the Moody-type frictional losses. These losses are associated with frictional losses along the inner wall of long, straight sections of pipe. The Moody Chart is a collection of semi-empirically obtained values of the frictional loss as a function of Reynolds number Re and pipe roughness factor $\frac{\varepsilon}{d}$. These losses are plotted in terms of the nondimensional Darcy friction factor f:

$$f = \frac{h_f}{\left(\dfrac{L}{d}\right)\left(\dfrac{v^2}{2g}\right)},$$

where L is the total length of the pipe.

The following steps should be taken to determine the head loss with known conditions:

(1) Evaluate the Reynolds number.

(2) Obtain the relative roughness, $\frac{\varepsilon}{d}$, from the Moody Chart.

(3) Obtain the friction factor, f, using the appropriate curve from the Moody Chart.

(4) Find the head loss using the friction factor.

The friction factor for laminar flow (Re < 2,300) can be approximated by:

$$f = \frac{64}{Re}.$$

The most common method of determining friction factor in turbulent flow is by use of the Moody Chart. It can also be approximated by the Swamee & Jain equation:

$$f = \frac{0.25}{\left[\log_{10}\left(\dfrac{\varepsilon/d}{3.7} + \dfrac{5.74}{Re^{0.9}}\right)\right]^2}.$$

The second component of h_{total} comes from the so-called "minor" losses. These are losses associated with parts of the piping system other than long, straight pipe sections, such as valves, bends, or elbows, sudden changes in pipe diameter, inlets, exits, etc. Again, empirical values of these losses can be obtained from tables or charts. The nondimensional minor loss coefficient K is typically the listed value, defined as:

$$K = \frac{h_m}{\dfrac{v^2}{(2g)}}$$

For a constant diameter section of the pipe system, $\sum h_m$ is simply $v^2 / (2g)$ multiplied by $\sum K$, the sum of all the minor loss coefficients along the pipe section.

For pipes of different diameters in *series*, volume flow rate Q must be the same along each section (for steady flow in the mean), and h_f and h_m must be summed independently in each section, then added to obtain the total head loss h_{total}. For pipes in *parallel*, however, the volume flow rate may be different in each parallel section. If the parallel pipes branch off at one point A and later rejoin at a point B as shown in Figure 14:

$$Q_A = Q_B = Q_1 + Q_2 + Q_3,$$

whereas the total head, along any of the three pipes from A to B must be identical since p_A and p_B are the same regardless of which pipe (1, 2, or 3) is under consideration. Another general rule is that the net volume flow rate into any junction must be zero, analogous to the statement that the net current into any junction in an electrical circuit must be zero. Also, the net head loss around any closed loop must be zero, just as in electrical circuits the net voltage drop around any closed loop must be zero.

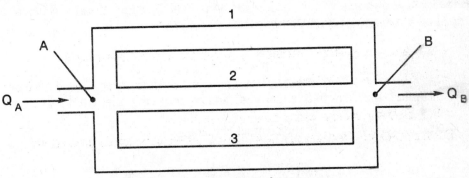

Figure 14. Parallel pipe system

For pipes with a cross section other than circular, the **hydraulic diameter** is defined as:

$$D_h = \frac{4A}{P},$$

where

A = The cross-sectional area,

P = The wetted perimeter.

Wetted perimeter refers to the portion of the perimeter in contact with the fluid. Once D_h has been calculated, the Moody Chart can be used to obtain friction factors based on D_h, which is basically the diameter of an equivalent round pipe that would give the same losses as the actual nonround pipe.

Open channel flow is somewhat similar to pipe flow except that, with a free surface exposed to atmospheric pressure, there can be no streamwise pressure gradient. The fluid flows due to gravity alone, with the flow rate determined by a balance between gravitational and frictional forces. A reasonable analysis can be obtained by using the Moody Chart for pipe flow with the hydraulic diameter of the channel. Instead, engineers prefer the **hydraulic radius**, defined as one-fourth of the hydraulic diameter:

$$\text{hydraulic radius} = R_h = \frac{A}{P} = \frac{D_h}{4},$$

where

A = The cross-sectional area of the fluid in the channel

P = The wetted perimeter (which does *not* include the free surface).

For uniform flow in long, straight, inclined channels of a constant shape, the average velocity v at any streamwise location remains constant, and the energy equation between two streamwise locations 1 and 2 reduces to:

$$h_f = y_1 - y_2 = L \sin \alpha,$$

where h_f is the frictional head loss which is exactly balanced by the change in surface height $y_1 - y_2$. In the equation above, L is the streamwise distance from 1 to 2 and α is the inclination angle with respect to the horizontal. Frictional head loss can also be expressed in terms of the Darcy friction factor f as:

$$h_f = f \frac{Lv^2}{8gR_h}.$$

Combining the last two equations, and introducing the Chézy constant $C = (8g / f)^{\frac{1}{2}}$, the velocity in the channel is found:

$$v = C(R_h S)^{\frac{1}{2}} = C(R_h \sin \alpha)^{\frac{1}{2}},$$

where S is the slope or hydraulic gradient. Typically, volume flow rate Q is the unknown in a channel flow problem. The procedure is to find the Chézy constant C for the given channel, and then to use the above equation to find v. Finally, $Q =$ volume flow rate $= vA$.

In some cases, direct empirical relationships for C are available, but in most cases, C is found by the Manning correlation:

$$C = \frac{1.49}{n} R_h^{\frac{1}{6}} \ (R \text{ in units of feet}),$$

when n is the Manning coefficient, a nondimensional roughness coefficient that can be obtained from empirical tables.

In practice, the calculation of Chézy's coefficient can be bypassed by combining the equation for velocity and the Manning correlation:

$$v = \frac{1.49}{n} R_h^{\frac{2}{3}} (\sin\alpha)^{\frac{1}{2}} \quad (R_h \text{ in ft, } v \text{ in ft/s})$$

or

$$v = \frac{1.0}{n} R_h^{\frac{2}{3}} (\sin\alpha)^{\frac{1}{2}} \quad (R_h \text{ in m, } v \text{ in m/s})$$

To find the volume flow rate, look up n for the channel (n depends greatly on the amount of roughness on the walls of the channel), calculate v using the appropriate equation above for inclination angle α and hydraulic radius R_h, and finally calculate $Q = vA$.

Another feature unique to open channel flows is the hydraulic jump. Here, a high-velocity supercritical flow (the Froude number is greater than one) is suddenly slowed to subcritical ($F_r < 1$) as the fluid depth increases and the velocity decreases across the hydraulic jump. The ratio of fluid heights downstream (y_2) and upstream (y_1) of a stationary hydraulic jump is:

$$\frac{y_2}{y_1} = \frac{1}{2}\left[\left(1 + 8F_{r_1}^2\right)^{\frac{1}{2}} - 1\right]$$

where F_{r_1} is the Froude number upstream:

$$F_{r_1} = \left[\frac{v_1^2}{(y_1 g)}\right]^{\frac{1}{2}}.$$

In problems where the hydraulic jump is not stationary as, for example, with a surge caused by a sudden gate closure, it is most convenient to transform the frame of reference to that of a stationary hydraulic jump, apply the fluid height ratio above, and then transform back to the original frame of reference.

EXAMPLE: A water transmission pipe having the diameter shown conducts water with flow rate of 0.5 m³/s. The relative roughness of the pipe, $\frac{\varepsilon}{D}$, is 3×10^3. Find the pressure loss over unit length of the pipe.

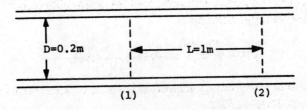

Figure 15. Dimensions of pipe

SOLUTION: Applying the general flow equation to compute the total head loss,

$$h_{total} = \frac{p_1}{\rho g} + \alpha_1 \frac{v_1^2}{2g} + h_1 - \frac{p_2}{\rho g} - \alpha_2 \frac{v_2^2}{2g} - h_2. \tag{1}$$

Assuming that the pipe is horizontal, then $h_1 = h_2$. With uniform internal pipe cross section, $v_1 = v_2$ since $A_1 = A_2$, and flow rate Q is constant throughout the pipe length. Substituting these conditions into (1), we obtain:

$$h_{total} = \frac{1}{\rho g}(p_1 - p_2) \tag{2}$$

Now,

$$h_{total} = f \frac{L}{D} \frac{v^2}{2g} \tag{3}$$

and, therefore,

$$p_1 - p_2 = \rho h g = f \rho \frac{L}{d} \frac{v^2}{2}. \tag{4}$$

The friction factor, f, is a function of the Reynold's number and the relative roughness, $\frac{\varepsilon}{d}$. The Reynold's number, Re, is obtained from:

$$Re = \frac{\rho v d}{\mu}. \tag{5}$$

The velocity of flow may be calculated from the flow rate equation:

$$v = \frac{Q}{A} = \frac{0.5}{\pi(.1)^2} = 15.95 \ \ m/s$$

The parameters $\rho = 9.99 \times 10^2 \ Kg/m^3$ and $\mu = 10 \times 10^{-4} \ Kg/m\text{-sec}$ can be assumed for water at 20°C. The Reynold's number, therefore, becomes

$$Re = \frac{(9.99 \times 10^2)(0.2)(15.95)}{10 \times 10^{-4}}$$

$$= 3.18 \times 10^6.$$

Referring to the Moody chart to obtain the friction factor when knowing the Reynold's number and the relative roughness, we find $f = .014$. Substituting values into Equation (4),

$$p_1 - p_2 = \frac{(.014)(9.99 \times 10^2)(15.95^2)}{(0.2)(2)}$$

$$= 8.89 \times 10^3 \, Pa.$$

EXAMPLE: What slope is required to produce a flow of 400 cfs at a uniform depth of 4 ft in a trapezoidal earth channel with a base width of 6 ft and side slopes of 1 vertical on 2 horizontal? Use a Manning coefficient of $n = 0.025$.

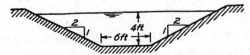

Figure 16. Dimensions of channel

SOLUTION: The Chézy formula is:

$$Q = AC\sqrt{R_h S} \qquad (1)$$

The Manning correlation is:

$$C = 1.5\frac{R_h^{\frac{1}{6}}}{n} \qquad (2)$$

The hydraulic radius is:

$$R_h = \frac{A}{P} = \frac{\frac{4(22+6)}{2}}{6+2\sqrt{4^2+8^2}} = \frac{56}{23.9} = 2.34 \text{ ft}.$$

From Equation (2),

$$C = 1.5\frac{R_h^{\frac{1}{6}}}{n} = 1.5\frac{2.34^{\frac{1}{6}}}{0.025} = 69 \text{ ft}^{\frac{1}{2}}/\text{sec}.$$

Hence, upon solving Equation (1) for S,

$$S = \frac{Q^2}{C^2 A^2 R_h} = \frac{400^2}{69^2 \times 56^2 \times 2.34} = 0.0046.$$

FLOW MEASUREMENT

There are two categories of devices that measure fluid flows: local velocity meters and volume flow meters. In addition, manometers are used to measure pressure in fluid flows as discussed previously in the fluid statics section.

Consider now the **pitot tube**, which is simply a slender tube with a hole in the front, aligned with the flow. The pressure at the nose of any body in an incompressible flow is the stagnation pressure p_o. Thus, p_o can be measured by connecting the opposite end of the pitot tube to a pressure meter (such as a manometer). The static pressure p can also be measured in a flow, either by a separate pressure tap or with additional holes in the pitot probe itself.

Bernoulli's equation is used to calculate the velocity from the pressure difference between p_o and p as follows (neglecting gravity):

$$p_o = p + \frac{1}{2}\rho v^2,$$

hence,

$$v = \left(\frac{2(p_o - p)}{\rho} \right)^{\frac{1}{2}}.$$

The pitot tube and pitot-static tube (which contains holes for both stagnation and static pressure) are local-velocity meters since they can easily be traversed through the fluid flow. Often it is only necessary to measure the volume flow rate in a pipe flow. The three most common volume flow meters are the orifice meter, the venturi meter, and the flow nozzle. All three work on the principle that pressure in an incompressible flow decreases as the velocity increases through a throat of a smaller area than the pipe's cross-sectional area. This is nothing more than Bernoulli's equation, and so the three devices are called Bernoulli obstruction devices.

In all three devices, a static pressure tap is located near the throat, and a second tap is located just upstream of the device. If there were no losses, the pressure difference between these two would give the average velocity (and, hence, the volume flow rate) by Bernoulli's equation and the integral conservation of mass. Of course, fluid that flows through any real device will not be inviscid, and frictional losses are taken into account through a **discharge coefficient** C_d which is always less than 1.0. (Empirical formulae for C_d can be obtained for any of the three devices.) The final expression for the volume flow rate is:

$$Q = C_d A_t \left[\frac{2(p_1 - p_2)}{\rho(1 - \beta^4)} \right]^{\frac{1}{2}},$$

where

p_1 and p_2 = The pressure upstream of the throat and near the throat respectively,

A_t = The throat (minimum) area, and

β = The ratio of the throat diameter to the upstream pipe diameter.

Three coefficients commonly used with volume flow meters are defined as follows. The coefficient of contraction is defined as:

$$C_c = \frac{A_v}{A_t},$$

where

A_v = Area contracted

A_t = Meter throat area.

The coefficient of velocity is defined as:

$$C_v = \frac{v_{2\,\text{actual}}}{v_{2\,\text{ideal}}},$$

where

$v_{2\,\text{actual}}$ = Actual mean velocity of the flow

$v_{2\,\text{ideal}}$ = Ideal velocity (no friction).

The coefficient of discharge is defined as:

$$C_d = \frac{Q_{\text{actual}}}{Q_{\text{ideal}}},$$

where

Q_{actual} = Actual flow rate

Q_{ideal} = Ideal flow rate,

also

$$C_d = C_v C_c.$$

Note: The value of the coefficients are determined experimentally.

A **weir** is an obstruction in the bottom of a channel over which the flow must deflect. The volume flow rate per unit width, q, is proportional to $H^{\frac{3}{2}}$ for a sharp-crested weir, i.e.:

$$q = \frac{2}{3} C_W (2g)^{\frac{1}{2}} H^{\frac{3}{2}},$$

where H is the height of the upstream flow above the crest of the weir, and C_W is the weir coefficient. The weir coefficient is an empirically determined coefficient that accounts for the losses associated with end effects, friction, etc. Typically,

$$C_W \approx 0.611 + \frac{0.075H}{Y},$$

where Y is the height of the crest of the weir. Once q is determined, the total discharge (or volume flow rate) Q is then simply q multiplied by the width of the weir.

For a broad-crested weir, i.e., one with a flat top, the surface of the fluid on top of the weir usually sinks to a height h above the crest, where $h \approx \frac{2H}{3}$. The discharge q per unit width then becomes:

$$q = c(2g)^{\frac{1}{2}} h^{\frac{3}{2}},$$

where c is the discharge coefficient of the weir, which must be empirically determined and is usually tabulated as a function of weir geometry (c also varies with *h*).

For flow through small orifices, the flow rate, Q, is given by

$$Q = C_D a (2gh)^{\frac{1}{2}}$$

where

a = The cross-sectional area of the orifice,

h = The difference in the head from one side of the orifice to the other, and

C_D = A discharge coefficient which again must be determined empirically.

EXTERNAL FLOW

The *no-slip* boundary condition requires that the velocity of a fluid immediately adjacent to a solid wall be equal to the velocity of the wall itself. In the usual frame of reference where the wall is stationary with fluid flowing over it, the fluid velocity right next to the wall must be zero. This no-slip condition leads to what is referred to as a **boundary layer**. A boundary layer is a thin fluid layer near the wall which experiences velocity variations. Inside a boundary layer, the fluid velocity goes from some finite value at the boundary layer edge to zero at the wall in a very short distance. Since viscous shear stress is proportional to viscosity μ and velocity gradient, the shear stress is quite large in a boundary layer, especially very close to the wall where the velocity gradient is steepest. This shear stress, by Newton's third law, imposes a frictional drag force on the wall in the same direction as the flow above the boundary layer.

Boundary layers can be either laminar (smooth and steady) or turbulent (quite unsteady and irregular). For uniform flow along a semi-infinite flat plate, the laminar boundary layer solution can be obtained exactly (albeit with the help of a digital computer), but turbulent boundary layers are too complex to solve exactly, even with the fastest computers. Turbulent boundary layer results are found empirically or semi-empirically.

For engineering analyses, the three quantities of most significance are the boundary layer thickness δ, the skin friction coefficient C_f, and the displacement thickness $\delta*$. $\delta*$ is usually defined as the distance from the wall where the velocity μ has increased to 99 percent of the freestream velocity v. Letting τ_w denote the shear stress acting on the wall by the fluid, C_f is the skin friction coefficient:

$$C_f = \frac{2\tau_w}{\rho v^2}.$$

Displacement thickness is defined as the distance to which streamlines outside the boundary layer are displaced away from the wall, and results from the fact that the fluid inside the boundary layer carries less mass flow than it would have in the absence of the wall. See Figures 17 and 18. The displacement thickness is:

$$\delta^* = \int_0^\infty \left(1 - \frac{u}{v}\right) dy.$$

For laminar flow on a flat plate, expressions for δ, C_f, and δ^* to be used for engineering calculations according to the Blasius solution are:

$\dfrac{\delta}{x}$	C_f	$\dfrac{\delta^*}{x}$
$\dfrac{4.91}{\text{Re}_x^{\frac{1}{2}}}$	$\dfrac{1.328}{\text{Re}_x^{\frac{1}{2}}}$	$\dfrac{1.73}{\text{Re}_x^{\frac{1}{2}}}$

where $\text{Re}_x = v\,x/v$ is the Reynolds number based on length from the plate leading edge.

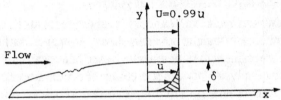

Figure 17. Boundary layer thickness

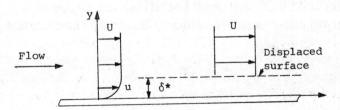

Figure 18. Displacement thickness

At a Reynolds number greater than about 300,000 based on v and x (the distance along the plate in the flow direction), the laminar boundary layer begins to oscillate and becomes turbulent. Empirical expressions for δ, C_f, and δ^* can be found in the literature for turbulent flow.

Skin friction is not the only source of drag on bodies—such as automobiles, baseballs, submarines, and airplanes—moving through a fluid. The uneven distribution of pressure forces along the body surface can produce significant (often dominating) drag forces as well. For nonstreamlined or blunt bodies in particular, the boundary layer along the body surface cannot remain attached and separates off the surface. This leads to a gross imbalance of pressure (pressure being very high on the front end and very low on the back end of the body) and a large pressure drag.

The total aerodynamic drag on a body usually must be found by experimentation.

Drag is expressed nondimensionally by a **drag coefficient**, C_D, defined as:

$$C_D = \frac{\text{Drag Force}}{\frac{1}{2}\rho v^2 A}$$

where

v = The freestream velocity

A = An area,

typically the projected frontal area, but sometimes (as in the case of airplane wings or flat plates) the platform area as listed in the last table. The drag coefficient has been determined to be:

$$C_D = \frac{1.328}{\sqrt{\text{Re}_x}}$$

for laminar boundary layer flow over a flat plate.

In general, a flow with a laminar boundary layer produces much less skin friction drag than a flow with a turbulent boundary layer. However, turbulent boundary layers are much more resilient to flow separation, and hence can lead to less pressure drag. In engineering analysis, one can sometimes force the flow to be turbulent in order to decrease the overall drag. The dimples on a golf ball are one such example. The dimples force the boundary layer to be turbulent, which delays separation and decreases the pressure drag. Since pressure drag dominates on bluff bodies such as spheres, the net effect is a decrease in total drag.

Lift, an upward force that is exerted on an object as it passes through a fluid, can be analyzed in much the same way as drag. Namely, a lift coefficient, C_L, is defined as:

$$C_L = \frac{\text{Lift Force}}{\frac{1}{2}\rho v^2 A}.$$

EXAMPLE: Air at 100°F is flowing over a flat plate 1 ft wide. Estimate the boundary layer thickness 1 ft from the leading edge. Also, determine the drag force. The air speed is 7.2 ft/sec.

SOLUTION: The boundary layer thickness δ can be estimated from the equation:

$$\frac{\delta}{x} = \frac{5}{\sqrt{\text{Re}_x}}$$

The kinematic viscosity of air at 100°F is $\nu = 1.8 \times 10^{-4}$ ft²/sec. The density is $\rho = 2.20 \times 10^{-3}$ slug/ft³. Therefore,

$$\text{Re}_x = \frac{v_0 x}{\nu} = \frac{7.2 \times 1}{1.8 \times 10^{-4}} = 4 \times 10^4$$

and

$$\delta = \frac{5 \times 1}{2 \times 10^2} \text{ft} = 0.025 \text{ft}.$$

In general, the drag coefficient is:

$$C_D = \frac{\text{drag}}{\frac{1}{2}\rho v^2 A}$$

where

v = The upstream velocity and

A = The cross-sectional area.

For laminar boundary layer flow over a flat plate, the drag coefficient has been found to be:

$$C_D = \frac{1.328}{\sqrt{\text{Re}_x}}.$$

Therefore:

$$C_D = \frac{1.328}{\sqrt{4 \times 10^4}}$$

$$= 6.6 \times 10^{-3},$$

and the drag force is:

$$\begin{aligned} \text{Drag} &= .5\rho \, C_D v^2 A \\ &= (.5)(2.2 \times 10^{-3})(6.6 \times 10^{-3})(7.2^2)(1^2) \\ &= 3.8 \times 10^{-4} \text{ lbf.} \end{aligned}$$

EXAMPLE: A parachutist weighs 175 lb and has a projected frontal area of 2 ft² in free fall. His drag coefficient based on frontal area is found to be 0.80. If the air temperature is 70°F, determine his terminal velocity.

SOLUTION: It is convenient to express the drag of a bluff body in terms of a nondimensional parameter, C_D, called drag coefficient:

$$C_D = \frac{\text{Drag}}{\frac{1}{2}\rho v^2 A}$$

or

$$\text{Drag} = C_D \frac{1}{2}\rho v^2 A,$$

with A the projected frontal area of the bluff body normal to the flow direction.

At terminal velocity, the parachutist's weight is balanced by his drag:

$$W = C_D \frac{1}{2}\rho v^2 A$$

The density of air at normal atmospheric pressure and 70° F is $\rho = 0.00233$ slug/ft³. Therefore,

$$v^2 = \frac{W}{C_D\left(\frac{1}{2}\right)\rho A}$$

$$= \frac{175}{0.80\left(\frac{1}{2}\right)(.00233)(2)}$$

and

$$v_{\text{terminal}} = 30 \text{ ft/s}$$

COMPRESSIBLE FLOW

There are several flow phenomena, such as choking, shock waves, etc., which occur only when a fluid flow is highly compressible. Compressibility becomes important when the **Mach number** becomes greater than about 0.3. The Mach number, defined earlier in the Force Ratio table, is the ratio of an object's speed to the speed of sound in the medium through which the object is traveling:

$$M = \frac{v}{a}.$$

For a perfect gas, the speed of sound a is:

$$a = (kRT)^{\frac{1}{2}}$$

where

k = The ratio of specific heats C_p/C_v,

R = The gas constant, and

T = The *absolute* temperature.

When M is less than one, the flow is subsonic, while supersonic flows are those with Mach numbers greater than one.

Most compressible flow problems encountered by engineers involve the flow of a gas in a duct. Of these, three different simplifications enable the analysis of three primary categories of compressible duct flow: isentropic flow in a duct of a changing area, adiabatic flow in a duct of a constant area (with friction), and frictionless flow in a constant area duct with heat transfer. When none of these simplifications can be made, the problem is much more complicated. Isothermal duct flow with friction is one case where there is heat transfer as well as frictional effects, but this can be analyzed.

For the case of adiabatic flow in a constant area duct with friction, the stagnation enthalpy of the fluid must remain constant since no energy is added and no work is done. This leads to the rather unique result that the Mach number always approaches unity toward the end of the duct. This applies to both subsonic flow (where the Mach number will increase toward one) and supersonic flow (where M will decrease toward one). Problems of this type are attacked by utilizing L^*, **the sonic length**, defined as the duct length required to develop from some initial Mach number to $M = 1$. The dimensionless parameter fL^*/D is tabulated as a function of the Mach number where f is the Darcy friction factor obtainable from the Moody Chart, and D is the diameter of a round duct or the hydraulic diameter of a nonround duct. For cases where the duct is not long enough to reach sonic conditions at its exit, the relationship between the duct length and the Mach numbers M_1 and M_2 at the inlet and outlet of the duct respectively is:

$$\frac{fL}{D} = \left(\frac{fL^*}{D}\right)_{M_1} - \left(\frac{fL^*}{D}\right)_{M_2}.$$

Constant area duct flow problems, where friction can be ignored but heat is added or subtracted, must be addressed using the integral (control volume) conservation laws of mass, momentum, and energy, where heat transfer rate $\dot{Q}$ appears in the energy equation. Ratios of temperatures, stagnation temperatures, pressures, stagnation pressures, and velocities can be found as functions of the Mach number and are tabulated for air and aid in the solution of problems.

The most common category of flows encountered is adiabatic, isentropic flow in a duct of a varying area. Thermodynamic relationships, combined with the integral conservation laws of mass and energy, lead to expressions of temperature, pressure, and density ratios as functions of the Mach number and the ratio of specific heats, k. These are tabulated in the form T/T_o, p/p_o, ρ/ρ_o, and A/A^* as functions of M, where the subscript o denotes total properties and the asterisk denotes the value at $M = 1$. The only way to attain supersonic flow in a duct of this kind is by first passing through a converging section, then a throat, followed by a diverging section. The flow upstream of the throat will be subsonic, and the flow downstream may reach supersonic conditions if the pressure drop is sufficient. The condition $M = 1$ can only occur at the throat, and the throat area is then defined as A^*. If the flow remains isentropic, one can find all desired quantities for a given A/A^* by using the isentropic tables or equations to find M, and then using M to find all other quantities.

TABLE 4
BASIC EQUATIONS FOR ISENTROPIC FLOW OF AN IDEAL GAS

Continuity:	$\rho_1 v_1 A_1 = \rho_2 v_2 A_2 = \dot{m}$
Momentum:	$R_x + p_1 A_1 - p_2 A_2 = \dot{m} v_2 - \dot{m} v_1$
First Law:	$h_1 + \dfrac{v_1^2}{2} = h_2 + \dfrac{v_2^2}{2}$
Second Law:	$s_1 = s_2$
Equation of State:	$p = \rho RT$
Process Equation:	$\dfrac{p}{\rho^k} = \text{constant}$

TABLE 5
ISENTROPIC FLOW RATIOS

Stagnation pressure:
$$\frac{p_0}{p}=\left[1+\frac{k-1}{2}M^2\right]^{\frac{k}{(k-1)}}$$

Stagnation temperature:
$$\frac{T_0}{T}=1+\frac{k-1}{2}M^2$$

Stagnation density:
$$\frac{\rho_0}{\rho}=\left[1+\frac{k-1}{2}M^2\right]^{\frac{1}{(k-1)}}$$

In addition, the ratio of gas velocity at some point to the sonic velocity in the throat is:

$$\frac{v}{a^*}=\left[\frac{\frac{(k+1)}{2}M^2}{\frac{(k-1)}{2}M^2+1}\right]^{\frac{1}{2}},$$

and the ratio A/A^* is:

$$\frac{A}{A^*}=\frac{1}{M}\left[\frac{1+\frac{k-1}{2}M^2}{1+\frac{k-1}{2}}\right]^{\frac{(k+1)}{2(k-1)}}$$

since,

$$\rho A v = \text{constant}.$$

If the downstream pressure is not low enough to attain supersonic flow throughout the entire length of the diverging nozzle section, a normal shock wave will form in the diverging section. A normal shock leads to a sudden rise in temperature and pressure, and the flow abruptly changes from supersonic to subsonic. Tables are available for the ratios of temperature, pressure, etc., across a normal shock wave. These, combined with the isentropic tables, are used to find the Mach numbers upstream and downstream of the shock. The flow is assumed to be isentropic everywhere except across the shock where entropy increases significantly, resulting in a great loss of stagnation pressure.

EXAMPLE: An airflow ($k = 1.4$) is expanded isentropically in a nozzle from $M_1 = 0.3$, $A_1 = 1.0$ ft^2, to a Mach number M_2 of 3.0. Determine (a) the minimum nozzle area, (b) A_2, (c) $\frac{p_2}{p_1}$ and (d) $\frac{T_2}{T_1}$ (see figure).

Figure 19. Airflow through nozzle

SOLUTION: (a) Since flow in the nozzle goes from subsonic to supersonic speeds, the flow must pass through a minimum area $A*$ at which $M = 1$. From Table 6, at $M_1 = 0.3$, $A_1/A* = 2.0351$ so that the minimum area = $1/2.0351 = 0.491$ ft^2.

		TABLE 6					
		$\dfrac{p}{p_t}$ when $k = 1.4$					
M	$\dfrac{p}{p_t}$	$\dfrac{T}{T_1}$	$\dfrac{A}{A_0}$	M	$\dfrac{p}{p_t}$	$\dfrac{T}{T_1}$	$\dfrac{A}{A_0}$
0	1.0000	1.0000	∞	0.30	0.9395	0.9823	2.0351
.01	.9999	1.0000	57.8738	.31	.9355	.9811	1.9765
.02	.9997	.9999	28.9421	.32	.9315	.9799	1.9219
.03	.9994	.9998	19.3005	.33	.9274	.9787	1.8707
.04	.9989	.9997	14.4815	.34	.9231	.9774	1.8229
.05	.9983	.9995	11.5914	.35	.9188	9761	1.7780
.06	.9975	.9993	9.6659	.36	.9143	.9747	1.7358
.07	.9966	.9990	8.2915	.37	.9098	.9733	1.6961
.08	.9955	.9987	7.2616	.38	.9052	.9719	1.6587
.09	.9944	.9984	6.4613	.39	.9004	.9705	1.6234
.10	.9930	.9980	5.8218	.40	.8956	.9690	1.5901
.11	.9916	.9976	5.2992	.41	.8907	.9675	1.5587
.12	.9900	.9971	4.8643	.42	.8857	.9659	1.5289
.13	.9883	.9966	4.4969	.43	.8807	.9643	1.5007
.14	.9864	.9961	4.1824	.44	.8755	.9627	1.4740

TABLE 7

$$\frac{T}{T_t} \text{ when } k = 1.4$$

M or M_1	$\dfrac{p}{p_t}$	$\dfrac{T}{T_1}$	$\dfrac{A}{A_0}$
2.90	0.3165^{-1}	0.3729	3.850
2.91	$.3118^{-1}$	.3712	3.887
2.92	3071^{-1}	.3696	3.924
2.93	$.3025^{-1}$	.3681	3.961
2.94	$.2980^{-1}$	.3665	3.999
2.95	$.2935^{-1}$	.3649	4.038
2.96	$.2891^{-1}$	.3633	4.076
2.97	$.2848^{-1}$	.3618	4.115
2.98	$.2805^{-1}$	.3602	4.155
2.99	$.2764^{-1}$	.3587	4.194
3.00	$.2722^{-1}$	.3571	4.235
3.01	$.2682^{-1}$	.3556	4.275
3.02	$.2642^{-1}$	.3541	4.316
3.03	$.2603^{-1}$	.3526	4.357
3.04	$.2564^{-1}$	.3511	4.399

(b) At $M_2 = 3.0$, $\dfrac{A_2}{A*} = 4.235$ so $A_2 = 2.08$ ft².

(c) For this isentropic flow, T_t and p_t are constants.

$$\frac{p_2}{p_1} = \frac{\dfrac{p_2}{p_{t_2}}}{\dfrac{p_1}{p_{t_1}}} = \frac{0.0272}{0.9395} = 0.0290 \text{ (see the Table for } \frac{p}{p_t})$$

d) $$\frac{T_2}{T_1} = \frac{\dfrac{T_2}{T_{t_2}}}{\dfrac{T_1}{T_{t_1}}} = \frac{0.3571}{0.9823} = 0.363 \text{ (see the Table for } \frac{T}{T_t})$$

REVIEW PROBLEMS

PROBLEM 1

The specific weight of water at ordinary pressure and temperature is 62.4 lbf/ft³ (9.81 kN/m³). The specific gravity of mercury is 13.55. Compute the density of water and the specific weight and density of mercury.

SOLUTION: Knowing that density and specific weight of a fluid are related as follows:

$$\rho = \frac{\gamma}{g} \text{ or } \gamma = \rho g$$

and that specific gravity s of a liquid is the ratio of its density to that of pure water at a standard temperature, we can calculate:

$$\rho_{water} = \frac{\gamma_{water}}{g} = \frac{62.4 \text{ lbf/ft}^3}{32.2 \text{ ft/s}^2} = 1.94 \text{ slugs/ft}^3$$

$$\rho_{water} = \frac{9.81 \text{ kN/m}^3}{9.81 \text{ m/s}^2} = 1000 \text{ kg/m}^3 = 1.00 \text{ g/cm}^3$$

$$\rho_{mercury} = s_{mercury}\rho_{water} = 13.55(1.94) = 26.3 \text{ slugs/ft}^3$$

$$\rho_{mercury} = 13.55(1,000) = 13,550 \text{ kg/m}^3$$

$$\gamma_{mercury} = \rho_{mercury}g = 26.3(32.2) = 846.9 \text{ lbf/ft}^3$$

$$\gamma_{mercury} = (13,550)(9.81) = 132.4 \text{ kN/m}^3.$$

PROBLEM 2

Oil with a specific gravity of 0.80 is 3 ft (0.91 m) deep in an open tank that is otherwise filled with water. If the tank is 10 ft (3.05 m) deep, what is the pressure at the bottom of the tank?

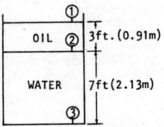

Figure 20. Tank schematic

SOLUTION: First determine the pressure at the oil-water interface staying within the oil, and then calculate the pressure at the bottom.

$$\frac{p_1}{\gamma} + z_1 = \frac{p_2}{\gamma} + z_2$$

where

p_1 = Pressure at free surface of oil

z_1 = Elevation of free surface of oil

p_2 = Pressure at interface between oil and water

z_2 = Elevation at interface between oil and water.

For this example, $p_1 = 0$, $\gamma = 0.80 \times 62.4$ lbf/ft³, $z_1 = 10$ ft, and $z_2 = 7$ ft. Therefore,

$$p_2 = 3 \times 0.80 \times 62.4 = 150 \text{ psfg.}$$

Now, obtain p_3 from:

$$\frac{p_2}{\gamma} + z_2 = \frac{p_3}{\gamma} + z_3,$$

where p_2 has already been calculated and $\gamma = 62.4$ lbf/ft³:

$$p_3 = 62.4\left(\frac{150}{62.4} + 7\right)$$

$$= 587 \text{psfg}$$

$$= 4.07 \text{psig.}$$

PROBLEM 3

A hydraulic turbine operates from a water supply with a 200-ft head above the turbine inlet, as shown in Figure 21. It discharges the water to atmosphere through a 12-in. diameter duct, with a velocity of 45 fps. Calculate the horsepower output of the turbine.

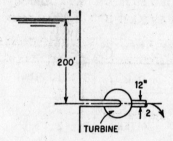

Figure 21. Hydraulic turbine

SOLUTION: If E is the energy extracted per pound of fluid flowing, then the energy equation may be written as

$$\frac{p_1}{\gamma} + \frac{v_1^2}{2g} + z_1 = E + \frac{p_2}{\gamma} + \frac{v_2^2}{2g} + z_2$$

where suffix 1 refers to a point upstream of the turbine, and suffix 2 to a point downstream of the turbine.

If the exit from the turbine is defined as the potential datum, then $z_2 = 0$ and:

$$\frac{p_1}{\gamma} + \frac{v_1^2}{2g} + z_1 = 200 \text{ ft}$$

where $p_1 = 0$ and $v_1 = 0$ at the surface.

Therefore:

$$200 \text{ ft} = E + \frac{p_2}{\gamma} + \frac{v_2^2}{2g} + 0.$$

Now since the discharge is to atmosphere, $p_2 = 0$. Hence:

$$200 = E + \frac{45^2}{2g}$$

Thus:

$$E = 200 - 31.3 = 168.7 \text{ ft.}$$

The rate of fluid flow is given by $Q = Av$ cfs, and the weight of fluid flowing by $Q\gamma$ lbf/s. Therefore, the rate of work done on the turbine is $EQ\gamma$ ft-lbf/s or:

$$\frac{EQ\gamma}{550} \text{hp} = \frac{(168.7)(\pi)(45)(62.4)}{(4)(550)} = 675 \text{ hp.}$$

PROBLEM 4

Consider the steady flow of water ($\rho = 1.94$ slug/ft^3) through the device shown in the diagram. The areas are $A_1 = 0.3$ ft^2, $A_2 = 0.5$ ft^2, and $A_3 = A_4 = 0.4$ ft^2. Mass flow out through section 3 is given as 3.88 slug/s. The volumetric flow rate in through section 4 is given as 1 ft^3/s, and $V_1 = 10\hat{i}$ ft/s. If properties are assumed uniform across all inlet and outlet flow sections, determine the flow velocity at section 2.

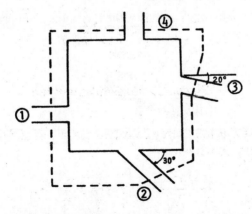

Figure 22. Control volume diagram

SOLUTION: The dashed lines in the figure represent a control volume. Equation (1) represents the control volume formulation of the conservation of mass.

$$0 = \frac{\partial}{\partial t} \int_{cv} \rho dv + \int_{cs} \rho v \times dA \qquad (1)$$

Since the flow is steady (not time dependent), Equation (1) becomes

$$\int_{A_2} \rho v \bullet dA = 3.88 \frac{slug}{s}$$

Since $v \bullet dv$ is positive at section 2, the flow is out.

Evaluating the integral at section 2, v_2 can be found:

$$\rho v_2 A_2 = 3.88 \frac{slug}{s}$$

$$|v_2| = \left(3.88 \frac{slug}{s}\right)\left(\frac{ft^3}{1.94\ slug}\right)\left(\frac{1}{0.6\ ft^2}\right) = 3.33 \frac{ft}{s}$$

$$v_2 = |v_2|(\sin\theta i - \cos\theta j)$$

$$v_2 = (1.66i - 2.88j)\frac{ft}{s}$$

PROBLEM 5

A submarine-launched missile, 1 m diameter by 5 m long, is to be studied in a water tunnel to determine the loads acting on it during its underwater launch. The maximum speed during this initial part of the missile's flight is 10 m s^{-1}. Calculate the mean water tunnel flow velocity if a $\frac{1}{20}$ scale model is to be employed and dynamic similarity is to be achieved.

SOLUTION: For dynamic similarity, the Reynolds number must be constant for the model and the prototype:

$$Re_m = Re_p,$$

$$\frac{v_m l_m \rho_m}{\mu_m} = \frac{v_p l_p \rho_p}{\mu_p}.$$

The model flow velocity is given by

$$v_m = v_p\left(\frac{l_p}{l_m}\right)\left(\frac{\rho_p}{\rho_m}\right)\left(\frac{\mu_m}{\mu_p}\right),$$

but $\rho_p = \rho_m$ and $\mu_p = \mu_m$. Therefore,

$$v_m = (10)(20)(1)(1) = 200m\ s^{-1}.$$

This is a high-flow velocity and illustrates the reason why a few model tests are made with completely equal Reynolds numbers. At high Re values, however, the divergences become of lesser importance.

PROBLEM 6

Water is discharged from a large reservoir through a straight pipe of 3 in. diameter and 1,200 ft long at a rate of 12 cfm. The discharge end is open to the atmosphere. If the open end is 40 ft below the surface level in the reservoir, what is the Darcy friction factor? Losses other than pipe friction may be ignored.

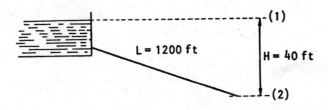

Figure 23. Flow into reservoir

SOLUTION: Applying the flow equation between the levels 1 and 2, as shown in the figure

$$z_1 + \frac{p_1}{\gamma} + \frac{v_1^2}{2g} = z_2 + \frac{p_2}{\gamma} + \frac{v_2^2}{2g} + h_f$$

Both ends of the system are open to the atmosphere, then:

$$p_1 = p_2.$$

Ignoring losses other than pipe friction, and using the Darcy equation:

$$h_f = 4f\left(\frac{L}{D}\right)\frac{v^2}{2g}$$

where v is the velocity in the pipe. It follows that $v_2 = v$ since

$$z_1 - z_2 = \frac{p_2}{\gamma} + \frac{v_2^2}{2g} + h_f - \frac{p_1}{\gamma} - \frac{v_1^2}{2g}$$

and

$$p_1 = p_2,$$

then

$$z_1 - z_2 = \frac{v_2^2}{2g} - \frac{v_1^2}{2g} + h_f.$$

as $v_1 = 0$ and $H = z_1 - z_2$:

$$H = \frac{v^2}{2g} + 4f\left(\frac{L}{D}\right)\frac{v^2}{2g}$$

$$H = \frac{v^2}{2g}\left(1 + 4f\frac{L}{D}\right)$$

$$Q = \frac{12}{60} = 0.2 \text{ cfs}$$

$$A = D^2\frac{\pi}{4} = \left(\frac{3}{12}\right)^2\frac{\pi}{4} = \frac{\pi}{64} \text{ ft}^2$$

$$v = \frac{Q}{A} = \frac{0.2}{\dfrac{\pi}{64}} = \frac{12.8}{\pi} \text{ fps}$$

Substituting into the equation for H:

$$40 = \frac{(12.8)^2}{(64.4)(\pi^2)}\left(1 + 4f\frac{1,200}{\dfrac{3}{12}}\right)$$

from which:

$$f = 0.008.$$

PROBLEM 7

A 4-in. by 1-in. nozzle, shown in the figure, is attached to the end of a 4-in hose line. The velocity of the water leaving the nozzle is 96 fps, the coefficient of velocity, C_v, is 0.96 and the coefficient of contraction, C_c, is 0.80. Determine the necessary pressure at the base of the nozzle. Use a specific weight of 62.3 lbf/ft³.

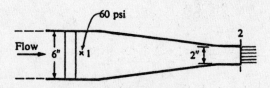

Figure 24. Flow in nozzle

SOLUTION: Figure 24 shows the nozzle attached to the 4-in. hose line. Since the coefficient of velocity, C_v, is $\dfrac{v_{\text{actual}}}{v_{\text{ideal}}}$,

$$v_{ideal} = \frac{96}{0.96} = 100 \text{ fps}.$$

The ratio of the area of the *vena contracta* to the area of the tip of the nozzle is called the coefficient of contraction. Incorporated with the continuity equation yields:

$$\frac{v_1}{v_2} = \frac{a_2}{a_1} = \frac{d_2^2}{d_1^2} = 0.8\frac{(1)^2}{4^2} = \frac{1}{20},$$

$$v_1 = \frac{100}{20} = 5 \text{ fps (ideal)}$$

Substituting these values into the Bernoulli equation, gives:

$$\frac{p_1}{\gamma} + z_1 + \frac{v_1^2}{2g} = \frac{p_2}{\gamma} + z_2 + \frac{v_2^2}{2g}$$

$$\frac{p_1}{0.433} + 0 + \frac{5^2}{64.4} = 0 + 0 + \frac{100^2}{64.4}$$

and

$$\frac{p_1}{62.3} = \frac{10,000 - 25}{64.4} = \frac{9,975}{64.4} = 155 \text{ ft}.$$

Therefore,

$$p_1 = 67.1 \text{ psi}.$$

PROBLEM 8

An aircraft is flying in level flight at a speed of 250 km/hr through air with standard conditions. The lift coefficient at this speed is 0.4, and the drag coefficient is 0.0065. The mass of the aircraft is 850 kg. Calculate the effective lift area for the craft.

SOLUTION: Apply the definition of lift coefficient.

$$C_L = \frac{F_D}{\frac{1}{2}\rho v^2 A_p}$$

Assume lift equals weight in level flight. Then:

$$F_L = mg = C_L \frac{1}{2}\rho v^2 A_p.$$

Solving for A_p:

$$A_p = \frac{2mg}{C_L \rho v^2}$$

$$A_p = \left(\frac{2}{0.4}\right)(850\,\text{Kg})\left(9.81\frac{\text{m}}{\text{s}^2}\right)\left(\frac{\text{m}^3}{1.23\,\text{Kg}}\right)\left(\frac{\text{hr}}{250\times10^3\,\text{m}}\times\frac{3,600\text{s}}{\text{hr}}\right)^2 = 7.03\,\text{m}^2$$

PROBLEM 9

Air flowing through a nozzle encounters a shock. The Mach number upstream of the shock is $M_x = 1.8$, and the static temperature downstream of the shock is $T_y = 800°R$. How much has the velocity changed across the shock? Assume $k = 1.4$.

TABLE 8

ONE-DIMENSIONAL NORMAL-SHOCK FUNCTIONS (FOR AN IDEAL GAS WITH CONSTANT SPECIFIC HEAT AND MOLECULAR WEIGHT, $k = 1.4$)

M_x	M_y	$\dfrac{P_y}{P_x}$	$\dfrac{\rho_y}{\rho_x}$	$\dfrac{T_y}{T_x}$	$\dfrac{P_{0y}}{P_{0x}}$
1.00	1.00000	1.00000	1.00000	1.00000	1.00000
1.05	0.95312	1.1196	1.08398	1.03284	0.99987
1.10	0.91177	1.2450	1.1691	1.06494	0.99892
1.15	0.87502	1.3762	1.2550	1.09657	0.99669
1.20	0.84217	1.5133	1.3416	1.1280	0.99280
1.25	0.81264	1.6562	1.4286	1.1594	0.98706
1.30	0.78596	1.8050	1.5157	1.1909	0.97935
1.35	0.76175	1.9596	1.6028	1.2226	0.96972
1.40	0.73971	2.1200	1.6896	1.2547	0.95819
1.45	0.71956	2.2862	1.7761	1.2872	0.94483
1.50	0.70109	2.4583	1.8621	1.3202	0.92978
1.55	0.68410	2.6363	1.9473	1.3538	0.91319
1.60	0.66844	2.8201	2.0317	1.3880	0.89520
1.65	0.65396	3.0096	2.1152	1.4228	0.87598
1.70	0.64055	3.2050	2.1977	1.4583	0.85573
1.75	0.62809	3.4062	2.2791	1.4946	0.83456
1.80	0.61650	3.6133	2.3592	1.5316	0.81268
1.85	0.60570	3.8262	2.4381	1.5694	0.79021
1.90	0.59562	4.0450	2.5157	1.6079	0.76735
1.95	0.58618	4.2696	2.5919	1.6473	0.74418
2.00	0.57735	4.5000	2.6666	1.6875	0.72088
2.05	0.56907	4.7363	2.7400	1.7286	0.69752
2.10	0.56128	4.9784	2.8119	1.7704	0.67422
2.15	0.55395	5.2262	2.8823	1.8132	0.65105
2.20	0.54706	5.4800	2.9512	1.8569	0.62812
2.25	0.54055	5.7396	3.0186	1.9014	0.60554
2.30	0.53441	6.0050	3.0846	1.9468	0.58331
2.35	0.52861	6.2762	3.1490	1.9931	0.56148
2.40	0.52312	6.5533	3.2119	2.0403	0.54015
2.45	0.51792	6.8362	3.2733	2.0885	0.51932
2.50	0.51299	7.1250	3.3333	2.1375	0.49902
2.55	0.50831	7.4196	3.3918	2.1875	0.47927
2.60	0.50387	7.7200	3.4489	2.2383	0.46012
2.65	0.49965	8.0262	3.5047	2.2901	0.44155
2.70	0.49563	8.3383	3.5590	2.3429	0.42359

SOLUTION: We seek $v_x - v_y$, which may be expressed as:

$$v_x - v_y = v_x\left(1 - \frac{v_y}{v_x}\right)$$

$\dfrac{v_y}{v_x}$, can be found from the table because M_x is given:

From the table and $M_x = 1.8$:

$$\frac{v_y}{v_x} = \frac{\rho_x}{\rho_y} = \frac{1}{\dfrac{\rho_y}{\rho_x}} = \frac{1}{2.36} = 0.425$$

$$\frac{T_x}{T_y} = \frac{1}{\dfrac{T_y}{T_x}} = \frac{1}{1.53} = 0.653.$$

v_x can be determined from:

$$v_x = M_x a_x = M_x\sqrt{kRT_x} = M_x\sqrt{kR\left(\frac{T_x}{T_y}\right)}\sqrt{T_y},$$

where $\dfrac{T_y}{T_x}$ is known as a function of M_x:

$$v_x = (M_x)\left(49.02\sqrt{\frac{T_x}{T_y}}\sqrt{T_y}\right) = (1.8)(49.02\sqrt{0.653})(\sqrt{800})$$

$$= 2,020 \text{ ft/s}.$$

Therefore:

$$v_x - v_y = 2,020 \text{ ft/s } (1 - 0.425) = 1,160 \text{ ft/s}.$$

Material Science/ Structure of Matter

Chapter 11

MATERIAL SCIENCE/ STRUCTURE OF MATTER

The key to the success of technology over the past century lies in our ability to make better use of materials obtained from nature and to develop new materials with starting, useful properties. It is difficult for us to imagine a world without plastics, lightweight metals for planes, and medical equipment that makes techniques for the treatment of illness possible. Our learning how to make such materials lies in the area of Material Science. In this chapter we explore the main ideas of this field in eight sections. Section 10.1 begins with the atomic structure of materials. Here we explore the forces that bind materials together and the various phases in which these materials are found. Similarly we classify materials on the basis of the structural arrangement of their molecules. In Section 10.2 we examine structures and structural defects of materials, a subject forming the basis for such areas as metals and ceramics. Section 10.3 is devoted to the mechanical and thermal properties of materials, forming the basis for materials processing. In Section 10.4 we turn to multicomponent materials and phase diagrams, describing the relationship between the composition of a material and its phase for a given temperature. We next turn (in Section 10.5) to diffusion and reactions in materials. Here we look at such processes as the movement of the atoms of one material through the solid, crystalline phase of another with subsequent implications for the strength of the material. In Sections 10.6 and 10.7 we examine the processes of corrosion and radiation alteration of materials while Section 10.8 concerns composite materials.

MATERIALS AND THEIR ATOMIC STRUCTURE

For our purposes the building blocks of all materials are the atoms of those elements making up the material. To understand the atom we must first recall the concept of electric charge. Forces exerted between bodies due to electrical effects are **electromagnetic forces**. Unlike gravity, which is always an attractive force tending to draw bodies nearer to each other, electromagnetic forces can be either attractive or repulsive, the latter driving the bodies further apart. We can focus only on **electrostatic forces**, which are those exerted between two electrically charged bodies at rest. The fact that both repulsive and attractive forces are involved leads us to postulate two kinds of charges: positive and negative. Bodies that are charged in the same way (both negative, or both positive) will be repelled by each other, while oppositely charged bodies are attracted to each other. In 1909 American physicist Robert Millikan found that the magnitude of an electric charge always appears as an integer multiple of a fixed value Ne, $N = 1, 2, 3,...$ The fundamental charge "e" is that of a single fundamental charged particle, the electron. Electrical charge is given in terms of the **Coulomb**, which is the amount of charge flowing past a point in a wire in one second when the current in the wire is one Ampere. Denoting the Coulomb by "C" we have $1 C = 1$ A/s. The charge of an electron, which by convention is taken as negative, is equal to $e = -1.602 \times 10^{-19}$ C while its mass is 9.11×10^{-31} kg.

The atom is essentially composed of three kinds of particles: electrons, protons, and neutrons—the latter two constitute the **nucleus** of the atom. A **proton** is a positively charged particle, the magnitude of whose charge is equal to that of the electron. Its mass, however, is approximately 1,840 times as large: proton mass = 1.67×10^{-27} kg. Since they have equal and opposite charges, the proton and electron can be in electrical balance; together they compose the atom of hydrogen, the lightest element. For the hydrogen atom, the nucleus is comprised simply of the proton itself, while the single electron is viewed as rotating about the proton. The **neutron** is an electrically neutral particle (neither positively nor negatively charged) whose mass can be taken as equal to that of the proton. While the hydrogen atom has no neutrons in its nucleus, the nucleus of the helium atom has two protons and two neutrons, with two electrons rotating around it. Regarding the nucleus as a sphere, its diameter is of the order of 10^{-14} m and at some distance from the nucleus we find the electrons rotating about it. Normally the number of electrons in orbit is equal to the number of protons in the nucleus. In this case the atom is electrically neutral. If one or more electrons are missing, then the resulting atomic structure is positively charged and is referred to as a **positive ion**. If, on the other hand, the atom has acquired one or more additional electrons, then it is called a **negative ion**.

All materials are composed of combinations of over 100 **elements**. Typical elements are hydrogen and helium, naturally occurring as gases; mercury, which we recognize readily in its liquid form in thermometers; and iron or aluminum, familiar to us in solid form. Some elements, such as radium and plutonium, are essentially not present in nature but can be created by means of suitable laboratory or industrial processes. An element is composed of atoms that uniquely correspond to it. The number of protons in the nucleus of an atom defines the **atomic number** of the element. Thus the atomic numbers of hydrogen, helium, sodium, aluminum, and oxygen are, respectively, 1, 2, 11, 13, and 8, that is, number of protons in any atom of the corresponding element. Elements are named (e.g., helium, hydrogen) and assigned symbols such as H (hydrogen), He (helium), Al (aluminum), O (oxygen), and Fe (iron). The number of neutrons for atoms of the same element may differ. Thus uranium (U) with an atomic number 92 (the number of protons in the nucleus) may have atoms with 143 neutrons or 146 neutrons. The various forms of an element differing in the numbers of neutrons in their atoms are referred to as **isotopes** of the element. The **atomic weight** of an atom is its mass expressed in "Atomic Mass Units" (AMU) where oxygen, with 8 protons and 8 neutrons, is taken to have a mass of exactly 16 Atomic Mass Units. The atomic **weight number** is the nearest natural number to the atomic weight in AMUs. Thus the atomic weight number of normally occurring hydrogen is 1; an isotope of hydrogen is deuterium, with an atomic weight number of 2, corresponding to the appearance of a neutron in the atomic nucleus. Normally occurring carbon has an atomic number of 6 and a mass number of 12 (with 6 neutrons). Carbon also has an isotope whose mass number is 13 (with 7 neutrons). The mass numbers of the two isotopes of uranium noted above are 235 and 238. Since the atomic weight and atomic weight number are very near in value, it is standard practice to use the rounded atomic weight number in place of the actual atomic weight, and to refer to it as the atomic weight of the element. This will be our practice as well (Figure 1).

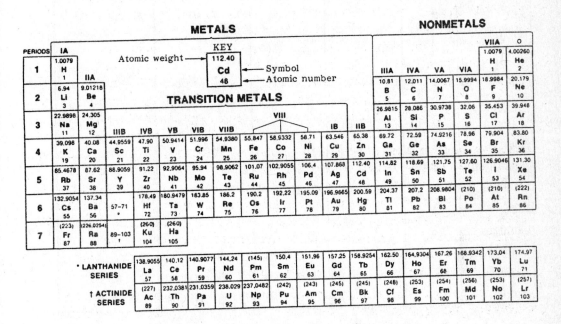

Figure 1. The periodic table of elements

The electrons in orbit around the nucleus of an atom have different energy levels. This has given rise to our viewing the atom as a nucleus surrounded by shells or groups of electrons having different energy levels. The lowest energy shell (**quantum shell**) has at most 2 electrons. The second shell has at most 8, the third 18, and the fourth 32. The maximum number of electrons in the n th shell is thus given by $2n^2$; n is called the **principal quantum number** of the shell. Thus, the hydrogen atom with atomic number 2 will have one electron rotating around its nucleus in the lowest energy shell, while oxygen, with atomic number 8, will have 2 electrons in its lowest energy shell and 6 in the next shell. It is important to note that the energies of electrons in different shells differ by fixed amounts. If an electron moves from a higher energy shell to the next lower energy shell, then an amount (quantum) of energy is released from the atom. This amount of energy is referred to as a **photon**. For reasons that are beyond the scope of this chapter, stability considerations favor the completion of the outer shell—that is, that there should be exactly $2n^2$ electrons in shell number n. We refer to electrons in the outer shell of an atom as **valence electrons**.

Shells are further subdivided into **orbitals** labeled, respectively, the s, p, d, and f orbitals. These will have maximum numbers of 2, 6, 10, and 14 electrons respectively. The shell and orbital structures of some familiar elements are shown in Table 1:

TABLE 1.
SHELL AND ORBITAL STRUCTURES

Shell	1	2		3			4				5			
Element	s	s	p	s	p	d	s	p	d	f	s	p	d	f
Hydrogen	1													
Helium	2													
Carbon	2	2	2											
Nitrogen	2	2	3											
Oxygen	2	2	4											
Sodium	2	2	6	1										
Aluminum	2	2	6	2	1									
Chlorine	2	2	6	2	5									
Iron	2	2	6	2	6	6	2							
Copper	2	2	6	2	6	10	1							
Cadmium	2	2	6	2	6	10	2	6	10		2			
Xenon	2	2	6	2	6	10	2	6	10		2	6		
Gold	2	2	6	2	6	10	2	6	10	14	2	6	10	

When atoms of two or more elements bind together in definite proportions to form a substance, this substance is referred to as a **compound**. Examples are water (two atoms of hydrogen and one of oxygen, H_2O) and salt (one atom each of sodium and chlorine, NaCl).

A molecule is the smallest particle into which a material can be subdivided without changing its physical or chemical properties. It is composed of atoms that are bound together by "binding forces." These forces may be weak or strong, and determine the nature of the material. We will now examine the various kinds of binding forces maintaining a molecule of a material.

We begin with ionic binding which is based on attraction between oppositely charged particles. Let us consider ordinary table salt, whose molecules consist of atoms of sodium (Na) and chlorine (Cl). Their atomic numbers are, respectively, 11 and 17. The shell makeup of sodium is 2 + 8 + 1 (that is, 2 on the inner shell, 8 on the next, and 1 on the outermost), while that of chlorine is 2 + 8 + 7. The two atoms would have an electrical propensity to "meld" together with the single outermost electron of sodium "leaving" the Na atom and "joining" the outermost shell of the Cl atom. The result is the formation of a positively charged Na ion being attracted to the negatively charged Cl ion, arising from the addition of the electron to its outer shell. **Ionic binding** is the binding stemming from mutual attraction of oppositely charged particles arising from the "sharing" of available electrons to complete shells in the atomic structure. Another example of ionic binding is that of magnesium chloride ($MgCl_2$). Magnesium, with an atomic number 12, has an electron structure $2 + 8 + 2$; chlorine has the structure $2 + 8 + 7$. Hence, the two magnesium electrons in the outer shell can move to two chlorine atoms, resulting in a positively charged Mg ion (charge $= +2e$) and two negatively charged Cl ions (charge $= -2e$) which then are bound to form magnesium chloride.

While ionic bonding is characterized by transfer of electrons from one atom to another, **covalent bonding** is that in which electrons are shared between the atoms. An example is that of the hydrogen molecule, consisting of two covalently bound hydrogen atoms (Figure 2). Roughly speaking, the two electrons available from the pair of atoms form a negatively charged "entity" binding the two positively charged nucleii. Another example of covalent binding is that of methane (CH_4) in which carbon (atomic number 6, shell structure $2 + 4$) has each of its four outer electrons paired with the single orbital electron for each of the four hydrogen atoms to form a strong negatively charged link between the hydrogen nucleii and the carbon nucleus. Other familiar examples of covalently bonded molecules are the oxygen molecule (O_2) (atomic number 8, shell structure $2 + 6$) in which a bridge of four electrons is formed, and the nitrogen molecule (N_2), (atomic number 7, shell structure $2 + 5$) in which a bridge of six electrons is formed. A third binding force is the **metallic bond**. In this structure the valence electrons form a negatively charged "cloud" surrounding and moving through a positively charged "core" of positively charged particles.

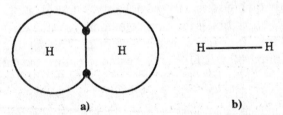

a)   b)

Figure 2 . Simplified representations of the covalent bond of hydrogen

Ionic, covalent, and metallic bonds are the primary mechanisms through which materials form structures. Other, weaker binding mechanisms are also present. These arise primarily from assymetries in the electrical structures of atoms and molecules, and are generally referred to as **Van der Waals** forces. An example of Van der Waals forces is given by the binding forces maintaining the physical structure of water in its liquid and solid (ice) phases.

Material structure rests on the fact that atoms assume spatial arrangements as a result of interactions between them. These interactions are the result of a combination of the above attractive forces, tending to draw atoms together, and repulsive forces, arising from the interactions of electrons. The sum of the attractive and repulsive forces is the net force between two atoms. The energy arising from this net force is said to be the **bonding energy** between the atoms. This energy represents the potential energy of the atomic pair. The atomic distance for which it is minimum coincides with the distance for which the sum of attractive and repulsive forces is zero. The addition of energy, such as that due to heating, can serve to drive atoms apart and break the bonds maintaining structure, while removal of energy will lead the atoms to assume their equilibrium position.

The molecular weight of a molecule is the total sum of the molecular weights of its atoms. Thus the molecular weight of water, is determined by adding that of two atoms of hydrogen (2×1) to that of one atom of oxygen (1×16) yielding the value 18. A **gram-mole** or **mole** of a molecule is that amount of the material whose weight in grams is equal to the molecular weight of the molecule. The **gram-molecular weight** is the mass of one mole of a compound equal in grams to the molecular weight. The number of molecules per gram-molecular weight is a constant, referred to as **Avagadro's number**, and equal to 6.02×10^{23}. Similarly every gram-atom contains Avagadro's number of atoms.

EXAMPLE: How many atoms are in 80 grams of aluminum?

SOLUTION: One gram-atom of aluminum contains the atomic weight of aluminum in grams, or 26.98 grams. Hence 80 grams of aluminum is equal to $80 \div 26.98 = 2.965$ gram-atoms, which in turn contains

$$2.965 \times 6.02 \times 10^{23} = 17.849 \times 10^{23} \text{ atoms.}$$

Most materials may be categorized into three classes: **metals, ceramics**, and **polymers**. Metals are composed of elements that easily give up electrons to form the "electron cloud" of a metallic bond. These include iron, copper, aluminum, nickel, and alloys or combinations of metals. Metals are characterized by high thermal and electrical conductivity, high strength and ductility. These properties will be further explained later in the text. Iron is the most commonly used metal; iron oxides, or pig iron are extracted from the iron ore in a blast furnace. The pig iron, which contains carbon and other impurities, is further processed to reduce the carbon content and obtain various grades of steel. Examples of processes include **Bessemer** and **oxygen processes**, also known as the L-D or **Linz-Donawitz processes**. Carbon steels are the simple grades and have carbon as the major non-ferrous element. Nickel, copper, manganese and other metals are alloyed with steel to obtain desirable properties. For example, adding chromium improves the steel and reduces corrosion, thus giving stainless steel.

Ceramics are materials containing compounds of both metals and nonmetals that are bound by both ionic and covalent bonds. Ceramics usually have high thermal and chemical resistances, and they are generally poor conductors of heat and electricity. Some simple examples include: magnesium oxide, beryllium oxide, silicon carbide, and silicon nitride. The structure may be a combination of highly ordered crystals and glassy regions. Ceramics may be roughly classified into four groups: **clays, refractories, cements**, and **glasses**. When wet, the clays can be easily blended and molded. Upon drying or **firing**, materials such as bricks, tiles, porcelain, and stoneware can be manufactured. The type of drying process alters the porosity and permeability of the substance. Refractories are designed to withstand high temperatures in industrial operations: gas-turbines, ram-jet engines, and nuclear reactors. Some examples include alumina-silica compositions, carbides, nitrides, carbon and graphite. Cements are characterized by their ability to set and harden after being mixed with water; a common type is Portland cement. Finally glasses, mainly made of silica, are materials that have been cooled to a rigid condition but have not been crystallized. For economic reasons, metal oxides are usually added to the glass mixture to reduce the melting temperature. Commercial products include soda lime or lime glass, lead glasses, borosilicate glasses, and high-silica glasses.

To define polymers we must recall the notion of an organic compound, one made up of molecules containing carbon. Examples are methane (CH_4) and propane (C_3H_8). Compounds containing only hydrogen and carbon are **hydrocarbons**. Plastics consist of hydrocarbon molecules that are linked together, primarily through covalent bonds, to form giant molecular chains. To see how this is done consider the methane molecule CH_4. The four atoms of hydrogen and one of carbon are bound covalently through the sharing of one electron of each of the hydrogen atoms, with the outer shell of the carbon atom. The resulting molecule of methane can be visualized as shown in Figure 3:

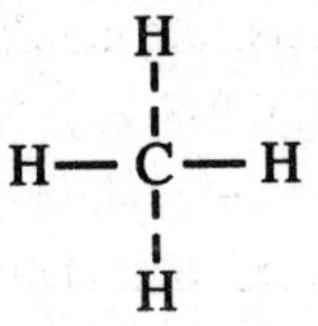

$$
\begin{array}{c}
H \\
| \\
H - C - H \\
| \\
H
\end{array}
$$

Figure 3. The methane molecule

The single line represents the covalent sharing of a single hydrogen electron. The molecule is electrically balanced with the four outer-shell vacancies in the carbon atom being compensated for by the "bridge" of four electrons. We can easily expand on our figure to a more complex hydrocarbon molecule by considering the case of two carbon atoms. For this case we can build a new hydrocarbon, bonded covalently, with the structure shown in Figure 4:

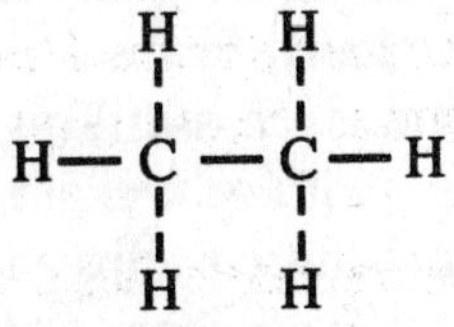

$$
\begin{array}{cc}
H & H \\
| & | \\
H - C - C - H \\
| & | \\
H & H
\end{array}
$$

Figure 4. The ethane molecule

Figure 4 describes the bonding structure of a molecule of the hydrocarbon ethane. In the same way we can successively add triplets of a single carbon atom and two hydrogen atoms to build long chain molecules with a "backbone" of carbon atoms.

Generally speaking the extension of this structure to one of n carbon atoms contains $2n + 2$ hydrogen atoms and has the chemical formula C_nH_{2n+2}; these compounds are referred to as **paraffins**. Since all electron shells of paraffins are filled there is no way for a paraffin molecule to chemically combine with any additional hydrogen atom. Such a molecule is referred to as saturated. A molecule that can admit additional hydrogen atoms is unsaturated. An example of an unsaturated molecule having a "double bond" between carbon atoms is ethylene, whose binding structure is shown in Figure 5:

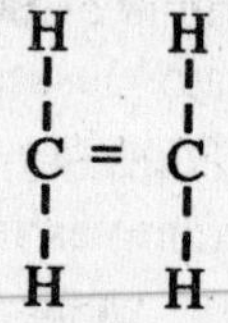

$$
\begin{array}{ccc}
H & & H \\
| & & | \\
C & = & C \\
| & & | \\
H & & H
\end{array}
$$

Figure 5. The ethylene molecule

Here as above, each line represents a covalent structural link of an electron pair; hence, between the carbon atoms we have a bridge of four electrons. Under appropriate processes ethylene molecules may be altered in such a way that the double carbon-carbon bond is replaced by a single carbon-carbon bond, and the now free remaining electrons can be used to create bonds with new carbon atoms in other former ethylene molecules, resulting in a structure of possibly enormous length, a typical segment of which would assume the following form:

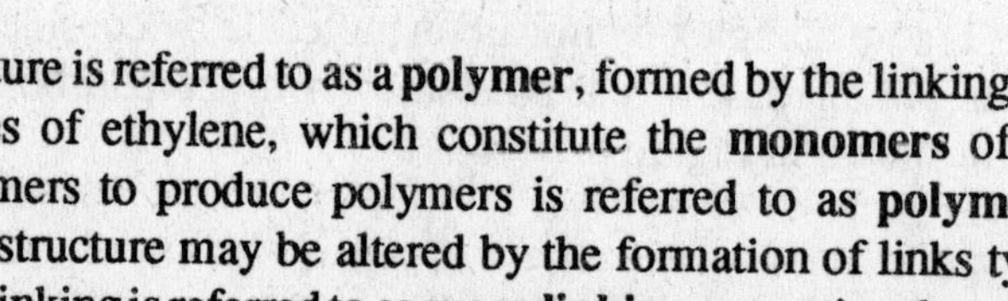

$$
\cdots -\overset{\displaystyle \overset{H}{|}}{\underset{\displaystyle \underset{H}{|}}{C}} -\overset{\displaystyle \overset{H}{|}}{\underset{\displaystyle \underset{H}{|}}{C}} -\overset{\displaystyle \overset{H}{|}}{\underset{\displaystyle \underset{H}{|}}{C}} -\overset{\displaystyle \overset{H}{|}}{\underset{\displaystyle \underset{H}{|}}{C}} -\overset{\displaystyle \overset{H}{|}}{\underset{\displaystyle \underset{H}{|}}{C}} -\overset{\displaystyle \overset{H}{|}}{\underset{\displaystyle \underset{H}{|}}{C}} -\overset{\displaystyle \overset{H}{|}}{\underset{\displaystyle \underset{H}{|}}{C}} -\overset{\displaystyle \overset{H}{|}}{\underset{\displaystyle \underset{H}{|}}{C}} -\overset{\displaystyle \overset{H}{|}}{\underset{\displaystyle \underset{H}{|}}{C}} - \cdots
$$

Figure 6. A polymer structure

Such a structure is referred to as a **polymer**, formed by the linking of two or more C_2H_4 molecules of ethylene, which constitute the **monomers** of our structure. Linking monomers to produce polymers is referred to as **polymerization**. The polymer chain structure may be altered by the formation of links tying the chains together. Such linking is referred to as **cross-linking**, occurring through connections between unsaturated carbon atoms within the chain. The result of cross-linking is to significantly restrict movement between adjacent chains and alter mechanical properties. One example is the aging of rubber. A second form of a polymer chain is branching. **Branching** occurs when a polymer chain bifurcates into two chains, resulting in a material with little movement possible between adjacent molecules.

EXAMPLE: Show the structure of a polyethylene monomer which incorporates a chlorine atom through substitution of a hydrogen atom, as well as the addition polymerization of this monomer.

SOLUTION: Substitution of a hydrogen atom by one of chlorine will produce the monomer with the following structure:

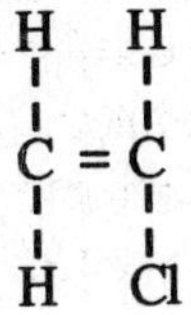

$$
\begin{array}{ccc}
H & & H \\
| & & | \\
C & = & C \\
| & & | \\
H & & Cl \\
\end{array}
$$

Figure 7. Substituted monomer

Addition polymerization of this monomer yields the material vinyl chloride, with the structure

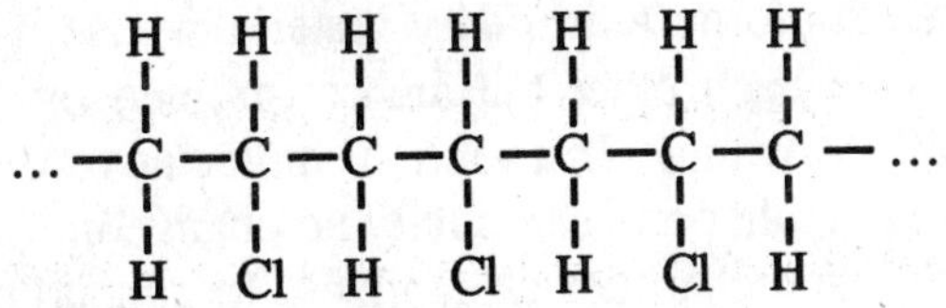

$$
\cdots - \underset{|}{\overset{|}{C}} - \underset{|}{\overset{|}{C}} - \underset{|}{\overset{|}{C}} - \underset{|}{\overset{|}{C}} - \underset{|}{\overset{|}{C}} - \underset{|}{\overset{|}{C}} - \underset{|}{\overset{|}{C}} - \cdots
$$

$$
\begin{array}{ccccccc}
H & H & H & H & H & H & H \\
H & Cl & H & Cl & H & Cl & H \\
\end{array}
$$

Figure 8. Polymer of chlorinated monomer

Organic materials occur in nature (e.g., wood) and are artificially produced (e.g., plastics). Our manufactured organics are made by creating large molecules such as the above via polymerization. We define the **degree of polymerization** as the ratio of the molecular weight of the polymer to that of the monomer. For most commercial plastics the degree of polymerization ranges from 75 to 750 mers per molecule. Polymerization is carried out by subjecting monomers to combinations of heat, pressure, light, or a catalyst, resulting in replacing the double carbon-carbon bond by a single one. The mechanisms for doing this are either **addition polymerization**, in which successive monomer bonds are broken down and the monomer is added to the molecule, or by **condensation polymerization** in which the polymer is a direct by-product of a batch-type process. The term **copolymerization** refers to the situation in which combinations of more than one monomer are used in producing the polymer.

The actual atomic arrangement of atoms in a molecule is not determined uniquely by the makeup of the molecule. Distinct structures of molecules having the same composition are referred to as **isomers**. Examples of two isomers of the compound H_8C_3O are given in Figure 9. The isomer on the left is normal propyl alcohol; the isomer on the right is isopropyl alcohol. Despite their identical composition these isomers have different physical properties.

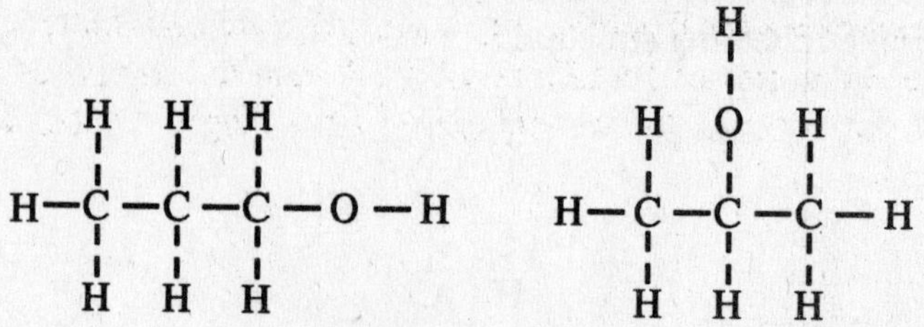

$$
\begin{array}{ccccccc}
 & H & H & H & & & \\
 & | & | & | & & & \\
H- & C- & C- & C- & O- & H & \\
 & | & | & | & & & \\
 & H & H & H & & &
\end{array}
\qquad
\begin{array}{cccc}
 & & H & \\
 & & | & \\
 & H & O & H \\
 & | & | & | \\
H- & C- & C- & C-H \\
 & | & | & | \\
 & H & H & H
\end{array}
$$

Figure 9. Isomers

The way in which the molecules of a material are arranged relative to each other determines the phase of the material. Familiar phases are gas, liquid, and solid. A **gas** is a material whose molecules are free to move independently of each other. The material has no structure or form. A **liquid** is a material whose molecules can change their position relative to each other but are constrained by attractive forces to maintain a relatively fixed volume. In a **solid** the molecules are constrained to fixed positions relative to each other and essentially no change in shape or volume will occur. Solids, in turn, can be further categorized. In crystalline solids we find a periodic structure in any spatial direction on the atomic scale: molecules are arranged in a definite ordering relative to each other, much like the corner points on a stack of identical boxes. Amorphous materials (solid or liquid) have no such ordering, and on the atomic scale one will see little or no ordering. Examples of crystalline solids include common metals; window glass is an example of an amorphous material.

STRUCTURES AND STRUCTURAL DEFECTS OF MATERIALS

As noted, materials can be structured in either an ordered, periodic form, or at the other extreme, be totally disordered. In this section we turn to a more detailed examination of the physical structure of materials.

We may regard a molecule as a collection of a small number of atoms, which are strongly bound together but whose bonds with other similar atomic groups is much weaker. An example is that of the water molecule H_2O which is covalently bonded. The bonds between water molecules arise from Van der Waals forces and are significantly weaker than the internal bonding forces. The relative strengths of the internal and external binding forces for molecular compounds such as water are manifested in such effects as: a) relatively low melting and boiling temperatures (as compared, e.g., with metals); b) maintenance of molecular structure in the liquid and gaseous phases; and c) solids formed from compounds are relatively soft and can be made to move with the application of relatively small forces. We note that the "small" number of atoms in a molecule may nevertheless be several thousand in number—particularly in the case of hydrocarbons.

Most solid materials used in engineering are crystalline, with their molecules consisting of structured, periodic arrangements of atoms in all directions (Figure 10).

The smallest unit, whose periodic repetition determines the solid, is called the **primitive lattice** or **unit cell** of the crystal. The geometric structure of the unit cell determines the categories of crystalline solids. A material composed of several or many crystals, in contrast to one composed of a single crystal, is said to be **polycrystalline**.

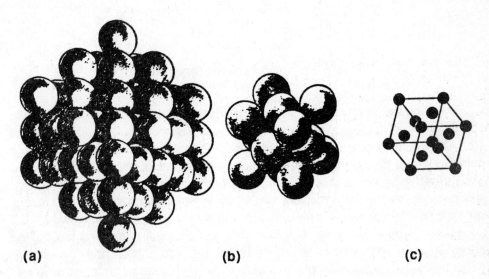

Figure 10. The crystal lattice of copper.
Copper Ions are shown as spheres: (a) Is the lattice, (b) Is the unit cell and (c) Is the representation of the unit cell.

Every atom has an **atomic radius** measured effectively at the radius of the electron cloud around the nucleus. The equilibrium distance between the centers of two adjacent atoms of a material can be regarded as the sum of their atomic radii, possibly increased by temperature or other factors. One such factor would be the presence of one or more additional atoms in their neighborhood. As more atoms are present the repulsive forces between atoms grow as do their equilibrium distance. The **coordination number** for an atom is the number of closest neighbors that an atom may have. Consider three atoms of radius R placed next to each other as the vertices of a triangle (Figure 11). Between the three there is a gap in which an additional atom could be placed. The largest ratio r of this atom that could physically

fit in this gap is found to be given by $\frac{r}{R} = 0.155$. For four atoms, this ratio is at most

0.225 while for six neighbors it would be 0.414. The ways in which these atoms are arranged in the solid determines the properties of the material. The density of atoms in the crystal is described by the **packing factor** and defined as the ratio of volume of atoms per unit cell to the colume of the unit cell. The volume of the unit cell is dependent on the crystal structure and is discussed below.

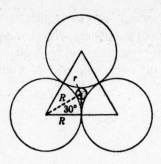

Figure 11. A gap between three atoms

The simplest structure is that of a **cubic primitive lattice**. In this case four atoms are placed, relative to each other, as the vertices of a cube, that is, at equal distances from three neighbors with perpendicular axes. Three lattice types are based on the cubic structure: the **simple cubic**, for which the atoms are located at the vertices and each primitive lattice has exactly eight atoms; **body-centered cubic**, for which eight atoms correspond to the vertices of a cube, while a ninth is located at the center of the cube, and **face-centered cubic**, in which in place of a single atom located at the cube center, we have six additional atoms located at the centers of the six faces of the cube. These structures are commonly referred to as "sc," "bcc," and "fcc" lattice structures, respectively. At room temperature the crystal lattice structure of iron is body-centered cubic. Similarly, copper atoms are arranged in a face-centered cubic lattice structure.

Moving from a cubic structure we encounter the **tetragonal lattice structure**, whose primitive lattice has two of three axis lengths equal, and all three axes at right angles to each other. For this configuration we find both simple and body-centered lattice structures; unlike the cubic lattice we do not find face-centered lattice structures.

The classification of all primitive lattices is shown in Table 2. In it a, b, and c denote the axis lengths for the lattice, while α, β, and Γ denote the angles between the axis pairs (a, b), (b, c), and (c, a), respectively. Note that with this notation the cubic lattice corresponds to the case $a = b = c$, and $\alpha = \beta = \Gamma = 90°$. The "options" shown refer to "simple" (atoms only at vertices), body-centered (an additional atom at the center), face-centered (additional atoms located on each face), and base-centered (an additional atom located on one face).

TABLE 2
PRIMITIVE LATTICE TYPES

System Name	Axes	Angles	Options
Cubic	$a = b = c$	$\alpha = \beta = \Gamma = 90°$	simple face-centered body-centered
Tetragonal	$a = b \neq c$	$\alpha = \beta = \Gamma = 90°$	simple body-centered
Monoclinic	$a \neq b \neq c$	$\alpha = \beta = 90° \neq \Gamma$	simple base-centered
Hexagonal	$a = b \neq c$	$\alpha = \beta = 90°$ $\Gamma = 120°$	simple
Triclinic	$a \neq b \neq c$	$\alpha \neq \beta \neq \Gamma$	simple
Orthorhombic	$a \neq b \neq c$	$\alpha = \beta = \Gamma = 90°$	simple face-centered body-centered base-centered
Rhombohedral	$a = b = c$	$\alpha = \beta = \Gamma \neq 90°$	simple

The crystal structure of salt (NaCl) is cubic, with each sodium atom surrounded by four chlorine atoms, and each chlorine atom surrounded by four sodium atoms arranged like the vertices of a cube. Iron at room temperature has the body-centered cubic (bcc) crystal structure; at 910°C iron undergoes a phase transformation in which its crystal structure changes to face-centered cubic (fcc). Such a phase transformation is referred to as **recrystallization**. At room temperature the crystal structure of copper is face-centered cubic (fcc), while that of sodium is body-centered cubic (bcc). Other metals having the fcc crystal structure include nickel and platinum.

The primitive lattice or unit cell of the hexagonal crystal can be considered as part of a structure in which we have three axes lying in a plane at angles of 120° from each other, in addition to a fourth axis perpendicular to the plane of the other three. Such a structure is found, for example, in magnesium, titanium, and zinc, and is referred to as **close-packed hexagonal (hcp)** (Figure 12).

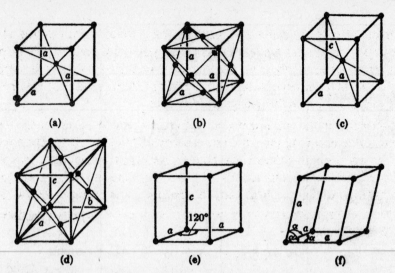

a) Body-centered cubic b) Face-centered cubic c) Body-centered tetrangonal d) Face-centered othorhombic e) Simple hexagonal, and (f) Rhombohedral

Figure 12. Unit cells

Directions and planes of a crystal are identified using a vector notation. Placing an origin at a vertex of the primitive lattice, the vector notation [hkl] indicates directions within the crystal, while (hkl) denotes crystal planes. The entities [100], [010], and [001] correspond to directions along the corresponding a, b, and c axes. Similarly [110] leads us to the diagonally opposite crystal vertex. On the other hand face-centered atoms are indicated by the value two in the corresponding face, whence [112] indicates the center of the top face of the unit cell. Crystal planes are indicated by the Miller indices (after W. H. Miller; Figure 13). Here the convention is that if h, k, and l are the reciprocals of the intercepts of the plane with the x, y, and z axes, then the Miller index representation of the plane is (hkl). If the intercept is negative, then the minus sign is replaced by a bar above the corresponding value.

Thus the plane given by $(1\bar{1}1)$ cuts the x, y, and z axes at 1, –1. and 1.

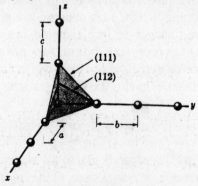

Figure 13. Miller indices. The (112) plane intersects the axes at distances

of 1, 1, and $\frac{1}{2}$.

A material with the same composition may assume various crystal structures. These crystal structures are referred to as **polymorphs**. An example of polymorphism is iron, assuming both the bcc and fcc structures. As a result the density may well change, as it does in iron, since the packing of the crystal lattice by the material atoms will change with the change of the lattice.

Most metals in the solid phase assume crystal structures. The most convenient method for determining the structure is **x-ray diffraction**. Diffraction occurs because the wavelength of the x-ray is on the same order of magnitude as the distance between atoms in the crystal, about 0.1 to 0.2 nm. Consider three layers (or planes) of atoms (Figure 14). The difference between the paths *DEF* and *ABC* is *GEH*, and *GEH* = 2 *GE*. From trigonometry, we have *GE* = *EH* = $d \sin \theta$. Bragg discovered that *GE* is an integer if

$$n\lambda = 2d\sin\theta$$

with $n = 1, 2, 3, \ldots$ (an integer) and λ = wavelength. This relationship can be combined with the Miller indices [*hkl*] to give:

$$\frac{(\lambda n)}{2\sin\theta} = \frac{a}{\sqrt{\left(h^2 + k^2 + l^2\right)}}$$

where *a* is the lattice constant of the crystal.

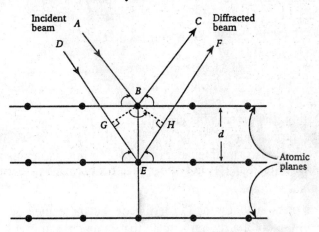

Figure 14. Diffraction from planes of atoms

The **density** of a crystal can be easily calculated if the structure is known. For a unit cell of a cubic crystal with side, *a*, and *n* atoms in the pattern, the weight of the unit cell is $\frac{nM}{N_A}$. *M* is the atomic weight, N_A is Avagadro's number. Density, ρ, is the weight divided by the volume, a^3.

$$\rho = \frac{nM}{a^3 N_A} = \frac{nM}{V N_A}$$

where V is the volume of the unit cell. Since for a face centered cubic (fcc), the lattice constant, a, is related to the atomic radius, ρ, by $a = \dfrac{4\rho}{\sqrt{2}}$, then $V = \dfrac{32r^3}{\sqrt{2}}$.

Ceramics are defined as compounds of metals and nonmetals. As for metals, ceramic structures are generally crystalline, but unlike the case of metals, these lattices are unaccompanied by clouds of free electrons. Ceramics generally have much higher melting temperatures than do metals; they are harder and resistant to chemical change. In addition they are ordinarily electric insulators, and are poor heat conductors. Because of their rather complex crystal structure, the response of ceramics to thermal events is slow. For this reason, cooling of molten ceramics generally occurs too rapidly for crystal lattice structures to form, hence the materials often solidify as **supercooled liquids** still containing their latent heat of melting.

Because polymer molecules can bind with each other only via the relatively weak Van der Waals forces, crystal structures are either imperfect or totally lacking. The actual structure of organics, which can serve to link neighboring organic chains, can be based on **cross-linking**, in which adjacent chains are bound together through a number of unsaturated atoms. An example of the use of cross-linking is in the rubber of an automobile tire, in which sulfur atoms are used for linking between neighboring chains. The addition of sulfur for this purpose is referred to as **vulcanization**.

By definition, **amorphous structures** are those that are not crystalline. These may be divided into a number of categories. Two examples of interest to us are liquids and glasses. A liquid, while not assuming a crystalline form, does nevertheless possess a local ordering for its molecules. To define a glass, let us examine the performance of a material as its temperature is reduced. Generally, with reduction in temperature, the energy of motion of the material molecules decreases; hence, the average density increases and the specific volume (reciprocal of the density) declines. During this process the crystalline structure is preserved. At the **fictive temperature**, representative of the material, the decline in specific volume may cease, corresponding to a maintenance of local ordering, and a breakdown in longer-range ordering of the material and of its crystal structure. The material is now said to be in its **glassy phase**.

Crystals express a long-range ordering of the molecules of a material. This ordering is ultimately marred on a scale that is macroscopic, by **crystal defects** which may be expressed as point defects, line defects, or boundaries (Figure 15). Let us now examine these concepts.

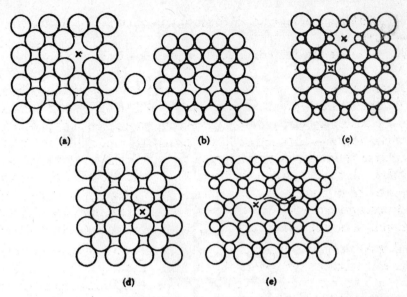

**a) Vacancies, b) Di-vacancy (two missing atoms), c) Schottky defect
(d) Interstitialcy, (e) Frenkel defect**

Figure 15. Kinds of defects

Point defects may arise when atoms in the crystal lattice are missing, displaced from their desired lattice locations, or when extra atoms appear in the lattice. A **vacancy** is a defect in which an atom is missing from where it should be in the crystal lattice. Vacancies occur because of imperfect packing of atoms when the crystal was formed or from vibrations at elevated temperatures. Similarly, they can be induced by bombardment by neutrons in situations involving radioactivity. **Schottky defects** are related to vacancies and involve the lack of a pair of oppositely charged ions from the lattice. Another kind of point defect is that of an **interstitialcy**, which occurs when an additional atom is found in the lattice structure. This may occur for a material of high specific volume and will generally result in a localized distortion of the lattice structure. Similarly, a **Frenkel defect** is one in which an atom is displaced from its site to a neighboring site where it constitutes an interstitialcy.

Line defects occur when an entire line of atoms is misplaced in the crystal lattice. An example is an **edge dislocation**, which arises from an edge or additional plane of atoms within the lattice. This, in turn, induces regions of compression and tension in the lattice. A **screw dislocation** is a special case of an edge dislocation in which the crystal lattice is placed much like a winding staircase corresponding to a dislocation (Burgers) vector. The dislocation appears very much like the steps of a screw located along the Burgers vector direction (Figures 16 and 17).

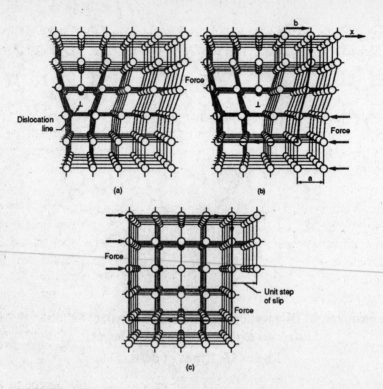

a) Misplaced atoms, b) Compression and tension in the lattice,
c) Resultant unit step of slip

Figure 16. Motion of an edge dislocation and production of a unit slip defect at a crystal surface

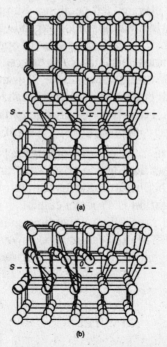

Figure 17. Production of a screw dislocation in a crystal

The **Burgers vector, b**, is parallel to the screw dislocation. It can be related to the shearing force, V, per unit length, L, and the shear stress, τ, by the equation:

$$\frac{V}{L} = \mathbf{b}t$$

These dislocations, when subjected to stress, can move, producing **slip** or **climb**. Slip refers to the motion of the dislocation, or **slip plane**, parallel to the crystalline plane. It is a plastic deformation. In climb, the edge dislocation moves perpendicularly to the slip plane. Climb occurs at much higher temperatures and is not as common as slip.

Boundaries may appear as external surfaces. At such surfaces the atoms are different from those on the interior of the lattice. Their energies are different, since they have fewer neighbors, and surface tension effects tend to bind them together along the surface. The comparable effect for liquid droplets where flow is possible is to tend toward a spherically shaped boundary. A **grain boundary** is a planar imperfection. It separates crystals (or grains) of different orientations. In this case there is a jump in the lattice orientation across a two-dimensional surface (the grain boundary itself) such that on each side the material has a lattice structure. Grain boundaries constitute lines of weakness of the material since binding forces across them are not as strong as those within the lattice structure.

Crystal lattice defects often occur in the process of crystallization of liquid material. If a homogeneous material (e.g., liquid copper) is gradually cooled, its molecules will gradually align themselves in such a way that we find a common orientation in a lattice arrangement on an ever-growing spatial scale. At the solidification temperature, an amount of energy is withdrawn from the material which is large enough to "fix" the atoms in the lattice. The amount of energy needed to do this is the **latent heat of melting**. If the cooling is slow, then the alignment will take place with local attractive and repulsive forces controlling the alignment. If, on the other hand, the cooling is rapid, then time may be lacking for the proper alignment to take place. This, in turn, has a high probability of resulting in misalignments of the crystal lattice and the formation of the defects discussed above. Once solid is formed it is generally subject to a variety of compressive and shear stresses. When these are sufficiently small, their application results in a distortion of the lattice whose extent is reversed upon removal of the stress. However, should the stress be sufficiently large, even when it is removed, the distortion remains permanent. Distortion of this kind may involve displacement of the lattice along planes, grains, or locally.

PROPERTIES OF MATERIALS

The chief aim of Materials Science is to learn how materials respond to processes being applied to them. These processes may be physical, such as compression or tension, electrical, thermal, or radioactive. As a result of these processes, the response of the material can be characterized quantitatively. In this section we turn to some of the key properties of materials in the light of applied processes.

We begin by examining **mechanical properties** of materials. The key properties of this type are strength, ductility, elasticity, creep, hardness, and toughness. Each of these properties measures some aspect of the response of the material to some form of mechanical force. The **tensile test** is performed on materials to determine their properties, for example: stiffness, elasticity, or fracture strength. Briefly, a specimen of loaded with tension and elongation (or deformation) is measured as the load increases.

Stress is a force per unit area applied to the material. **Strain** is a measure of the deformation of the material. It is generally measured as the relative length of deformation of the material. An example of stress-strain curve for a ductile sample is shown in Figure 18. From point O to P, Hooke's law applies and a linear relationship exists between stress and strain. The slope of the curve is an index of the **stiffness** of the material. Sometimes it is called the **modulus of elasticity** or **Young's modulus**. Point E is the **elastic limit** of the material. Up to this point, the material returns to its original state after the tension has been removed. However, beyond this point, all deformations are permanent and the material does not recover. Point Y, the **upper yield point** (or strength) is the stress necessary to free dislocations. The dislocations are moved through the lattice at the **lower yield point** L. The **ultimate strength** of the material is shown at point U. Point R represents the **breaking point** or **fracture strength**.

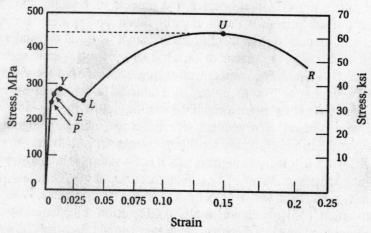

Figure 18. Stress-strain diagram for mild steel in tension

For plastics, the stress-strain curve has a slightly different shape as seen in Figure 19. After the yield point Y, the sample elongates and the diameter decreases. A phenomenon known as **necking** occurs when the deformation becomes concentrated, altering the shape of the curve. Brittle materials such as glass, cast iron, and ceramics can only support small stresses, and approach failure (or fracture strength) rapidly.

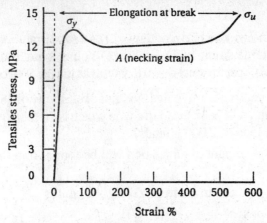

Figure 19. Stress-strain curve for a plastic

A stress-strain curve predicts a fracture stress that is lower than the ultimate strength of the material. This error is due to the fact that the stress is calculated on the basis of the original cross-sectional area. The **true stress-strain curve** calculates the instantaneous deformation. A differential strain element $d\varepsilon$ is the ratio of the instantaneous length, dl to the original length, l.

$$d\varepsilon = \frac{dl}{l}$$

The total strain during deformation is then the integral

$$\varepsilon = \int_{l_o}^{l} \frac{dl}{l} = \ln\left(\frac{l}{l_o}\right)$$

Figure 20 compares the true stress-strain curve (corrected) to the nominal curve (uncorrected). In addition, the part of the true stress-strain curve from Y to R can be approximated by the **Power law equation**:

$$\sigma = K\varepsilon^n$$

where σ is the true stress, K is the strength coefficient of the material, and n is the strain hardening coefficient.

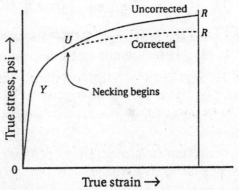

Figure 20. True stress-strain curve

For large deformation Young's modulus will be a small value, while for inelastic materials it will be large. Some typical values of Young's modulus are 10×10^6 psi for aluminum, 30×10^6 psi for steel down to values as low as 500 for certain rubbers.

EXAMPLE: For a steel with an average modulus of elasticity of 28,000,000 psi how much of an elongation will be induced in a wire which is 20 feet long and 0.2 inches in diameter under a load of 2,000 pounds?

SOLUTION: The amount of elongation will be found from the strain which is given by

$$\text{Strain} = \text{Stress/Young's modulus}$$

From our conditions

$$
\begin{aligned}
\text{Stress} \quad &= \text{load/cross-sectional area} \\
&= 2{,}000/[(\pi/4)(0.2)^2] \\
&= 63{,}662 \text{ psi}
\end{aligned}
$$

whence

$$\text{Strain} = 63{,}662/28{,}000{,}000 = .00227 \text{ in/in}$$

and so the elongation of the 20 ft (240 in) rod is given by

$$\text{Elongation} = .00227 \times 240 = .5548 \text{ in.}$$

Normally as an applied stress increases the responding strain will increase in direct proportion, with the reciprocal to Young's modulus serving as proportionality constant. At some point (the **elastic limit**) the increase ceases to be given by the (constant) Young's modulus value, but rather the material suddenly "gives way," allowing much greater strains in response to stress increases, to the point of total breakdown of the cohesive structure. At this point, referred to as the **breaking point** of the material, the plastic deformation is referred to as the **ductility** of the material. **Ductility** is defined as the ratio of the ultimate failure strain to the yielding strain. The percent of elongation at failure is calculated by

$$\text{percent elongation} = \frac{\left(L_f - L_o\right)}{L_o} \times 100\%$$

$$= e_f \times 100\%$$

where L_o and L_f are the initial and final lengths of the sample, respectively, and e_f is the final elongation. Likewise, the reduction in area at the point of failure is expressed as:

$$\text{reduction in area} = \frac{\left(A_o - A_f\right)}{A_o} \times 100\%$$

Typically ductile materials have reductions around 50% or greater, and for brittle materials, less than 10% reduction is expected.

Another term for the plastic deformation at the elastic limit is the **yield**. If we divide the stress inducing the elastic limit by the cross-sectional area of the body being deformed, then the result is the **yield strength** of the material. The yield strength indicates the ability of the material to resist plastic deformation. For certain materials there is a definite yield strength, to which we refer as the **yield point**. In others the transition is somewhat more gradual. Under elastic deformation arising from an applied stress, the strain is essentially proportional to the stress. A compression or elongation of a body in one direction due to the applied stress produces a change in the dimensions of the body in the direction perpendicular to the applied stress. Thus, generally, a compression in the stress direction will produce a broadening of the body in the normal direction, with the opposite holding for elongation. The negative of the ratio of the strain in the normal direction, or **lateral strain** to the strain in the stress direction, is **Poisson's ratio**. This value is normally between 0.25 and 0.5.

When a stress is applied to part of the surface of a body, a **shear stress** is created, producing a displacement of one plane of atoms relative to its neighboring plane. The **shear angle**, α, is the angular displacement produced in this way; for no displacement, this angle would be zero. Its tangent, $\Gamma = \tan(\alpha)$, defines the **elastic shear strain**. The **shear modulus** is the ratio $G = $ Shear Stress$/\Gamma$, which corresponds to Young's modulus for the induced strain in the direction of the applied stress. We find that Young's modulus E is related to the shear modulus G and the Poisson ratio v as:

$$E = 2G(1 + v).$$

An applied pressure δP to a body of volume V results in a change in volume δV. The **bulk modulus** of a material is defined as

$$K = \frac{V\delta P}{\delta V}.$$

This is related to the modulus of elasticity and the Poisson ratio as

$$K = \frac{E}{3(1 - 2v)}.$$

The maximum stress applied to a body before failure occurs is a known, measurable value. Dividing this by the original cross-sectional area of the body results in the **tensile strength** of the body. Dislocations in the crystal structure of a material make failure possible at reduced stresses, compared with the theoretical limits for perfect crystalline structures.

These concepts refer to the strength of the material arising from the molecular and atomic bonds of the body. We now turn to a property which relies partly on the surface nature of the body. We refer to the ability of a body to resist penetration by an indenting body as its **hardness**. Two standards for measuring hardness are the **Brinell hardness number** (BHN) and the **Rockwell hardness** (R). There is a linear correlation between the hardness and the tensile strength, as might be expected.

Energy, or work, is the product of force and distance. The ability of a body to withstand breaking under an imposed energy is its **toughness**. This property is found by using standard tests such as the **Charpy**. Another term for toughness is **impact strength**.

In general the Young's modulus of polymers is smaller than that of metals (by one or more orders of magnitude), corresponding to the fact that a given stress will induce a greater strain. A polymer with a substantial degree of elasticity is referred to as an **elastomer**.

Thermal properties of materials relate to their response to the injection or removal of thermal energy. These include the **density**, which normally decreases with increasing temperature; the **specific heat**, which is essentially the amount of heat needed to raise the temperature of the material through a single degree; and the **thermal conductivity** (measuring the ratio of the heat flux generated by a temperature gradient to the gradient itself). The **thermal diffusivity** is the ratio of the conductivity to the product of density and specific heat, and it represents the ratio of the rate of temperature increase to spatial change in the gradient. Normally materials have a characteristic temperature at which they melt, and at which they vaporize. The **heat of fusion** (latent heat) is the amount of heat inducing melting of the solid phase of the material at the melting temperature; similarly, the **heat of vaporization** is the amount of heat inducing the vaporization of the liquid material at the vaporization temperature. The **thermal expansion coefficient** is the relative change in volume of a material induced by a unit temperature rise. In SI units these properties are given by the following:

thermal conductivity	$\dfrac{kJ}{m-s-°C}$
specific heat	$\dfrac{kJ}{kg}$
density	$\dfrac{kg}{m^3}$
diffusivity	$\dfrac{m^2}{s}$
latent heat of fusion	$\dfrac{kJ}{kg}$
heat of vaporization	$\dfrac{kJ}{kg}$
expansion coefficient	$1/°C$

We recall that a voltage V (1 Volt = 1 Watt/A = 1 Joule/s – A) across a wire is equal to the product of the current I (A) through the wire and the resistance R (ohms), $V = IR$. The resistance of the wire is in turn equal to the product of the resistivity and the length of the wire, whence

Resistivity = Resistance × Area/Length (ohm – m).

EXAMPLE: The resistivity of copper is 1.7×10^{-6} ohm-cm. What is the resistance of a 200 meter length of wire whose diameter is 1 cm in diameter?

SOLUTION: By our relation,

$$\text{Resistance} = \text{Resistivity} \times \text{Length/Area}$$
$$= 1.7 \times 10^{-6}\ 200 \times 10^2/\pi(.5)^2$$
$$= .022\ \text{ohms}$$

Electrical conductivity is the reciprocal of the resistivity. **Conductors** are materials with high conductivity such as copper and aluminum. A typical resistivity value is that of copper, which is 1.7×10^{-8} ohm – m. **Dielectrics** are electrical insulators having very low electrical conductivity. Certain materials lose all electrical resistance at a sufficiently low temperature at which point they become **superconductors**. In recent years certain ceramic materials have been found to go through this change of phase at relatively moderate temperatures.

Semiconductors are materials having resistivities lying between those of conductors and insulators. Examples are silicon, germanium, and diamond. While semiconductors are poor conductors of electricity, their conductivity can be enhanced by the addition of electronic imperfections. If, for example, in a silicon lattice a single silicon atom is replaced by an atom of another element for which an electron is missing which would otherwise be there if the atom of silicon were in place, then the resulting "hole" will tend to be filled by electrons coming from a negative side of an imposed electric current. Thus, the hole will migrate toward this negative source, effectively acting like a migrating positively charged particle. Electric current in this case is referred to as **p-type semiconduction**. If, on the other hand, the electronic imperfection is due to the presence of an atom with a surplus valence electron, then this excess electron cannot fit into the usual bonding structure and will instead migrate to a positive source. This current flow is referred to as **n-semiconduction**. Binding a semiconductor with an electron "hole" (p-type) to one with a surplus electron (n-type) results in a **p-n junction**.

Diffusion (or movement of atoms within a solid matrix due to random movements on an atomic scale) is the movement of atoms relative to each other in a homogeneous (single phase) body, and the movement of atoms of one material (solute) within a matrix of a second material (solvent). In each case the **diffusion coefficient** is the ratio of the flux of the diffusing material to its concentration gradient. The diffusion coefficient is generally a function of temperature as well as of the materials involved.

A number of metals including iron, cobalt, and nickel are magnetic, in the sense that in response to a magnetic field their atoms will become magnetically aligned, resulting in the material itself becoming magnetic. We refer to the magnetic properties in this case as **ferromagnetism**. If the magnetism is permanent, then the material is **magnetically hard**; if not, it is then **magnetically soft.**

Luminescence of a material is the capability of the electrons in the atoms of the material to be raised to higher energy levels in response to energy input; in certain cases, upon halting the energy input, the electrons move back to lower energy levels, releasing the excess energy in the form of photons and in this way emitting light.

Having examined the various properties of materials, we now turn to the manner in which a body will deform and mechanically fail, in the sense that it has been permanently damaged and is more likely to break apart. Permanent deformation is referred to as **plastic deformation**. The ability of a material to be permanently deformed without breaking apart is referred to as **plasticity**.

We define **slip** as the permanent displacement of the crystal planes relative to each other due to shear stress. The **critical shear stress** is that value of the shear stress which is sufficient to induce slip. Slip will be favored in certain directions over others, dependent on the crystal structure of the material. Solid solutions of metals, wherein one metal is dissolved in another, are less vulnerable to slip than are pure metals. Similarly, grain boundaries interfere with slip, so that materials having much reduced grain size (e.g., polycrystalline materials) are much less vulnerable to slip.

We refer to cold working as the shaping of a metal by mechanical operations while at moderate (room) temperature below the melting point. The term **cold work** is a measure of the distortion resulting from a decrease in cross-sectional area during plastic deformation,

$$\text{Cold Work} = 100 \times \frac{\delta A}{A_0}$$

for the original area A_0. One result of cold working is that small amounts of slip induce less ordering, and hence reduced vulnerability to additional slip and an increase in material hardness. The increase in hardness arising from plastic deformation is **strain hardening**. The dependence of properties on temperature for a cold-worked material is seen in Figure 21.

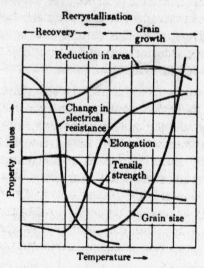

Figure 21. Property variation with temperature for a cold-worked metal

EXAMPLE: To have a tensile strength of more than 60,000 psi in its final form, we may subject a brass rod to cold work greater than 15%. What is the final diameter of a brass rod with an initial diameter of .3 inches and cold work of 25% as its drawing process specification?

SOLUTION: From the definition of the cold work if d is the final diameter of the rod then

$$.25 = [(.3)^2\, \pi/4 - d^2\pi/4]/(.3)^2\, \pi/4$$

$$d = .2598 \text{ inches.}$$

Annealing is the heating of a plastically deformed material to a temperature sufficient to move the lattice atoms into a crystalline array stronger than their original form, undergoing a process of **recrystallization**. **Quenching** and **tempering** are heat treatment methods for hardening and toughening materials. When applied to glass, tempering involves heating to a high temperature followed by immersion in oil to subject the surface to a permanent compression.

Creep is the slow deformation of a material under stress in the period of steady elongation due to applied stress. This period begins after initial adjustments of grain boundaries, flaws, etc., and ends when significant reduction in cross-sectional area begins. After the area reduction begins, the material elongates at a rapid rate and eventually fails or ruptures. The **creep rate** is the ratio of strain to time during the period of steady creep. The creep rate increases with temperature and stress, which in turn, reduce the time to failure.

The actual failure may involve a continuous reduction of cross-sectional area to zero, referred to as **ductile fracture**, or separation of the material into distinct parts, referred to as **brittle failure**. **Fracture** is the failure of the material in each of these ways.

Under cyclic stress loading, a material is more vulnerable to failure than under static loading. For sufficiently low stress the number of cycles leading to failure will be unlimited. At some stress value, referred to as the **endurance limit**, this ceases to be the case, and for stresses beyond this value the number of cycles leading to failure decreases rapidly. We refer to the failure of a material under cyclic loading as **fatigue**.

The term "plastics" is used because these materials undergo plastic deformations and have a high degree of plasticity. This is manifested by the fact that one can subject them to pressure and high temperature, pour them into a mold and upon "setting," cooling, and removal of the pressure, one will have a permanently shaped object. Materials whose plasticity increases with temperature are referred to as **thermoplastic resins**. In contrast, **thermosetting resins** set essentially as one large covalently bound molecule simply upon mechanical stress, with no slippage possible. Generally this type of resin is stronger than the thermoplastic type. The intermolecular forces of thermoplastic polymers are overcome at high temperatures while those of the thermosetting plastics are not.

MULTICOMPONENT MATERIALS AND PHASE DIAGRAMS

We now turn to the makeup and behavior of mixtures of pure materials. We refer to such mixtures as **solutions** or **alloys**. To gain an understanding of the behavior of a solution we begin with two examples.

Consider a quantity of pure water at some temperature T_0. We know that should we reduce the temperature to $T_1 = 0°C$ the water will freeze and pure ice will form. Let us now add a small amount of salt to the water stirring the mixture. We know that the small amount of salt added should totally dissolve in the water, forming liquid **brine**. Should we now reduce the temperature of the brine, we would find that at some temperature $T_2 < T_1$, which is less than the freezing temperature of pure water, we obtain a mixture of ice and brine; further reduction to a temperature of $-21°C$ results in a separation of salt from the brine, freezing of the water in the brine, and a mixture of ice and salt only. Adding more salt to our water at the initial temperature T_0 results in a brine solution, which upon cooling will produce ice and brine at yet a temperature below T_2 and whose further cooling to $-21°C$ results only in pure ice and salt. If we now successively add more and more salt to the water, we eventually obtain a solution which, upon reducing the temperature, remains a brine solution until we reach $-21°C$ at which point pure ice and salt, with no brine present, is formed. Adding additional salt to the water at T_0 results now in different behavior: if its temperature is reduced from this initial value, we obtain salt and brine, instead of ice and brine at some temperature which is now above $-21°C$! Further reduction of the temperature to $-21°C$ results again in ice and salt. We thus see that our mixture of water and salt can result in four distinct physical phases, depending on the amount (or concentration) of salt and the temperature. These phases can be listed as: a) brine (pure liquid consisting of water and dissolved salt); b) ice and brine (that is, liquid brine with salt crystals); c) salt and brine (that is, liquid brine with salt crystals) and d) ice and salt (that is, mutually exclusive bodies of salt and ice). The minimum temperature of $-21°C$ reached by the demarcation points between ice + brine and brine or salt + brine and brine is the **eutectic temperature** of a salt/water solution; the corresponding salt concentration is its **eutectic composition**. The diagram indicating the phase of the solution in the temperature/composition plane is referred to as a **phase diagram** for the solution (Figure 22). By means of the phase diagram we know what is the physical state of the solution for any combination of values of salt concentration and temperature. The curves demarcating the brine state and those of salt + brine and ice + brine are the solubility curves for the solubility of ice in brine and salt in brine, respectively. We note that this phase diagram is an equilibrium phase diagram, in the sense that it reflects the phases that would be formed after a time which is long enough for our system to reach equilibrium and undergo no transient changes.

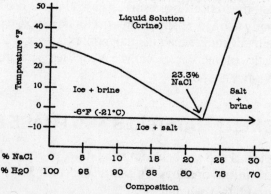

Figure 22. Phase diagram for water and salt mixture

An example of a metallic phase diagram is given by the solution of copper and nickel (Figure 23). It is found that the atoms of these metals occupy about the same volume, while in crystal form each has a face-centered cubic structure. Hence, each can physically replace the other in a crystal lattice. Pure nickel is known to melt at 1,455°C, while the melting point of pure copper is 1,083°C. Hence, if we begin with pure nickel at a temperature above 1,455°C and begin to reduce its temperature, the nickel will crystallize (freeze) at this temperature, and below it we will find solid nickel. Let us now add a small amount of copper to the liquid nickel at the initial temperature above 1,455°C. Reducing the temperature of this solution now produces, at a temperature T_1, which is less than 1,455°C, a semi-solid phase, comprised of intermixed solid and liquid particles; this phase is maintained as we reduce the temperature until we reach a value $T_2 < T_1$ at which point we find a solid solution of copper and nickel. The values of T_1 and T_2 depend on the relative concentrations of copper and nickel. As more copper is added and its concentration rises, their values decrease. Moreover, after an initial increase in their difference, we find that they tend to be closer to each other, agreeing at the value 1,083°C where we have pure copper only. The curve in the plane determined by temperature/copper composition obtained from the values of T_1, beyond which all compositions are liquid, is referred to as the **liquidus curve**; similarly that curve determined by T_2, beyond which all compositions are solid, is the **solidus curve**. The resulting phase diagram is again an equilibrium phase diagram corresponding to equilibrium conditions in which enough time has elapsed to allow movement of atoms and heat to produce uniform conditions in our sample. In an actual slow freezing or casting process for a copper/nickel alloy, we would expect to avoid the semi-solid state, obtaining, for uniform temperature, bodies of material at concentrations determined by the liquidus and solidus curves for that temperature. Equilibrium diagrams are often referred to as **constitutional diagrams**. The significance of the phase diagram is that it makes it possible for us to extract such information as what will be present and in what percentage for a given state. For the simple Cu-Ni phase diagram, any point below the solidus curve is a solid Cu-Ni alloy system. For a point in the semi-solid zone between solidus and liquidus curves, we have a mixture of solid and liquid phases whose compositions can be found from the so-called **Lever rule**; what we do is to find the points of intersection of a drawn through the point for constant temperature with the solidus and liquidus curves. These points of intersection determine the states of the two phases present.

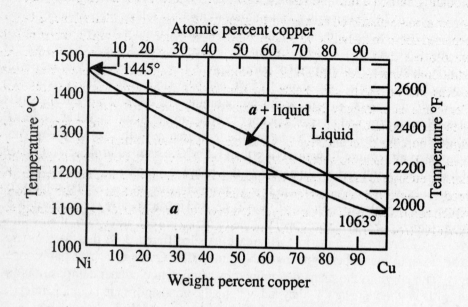

Figure 23. Copper-nickel phase diagram

For Figure 23, the fractions of solid and liquid at point A can be found by the lever rule and the definitions:

$$\text{fraction of solid} = \frac{s}{t} = 1 - \text{fraction of liquid}$$

$$\text{fraction of liquid} = \frac{l}{t} = 1 - \text{fraction of solid}$$

In this case, at a temperatue of 1,200°C and a 70 weight percent,

$$s = 70 - 60 = 10$$
$$l = 79 - 70 = 9$$
$$t = 79 - 60 = 19$$

$$\text{fraction of solid} = \frac{9}{19} = 0.47$$

$$\text{fraction of liquid} = \frac{10}{19} = 0.53$$

Probably the most important metallic alloy is steel, which is an alloy of iron and carbon. Pure iron undergoes recrystallization at the temperatures $T_0 = 910°C$ and $T_1 = 1,400°C$. Below T_0 it is body-centered; between T_0 and T_1 it is face-centered. Beyond T_1 it returns to the body-centered structure. The corresponding materials are respectively, ferrite (α-iron), austenite (Γ-iron), and δ-iron (Figure 24). **Ferrite** is familiar iron at room temperature. It is soft, ductile, and highly ferromagnetic. **Carbon** is essentially insoluble in ferrite which has no space available for accommodating carbon atoms. **Austenite** is soft and ductile, not ferromagnetic, but can accommodate dissolved carbon for forming steel. The δ-iron, like ferrite, is body-centered cubic in its lattice structure, but because of the high temperature at which we process it, it can accommodate substantially more carbon than ferrite. An additional carbon-iron alloy form is **cementite**, or **iron-oxide**, which arises when excess carbon atoms are bonded to iron atoms in an orthorhombic unit cell. Cementite is extremely hard and substantially strengthens steels in which it is present. A metallurgically important form of steel is **pearlite**, which results when slowly cooling steel undergoes a eutectic phase transformation at 723°C. It is composed of alternate sheets of iron and iron-carbide, Fe_2C. If steel is quenched rapidly enough to make it undergo a phase transformation at a low temperature, the product is referred to as **martensite**. Martensite is supersaturated with carbon atoms which could alloy with iron at a higher temperature but which, due to quenching, are literally "frozen" into place at the lower temperature (Figure 25).

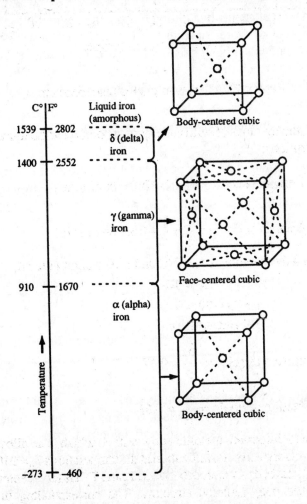

Figure 24. Crystal forms of iron

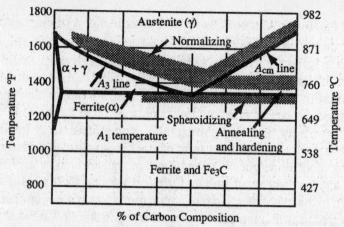

Figure 25. A portion of the iron-carbon phase diagram

Most steels contain elements in addition to iron and carbon; these may include manganese, chromium, or nickel for additional features. Such alloy systems contain more than two components, giving rise to **multicomponent phase diagrams.** For three components we would thus find **ternary phase diagrams.** We again obtain eutectic compositions and temperatures, which may now be planes or higher dimensional surfaces.

The **phase rule** for multicomponent alloy systems takes the form

$$P = C + E - V.$$

Here P is the maximum number of phases that could coexist under equilibrium conditions; C is the number of components; E is the number of environmental factors being taken into account (e.g., temperature, pressure, electric fields, magnetic fields, composition); and, V is the number of unassigned variables or degrees of freedom of the system.

DIFFUSION AND REACTIONS IN MATERIALS

Diffusion is the movement of atoms or molecules in a material. Fick's law describes this phenomenon; in one-dimension, steady-state diffusion is written as:

$$J = -D\frac{dC}{dx}$$

where J is the amount of the atoms or molecules moving per unit time, C is the concentration of the atoms, and x is the distance along which the diffusion occurs.

Atoms or molecules will tend to diffuse from points of higher to points of lower concentration. As the temperature rises the diffusion coefficient will tend to rise; similarly, the diffusion coefficient depends on the materials present as well as ambient conditions (pressure, etc.). An interesting example of diffusion is the case where the director of the British mint clamped together two well-cleaned blocks of gold and of lead and kept them in this way for four years. At the end of this period gold could be detected at a depth of 5/16 inches into the lead block. The diffusion of a metal's own atoms through its lattice is referred to as **self-diffusion**. This form of diffusion is examined by observing the movement of radioactive isotopes of a metal in the solid lattice of nonradioactive isotopes of the same metal. An industrial process based on diffusion of carbon atoms in steel is that of **carburizing**. In this process a low carbon steel is exposed to carbon atoms at a high temperature; the small carbon atoms diffuse into the subsurface of the solid metal to form a high-carbon coating. If the body is then quenched, we obtain a hard, wear-resistant, martensite surface with a ferrite core; such materials are of particular use in machining processes. Atoms of gases can also diffuse in metals. One undesirable instance of such a process is that of **hydrogen embrittlement**, wherein atoms of hydrogen arising, for instance from radiation sources, penetrate steel, diffuse to regions of greater stress, and induce cracks and premature fracture. This is of particular concern in nuclear reactors.

A variety of reactions that produce new solid phases are of major engineering interest. The simplest such reaction is that of a **polymorphic transformation**. An example would be that of recrystallization of iron at its appropriate transition temperatures. As we have seen earlier, this is accompanied by changes of density and other material properties. In studying phase diagrams, we noted that upon cooling an alloy system at a slow rate, we may obtain two or more distinct materials whose concentrations experience jumps. Such reactions are referred to as **eutectic reactions**. An example of a eutectic reaction is the pearlite formation reaction referred to earlier. Eutectic reactions must proceed slowly, since they require diffusion of atoms of the component materials with respect to each other to achieve the demands of the phase diagrams.

Other reactions exist at different phases and states while a **eutectic reaction** describes a liquid cooled to another liquid A and a solid B. (When a solid is cooled to a liquid A and a solid B, the term eutectic is replaced by **eutectoid**.) However, if the liquid and a solid A is cooled to a solid B, the reaction is **peritectic**, and liquid cooled to liquid B and solid A is **monotectic**. Again, **peritectoid** and **monotectoid** are used for the cooling of the solids.

Of major importance is the rate of a reaction. If the reaction requires, for example, that we have simultaneous diffusion and heat transfer, then rapid cooling may prevent the needed diffusion from occurring, and hence the material will enter into a state which is not equilibrium and which from the point of view of least potential energy, is unstable. Such a state is referred to as **metastable**. An example of such a metastable state is that of supercooled water, obtained when water, in a clean beaker, is cooled to a temperature below 0°C without freezing. In the same sense, an alloy system in its liquid phase may be cooled so rapidly that diffusion to accommodate the demands of the phase diagram cannot take place. In this case the material may enter the metastable constitutionally supercooled state. In general, to move from a metastable to a stable state may require some additional input of energy, an example being vibration of supercooled water, which immediately initiates freezing. The martensitic phase of steel is another example of a metastable phase.

By altering the microstructure of materials we can, in principle, incorporate desirable properties into the materials. These changes can be induced via plastic deformations, by recrystallization, by selection of appropriate solvent and trace components of our materials, and through appropriate crystal orientation. In turn the microstructure determines the thermal and mechanical properties (e.g., density, thermal conductivity, hardness, etc.) that are of importance to use of these materials. Some typical heat treating processes are **annealing**, in which we heat the material to remove strains (for cold worked materials and glass) and to soften (for steel); **quenching**, for hardening; **tempering** via quenching, to toughen materials such as steel and glass; **age-hardening**, in which we rapidly cool an alloy, reheat partly, and then cool again, for hardening of aluminum alloys; and **firing**, in which we heat a solid solution to form glassy bonds for producing bricks.

Treatments that help strengthen materials are of crucial importance. One approach is to reduce the grain size resulting in a reduced ability to encounter slip. For alloy systems, the low concentration (solute) atoms may tend to cluster around a dislocation in the high concentration (solvent) atoms raising the stress needed to cause failure. The hardening method based on this is referred to as **alloy hardening** and is found in all alloy systems. In some systems the method of **precipitation hardening** or **age-hardening**, occurring as a result of precipitation from a supersaturated solution, results in greatly improved material strength. Steel can be made particularly strong, as we have seen, as a result of the richness of hardening treatments such as the production of the martensitic state.

CORROSION AND RADIATION DAMAGE

Corrosion is the destructive attack of a metal by chemical or electrochemical reaction with its environment. This is in contrast with erosion, galling, or wear, which are physical processes. **Rusting** refers to the corrosion of iron or iron-based alloys. Thus, by definition, nonferrous metals may corrode but do not rust.

Corrosion generally occurs as the result of the processes of solution and oxidation. **Solution** is the process wherein a material dissolves in a solute. This may take the form of dissolving as molecules, as in the case of sugar in water, or as electrically charged ions, as in the case of salt, creating sodium and chloride ions. Generally, a solute dissolves more easily in a solvent if a) the solute molecules or ions are small; b) the solute and solvent are similar in structure; c) the temperature increases; and d) multiple solutes are present. **Oxidation** is the process of removing electrons from an atom. An example is the removal of electrons in two stages from an iron atom:

$$Fe \rightarrow Fe^{2+} + 2e^-$$

$$Fe^{2+} \rightarrow Fe^{3+} + e^-$$

The first reaction involves the oxidation of iron to form ferrous ions, while the second is the oxidation of ferrous ions to form ferric ions. In the presence of water and oxygen, rust, whose chemical formula is $4Fe(OH)_3$, can now be produced according to the reaction

$$4Fe + 3O_2 + 6H_2O \rightarrow 4Fe(OH)_3$$

which requires the initial dissolving of the iron in the water to produce iron ions. Thus iron will not rust unless both water and oxygen are present. This form of corrosion is referred to as **solution corrosion**. A second type of corrosion is **galvanic corrosion**; if iron is placed in water, it will rapidly produce electrons and ions. The presence now of negatively charged electrons and positively charged ions results in an electrical potential, the **electrode potential**. Hydrogen atoms, too, will dissolve in water, producing an electron and the positively charged hydrogen ion. The potential difference between a source of hydrogen and a plate of iron in water will be 0.44 volts, promoting an electric current from the iron to the hydrogen source. The pair consisting of the hydrogen source and the iron plate constitutes a **galvanic couple** or **cell**. We refer to each as an **electrode**. In general, the electrode supplying electrons to the circuit is referred to as the **anode**, while the electrode receiving the electrons is the **cathode**. The flow of electrons to the cathode, or hydrogen source results in the generation and release of hydrogen gas at the cathode, resulting in a greater tendency for oxidation of the iron at the anode to produce additional electrons. Hence iron at the anode continues to dissolve into the solution, releasing additional electrons to the cathode, and hence promoting the process. The iron plate will, in this way, corrode through the process of **galvanic corrosion**. The removal of hydrogen from the water results in a greater concentration of hydroxide ions $(OH)^-$ whose presence now promotes the production of rust through the reaction

$$Fe^{3+} + 3(OH)^- \rightarrow Fe(OH)_3.$$

The rust is essentially insoluble in water, whence it precipitates and allows the galvanic corrosion reaction to continue.

Galvanic cells may be formed between any two distinct materials. The material producing the greater flux of electrons (having the greater **electromotive potential**) will serve as the anode, while the second material will serve as the cathode. Examples include tools containing both steel and brass, steel pipes connected to copper plumbing, and others. Some methods for attempting to protect steel are based on this principle. Thus galvanized steel is made by coating the steel with a zinc coating. So long as the latter is not scratched, the zinc will indeed protect the steel; if, however, the steel is exposed at a point, then this point can serve as the source of corrosion arising from the zinc/steel galvanic cell, with zinc acting as the anode, and steel as the cathode.

Methods for protecting the surface of an object from corrosion are based on three approaches: a) isolation of electrodes via protective surfaces; b) avoidance of galvanic couples; and c) use of galvanic cells for protection. One approach to protecting steel by isolation is through the plating of the steel with ions of $(CrO_4)^{2-}$, with the chromium serving the role of introducing an electrically passive surface for the steel. Avoidance of galvanic couples is achieved, e.g., by the use of a single component. On the other hand, the galvanic reaction can be used to protect the material. An example is the use of zinc coating on a steel surface to serve as a sacrificial material for the galvanic reaction. Often corrosion can be inhibited by the formation of a protective oxide layer, an example being given by aluminum.

RADIATION ALTERATION OF MATERIALS

Radiation can be classified as **electromagnetic** and **particulate**. The former include radio waves, infrared radiation, light, x-rays, and gamma-rays; the latter include accelerated protons, electrons (β-rays), helium nuclei (α-rays), and neutrons. In the nuclear industry the principal particles are gamma-rays and neutrons.

Radiation results in transferring energy to the material on which it impacts, resulting in the breaking of bonds and the rearrangement of the atoms of the material. The nature of the impact depends on the particle. For charged particles such as α and β rays, proximity of the particle to the atoms of the material results in interraction and possible structural alteration. On the other hand, the neutron is a massive, uncharged particle that can only interract upon the occurrence of a collision. This collision will usually occur in the interior of the material and result in an interior vacancy. The displacement of an atom in a material can result in a number of possible changes: the material may become activated and available for further reactions; it may be degraded, resulting in loss of material strength; or it may become distorted via displacement. Some typical effects of neutron bombardment include coloring of glasses, loss of elasticity in natural and butyl rubber, hardening of natural rubber, loss of ductility of carbon steels, and rendering plastics unusable as structural materials, depending on the energy of the neutrons. A positive effect for certain stainless steels is an increase in tensile strength and hardness due to dislodging of atoms and restriction of slip. The effects of radiation damage can generally be repaired via annealing at temperatures which may be relatively moderate.

COMPOSITE MATERIALS

It is possible to prepare materials which are composites of various kinds of materials having desirable but complementary properties. An obvious example would be the coating of metal by glass to inhibit corrosion of the metal. Composite materials can be divided into three categories: agglomerated materials, surface coatings, and reinforced materials. We shall now examine each in turn. **Agglomerated materials** consist of a matrix material, "filler" material for the pore spaces in the matrix, and a "glue" to bind the rest of the structure together into the form of a single "brick." A typical glue or binding material would be water in a hydrated solid, linking together distinct molecules. Such is the case for concrete: **concrete** is gravel, with a mixture of sand to fill the pores of the gravel. The remaining space is then filled with a paste of cement and water. The cement hydrates to form a cement paste binding the agglomerate together. The actual chemical constitution of Portland cement is (before binding) a mixture of calcium aluminate and calcium silicate.

Sintering is a process of heating to agglomerate small particles into a large bulk structure. The exact relations may involve melting of a binding material, followed by its crystallization as part of a large, hardened structure. An example of this process is vitreous sintering which is used in brick manufacturing.

We often wish to coat a material surface with a hard surface for resisting wear. Methods for doing this include welding a hard metal coat onto the surface, spraying a special coating onto the surface, or alteration of the surface for this purpose. **Surface alteration** is done usually via heat treatment of the surface, using such techniques as **induction** and **flame hardening**, the former using high frequency currents at the metal surface, and the latter using localized heating. An alternative method of surface alteration rests on the diffusion of elements into the surface, an example being **nitriding**. Here suitable chemical reactions are used to produce a concentration of aluminum nitride at the surface, which forms a war-resistant surface layer. An alternative approach is **carburizing**, wherein suitable processes result in a high-carbon "case" immersed in the surface of the material. Compressive forces can be applied to the surface of the material by suitable volume changes induced by thermal or compositional changes. These too can result in desired surface layer changes.

Reinforcement of materials can be carried out by introducing small particles into the material which are substantially harder than their surroundings. With this aim we develop a variety of dispersion-strengthened alloys which are substantially stronger than their original forms. In place of particles it is sometimes advantageous to introduce high temperature "whiskers" of, e.g., Al_2O_3 to add strength to a ductile metal which is maintained at elevated temperatures. The same concept is the basis for such materials as glass-reinforced plastics and steel reinforced concrete.

Properties of composites are a reflection of their components and the purposes for which they have been designed. Thus for glass-reinforced plastic the glass provides tensile strength and dimensional stability while the plastic gives us coherency and is not porous. The determination of gross mechanical properties can be carried out in much the same manner as for noncomposite materials. Thus, for example, the determination of the heat transfer properties of a composite material would have to be found by means of standard heat flux and temperature experiments.

REVIEW PROBLEMS

PROBLEM 1

Discuss the difference between extensive and intensive properties and name the properties inherent to a given material.

SOLUTION: Any material properties can be described as either extensive or intensive. An extensive property is one which depends on the amount of material present. An intensive property is one which depends upon each specific material. Density, the yield point, Young's modulus, and Poisson's ratio are all intensive properties and are inherent to a given material. Flexural strength depends on the cross-section of the element and the direction in which the force is applied. Therefore, it is an extensive property and is not inherent to the material.

PROBLEM 2

Explain how the electrical conductivity of an *n*-type semiconductor crystal is dependent upon the number of donor atoms and the electron mobility.

SOLUTION: In an *n*-type semiconductor crystal, the density of electrons in the conduction band is essentially the density of donor atoms in the crystal, and hence the electrical conductivity is given by the relation:

$$\sigma_n = Ndq\mu_n$$

where

σ_n = Electrical conductivity of an n-type semiconductor

Nd = Number of donor atoms per cubic meter

μ_n = Electron mobility

q = Charge of an electron, which is constant

Hence the conductivity closely depends on the number of donor atoms and the electron mobility.

PROBLEM 3

The measure of specific heat of a material relates the quantity of heat required to change the temperature of a given mass of that material. At ordinary temperature ranges for a small temperature range the specific heat, C, may be considered constant. Explain the significance of aluminum having a greater specific heat (0.217 cal/gm°C) than copper (0.093 cal/gm°C).

SOLUTION: For a solid:

$$q = C\Delta T$$

Rearranging,

$$\Delta T = \frac{q}{C}$$

For the same ΔT, as C decreases, q must decrease. Likewise, as C increases, q must increase. Thus it takes more energy in the form of heat to change the temperature of aluminum than copper of equal mass.

PROBLEM 4

The viscoelastic behavior of an amorphous polymer will change with increasing temperature form its rigid structure at low temperatures. The polymer's behavior may proceed through the following stages: leathery, rubbery, and viscous. Explain the properties of these stages and their order.

SOLUTION: The polymer goes from a leathery stage, to a rubber and then a viscous stage. In the leathery stage the polymer can be deformed readily but it cannot regain its shape quickly if the stress is removed. In the rubbery stage the polymer can regain its original shape quickly. In the viscous stage the polymer deforms extensively by viscous flow.

PROBLEM 5

Molybdenum has a body-centered cubic lattice structure. What are the number of atoms per unit cell?

SOLUTION:

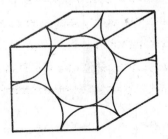

Figure 26. Body-centered lattice structure

$$1 \text{ atom centered} + 8 \times (\frac{1}{8} \text{ atoms in each corner}) = 2 \text{ atoms}$$

PROBLEM 6

Aluminum has a higher atomic packing factor that molybdenum by about 6%. What does this mean?

SOLUTION: Aluminum is heavier than molybdenum.

PROBLEM 7

Miller indices provide a system of notation for describing crystallographic planes and directions. Write a set of indices that is parallel to (101).

SOLUTION:

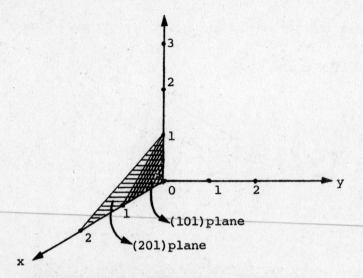

Figure 27. Parallel Miller indices and planes

Miller indices specify a plane by the integral common denominator of the values of the three coordinates of its location. The planes (101) and (201) overlap and thus are parallel.

PROBLEM 8

Using Miller indices and the lattice shown in Figure 29 describe the direction of [111].

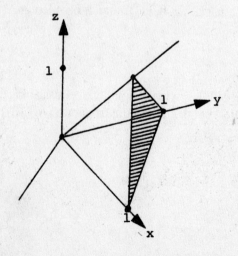

Figure 28. Miller indices and direction of [111]

SOLUTION:

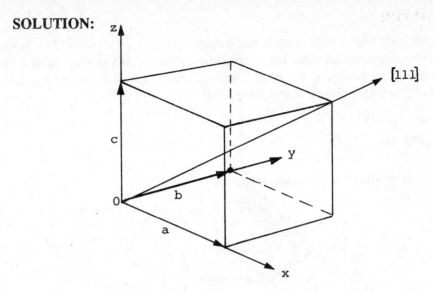

Figure 29. Plane (111) perpendicular to [111]

The plane (111) is shown in Figure 30; it can be seen that the direction [111] is perpendicular to it. This is true in general for Miller indices. But Miller indices apply only to cubic lattice structures.

PROBLEM 9

Compare the planes shown in Figure 31.

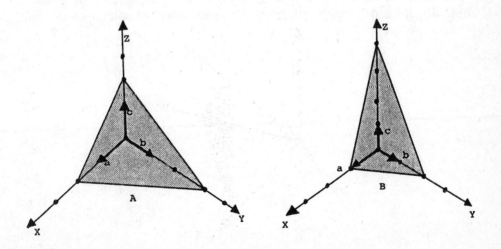

Figure 30. Miller indices of two planes

SOLUTION: Miller indices for A and B respectively are (232) and (124). Since Miller indices apply to all cubic lattice structures, plane A does not represent only body-centered crystal lattice structures. The planes are not parallel, and the direction vector [232] is perpendicular to the plane (232).

PROBLEM 10

Draw a phase diagram for a eutectic system.
SOLUTION:

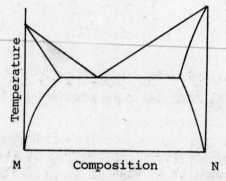

Figure 31. Phase diagram of a eutectic composition

A eutectic system is made up of two components and must satisfy the condition that at one point, the eutectic point, the liquid state must be completely insoluble in the solid state. In order to accommodate partial solubility in the solid state, a solid solution should exist above the eutectic point along with the liquid solution for each component. This solid solubility should increase with decreasing temperature above the eutectic point and decrease with decreasing temperature below the eutectic point.

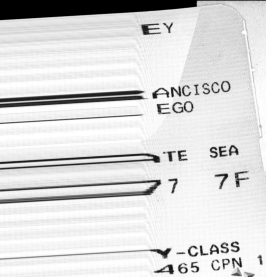

Engineering Economics

Chapter 12

ENGINEERING ECONOMICS

INVESTMENT

Money invested generates more money, and engineering economics gives the basis for making a systematic evaluation of investment alternatives for economic decisions to allocate the corporation's capital for a maximum return on these investments. An economic evaluation of a proposed investment includes determining the expected profit and capital expenditures, and these alternatives can include construction of a new plant and expansion of existing facilities.

New plants, products, and technology require new capital, and most firms have limited resources. Consequently, investment decisions require capital budgeting. This is the evaluation and selection of the best investments from a set of alternatives. Methods for evaluating investments include net present value and rate of return, among others, for private companies and benefit-cost analysis for public projects.

Risk is a part of the decision process, also. The analysis of projects must incorporate the level of risk to be able to compare projects with high returns and high risks with those having lower returns and more certain outcomes.

Engineering economics provides the framework for the preparation of economic feasibility studies as part of a company's on-going planning process. The fundamental concepts used in these evaluations include profit, capital, expenses, and cash flow. These require an understanding of the time value of money which is discussed in the next section.

TIME VALUE OF MONEY

Investment means committing funds in the present with a certain amount of assurance for a greater return of this money in the future. This growth in money is called the **time value of money**.

Interest is the cost of borrowed money, or it is the reward for lending money. Also, interest is the return which can be obtained when money is put to productive use.

Simple interest means that only the principal, P, is used in calculation of interest, I, due. Thus:

$$I = iPn \tag{1}$$

where n is the number of interest periods. It is understood that n and i refer to the same unit of time (year, month, etc.).

If $1,000 is borrowed for five years at 8% simple interest rate, the total interest would be:

$$I = (0.08)(\$1,000)(5) = \$400$$

and the total amount, F, due at the end of the loan period is equal to:

$$F = P + I = \$1,000 + \$400 = \$1,400$$

Simple interest is rarely used, and usually interest is compounded. **Compound interest** means that interest which has been accrued over the interest period is also subject to the interest rate in the next period. If $1,000 was borrowed for five years at 8%, compounded annually, the amount of principle plus interest, F_1, due after the first year could be calculated just as simple interest, since there is no compounding in the first year.

$$F_1 = P + iP = P(1 + i)$$
$$F_1 = \$1,000(1 + 0.08) = \$1,080$$

In the second year the interest rate i ($= 8\%$) is then applied to F_1 ($= \$1,080$) if no payment is made at the end of the first year:

$$F_2 = iF_1 + F_1$$
$$F_2 = iP(1 + i) + P(1+i)$$
$$F_2 = P(i + 1)^2$$
$$F_2 = \$1,000(1 + 0.08)^2 = \$1,166$$

In the third year, the procedure is repeated:

$$F_3 = iF_2 + F_2$$
$$F_3 = iP(i + 1)^2 + P(1 + i)^2$$
$$F_3 = P(i + 1)^3$$
$$F_3 = \$1,000(1 + 0.08)^3 = \$1,260$$

The amount to be repaid in years four and five would be:

$$F_4 = P(i + 1)^4 = \$1,361$$
$$F_5 = P(i + 1)^5 = \$1,469$$

In general, the formula to compute the amount F_n to be paid at the end of n time periods with an interest rate of i is given by the following equation

$$F_n = P(1 + i)^n \tag{2}$$

Compounding can be calculated more than once a year, e.g., quarterly, monthly, even daily. However, interest rates are usually quoted as an **annual nominal interest** rate, r. If m is the number of times the interest is compounded between payment, then the **annual effective interest rate**, i_e, is given by using equations (1) and (2), and the result is the following equation.

$$i_e = (1+\frac{r}{k})^k - 1 \tag{3}$$

For example, the effective interest rate is $i_e = 0.0824$ for a nominal interest rate $r = 0.08$ compounded quarterly ($k = 4$) using equation (3). The value of i_e is used in equation (2) to evaluate F_n given P; and for P equal to $1,000, the value of F_5 is $1,486 for five years.

As shown by equation (2), money, having an ability to earn interest, has its value increase over time, thus the term **time value of money**. The **future worth**, F_n of an amount of money P is given by equation (2), and P is called the **present value** of an amount of money whose future worth is F_n available in n time periods at an interest rate of i. An amount of money F_n, available in n time periods in the future is worth less than the same amount of money available in the present. That the amount F_n decreases in value is shown by rearranging equation (2) to have:

$$P = \frac{F}{(1+i)^n} \tag{4}$$

In the example, $1,469 received in five years is only worth $1,000 if the interest rate available is 8%.

INFLATION

The cost of goods and services increases with time as a result of inflation. The inflation rate is usually given as a percentage that is compounded annually. For a constant inflation rate, f, expressed as a fraction over a period of n years the future cost of a commodity, F_c, increases in relation to the present cost, P_c, by the following equation

$$F_c = P_c(i + f)^n \tag{5}$$

Also, the future worth, F, of money decreases in relation to the present value, P; and the devaluation is given by the following equation for a constant inflation rate for n years.

$$F = \frac{P}{(1+f)^n} \tag{6}$$

However, if P is invested at a constant interest rate i, the future worth F would be given by the following equation using equations (2) and (6).

$$F = \frac{P(1+i)^n}{(1+f)^n} = P\left[\frac{1+i}{1+f}\right]^n$$

For example, if the inflation rate was 6% ($f = 0.06$) for the first five year period

($n = 5$) when $P = \$1,000$ was invested at an interest rate of 8%, the future worth is only $1,098 compared to $1,469 not considering inflation.

TAXES

Interest earned in a given year is subject to taxes for that year. If the interest period is the same as the tax period, then the interest earned during this period is iP, and the tax due on this earned interest is tiP where t is the tax rate. For companies, t can be as much as 0.38, and the net return after taxes is interest minus taxes, $I - T = iP - tiP = (1 - t)iP$. Consequently, the future worth, F, after one year including taxes is:

$$F = P + (1 - t)iP = [1+(1 - t)i]P$$

and if inflation proceeds at a rate f during this year, the above equation can be written as:

$$F = \left[\frac{1+(1-t)i}{1+f}\right] \qquad P = \left[1+\frac{(1-t)i-f}{1+f}\right]P$$

Then this equation can be put in the form of equation (2) by defining a composite interest rate that includes inflation and taxes as:

$$i_c = \frac{(1-t)i-f}{1+f} \tag{7}$$

Now, the effects of interest, inflation, and taxes can be included in the evaluation of the future worth, F, knowing the present value P by the following equation.

$$F = P(1 + i_c)^n \tag{8}$$

For a tax rate of 0.20 with an inflation rate of 0.06, the future worth of $1,000 invested at an interest rate of 8% for five years is now only $1,019 compared to $1,469 not considering inflation and taxes.

The intervals for the interest, inflation, and tax rates in equation (8) must all be the same, and the rates must be constant. If this is not the case, the same procedure is used, but each interval has to be evaluated separately and sequentially.

RISK

The effect of risk can be approximated in this analysis by replacing the interest rate i in equation (7) with the sum $(i + i_r)$. Here i_r represents an addition to the cost of financing a project when there is more risk involved than is normally expected.

CASH FLOW

To this point we have been discussing the results of investing a single sum of money. However, money received from the sale of products and spent on manufacturing costs by a company occurs on nearly a continuous basis, and income is reinvested daily by the company. To be able to analyze the income and expenditures for a company, it is convenient to select an interval, typically a year, to perform the evaluations. The **cash flow** for the period is the difference between all of the funds received and all of the funds disbursed. It is convenient to represent this net annual cash flow on a diagram, and this is especially useful for project evaluations. A **cash flow diagram** will show what is expected to take place over the life of a project. The horizontal axis represents time intervals, and the vertical axis represents the amount of cash flow in each of the intervals. A simple cash flow diagram is shown in Figure 1, where the negative cash flows in the first two years represent a net loss for the project. After that, all flows are positive, meaning a net gain in each of those three years. This diagram could apply to a company that planned to purchase a new generator which would reduce utility bills by $4,000 per year for six years, and the generator would be purchased with two payments of $9,000 in the first two years. Then at the end of the sixth year the generator could be sold at a salvage value of $1,000 which is added to the income received in the sixth year.

This example illustrates the **end of year convention** which assumes that income and disbursements are made at the end of each year. Although receipts and payments are usually made throughout the year, the end of year convention greatly simplifies calculation, and poses no problem, since it will not lead to errors in choices between alternatives.

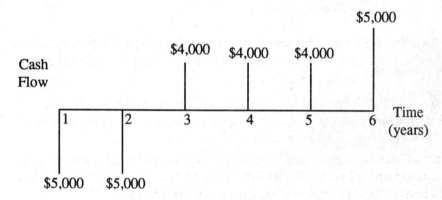

Figure 1. Simple cash flow diagram

A more realistic cash flow diagram is shown in Figure 2 for a typical projection of the net annual income for the estimated life of a new plant. To construct this diagram many items must be estimated. These include demand for product, plant capacity to meet the demand for the company's planned penetration into the market, selling price, cost of raw materials, direct fixed capital for design and construction of the plant, allocated and working capital, land costs, and out-of-pocket expenses.

As shown in the diagram, the new plant is in the negative cash position through the sixth year, and in the positive cash position from the eighth year through the end of the plant's planned life in the fifteenth year. The **break-even point** is shown at the seventh year when the firm's cumulative cash flow reaches zero. **Break-even ca-**

pacity is the production rate at which all of the costs, excluding depreciation, are equal to the sales realized.

The break-even point and the payback period (discussed later) are two ideas that are used in **break-even analysis** to compare different projects. Other methods used by companies to select the best alternative include determining the useful life for alternate pieces of equipment and the capacity utilization of alternative pieces of equipment in terms of time used per year.

INTEREST FACTORS

There are several interest factors that are routinely used in engineering economics calculations to evaluate the present value P, future worth F, uniform payment (receipt) A, and uniform gradient G. These equations require a constant interest rate i over a series of uniform time intervals n. Values for these factors can be found in tables of compound interest factors. Also, some electronic calculators incorporate these calculations with special keys. The cash flow diagrams for these factors are shown in Figure 3.

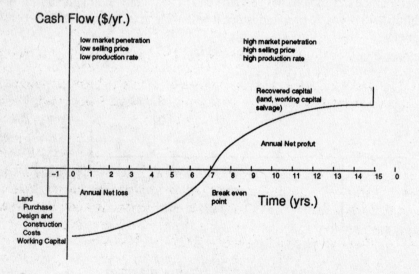

Figure 2. A typical diagram for the economic life of a proposed plant with a 15 year economic life

Single-Payment, Compound-Amount Factor, $\dfrac{F}{P}$, is equation (2) written in the following form.

$$\frac{F}{P} = (1+i)^n \qquad (9)$$

Values of the right hand side of equation (9) can be computed by specifying i and n, and these can be used to multiply the present value P to give the future worth F.

For example, with interest rate of 0.08 and five years, $n = 5$, the factor is $(1 + .08)^5$ = 1.469, and the future worth of $1,000 is $1,469.

There is a standard notation to represent the equations for these factors, and the notation $(\frac{F}{P}, i\%, n)$ is used to represent equation (9). Thus, for the example $(\frac{F}{P}, 8\%, 5) = 1.469$.

Single-Payment, Present-Worth Factor: This factor $\frac{P}{F}$ is the reciprocal of the single payment compound-amount factor and is given by the following equation.

$$\frac{P}{F} = (1+i)^{-n} \tag{10}$$

The quantity $(1 + i)^{-n}$ is also called the **discount factor**. The future worth F can be multiplied by this factor to determine the present value P. For example, with $i = 0.08$ and $n = 5$, the value of the factor is 0.68058, and the present value of $F = \$1,469$ is $1,000. Here $(\frac{P}{F}, 8\%, 5) = 0.68058$.

Uniform-Series, Compound-Amount Factor: This factor $\frac{F}{A}$ gives the future worth, F, of a uniform series of equal payments or receipts, A, that are made over n years earning an interest rate i. The equation for this factor is given below.

$$\frac{F}{A} = \frac{(1+i)^n - 1}{i} \tag{11}$$

A uniform annual series of payments can be multiplied by this factor to determine their future worth F. For example, if $1,000 is invested annually for five years, $n = 5$, at an interest rate of $i = 0.08$, the value of this factor is 5.867; and the future worth of these funds is $5,867. Here $(\frac{F}{A}, 8\%, 5) = 5.867$.

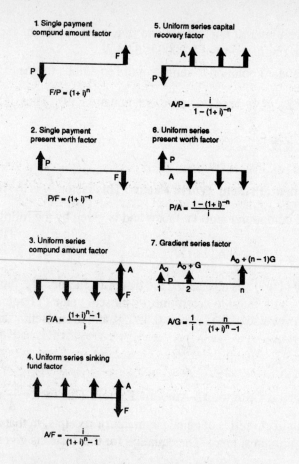

Figure 3. Cash flow diagrams for interest factors

Uniform-Series, Sinking-Fund Factor $\left(\dfrac{A}{F}\right)$, is the reciprocal of the uniform-series, compound amount factor and is given by the following equation.

$$\frac{A}{F} = \frac{i}{(1+i)^n - 1} \tag{12}$$

This factor provides a means to compute the value of a uniform series, A, to have a total amount F accumulated after n years. For five years $(n = 5)$ and an interest rate of $i = 0.08$, the factor is 0.17045; and using $F = \$5,867$ from above, then $A = \$1,000$. Here $(\dfrac{A}{F}, 8\%, 5) = 0.17045$.

Uniform-Series, Capital-Recovery Factor $\left(\dfrac{A}{P}\right)$, gives the uniform series value A that depletes an amount of money P over n years with an interest rate of i. This equation can be obtained by multiplying equations (9) and (12) i.e., $\dfrac{A}{P} = \left(\dfrac{F}{P}\right)\left(\dfrac{A}{F}\right)$, and the result is:

$$\frac{A}{P} = \frac{i}{1-(1+i)^{-n}} \qquad (13)$$

For five years ($n = 5$) and an interest rate of $i = 0.08$, this factor is 0.25046, and for an amount $P = \$3,993$, the value of A is $\$1,000$, i.e., five payments of $\$1,000$ could be distributed. Here ($\frac{A}{P}$, 8%, 5) = 0.25046.

Uniform-Series, Present Worth Factor is the reciprocal of the uniform-series, capital-recovery factor, is used to compute the principal needed to assure a uniform series of payments for n years at interest rate i. The equation is:

$$\frac{P}{A} = \frac{1-(1+i)^{-n}}{i} \qquad (14)$$

This equation has a value of 3.993 for $n = 5$ and $i = 0.08$; and for $A = \$1,000$, the value of P is $\$3,993$. Here ($\frac{P}{A}$, 8%, 5) = 3.993.

Gradient Series Factor is for an initial series value of A_O and each succeeding year A_O is increased, first by an amount G then $2G$ as shown in Figure 3. At year n the series value is $A_O+(n-1)G$. The amount accumulated after n years is F, and that amount can be converted into a series of uniform payments A by the following equation.

$$\frac{A}{G} = \frac{1}{i} - \frac{n}{(1+i)^n - 1} \qquad (15)$$

This equation is used with the **gradient-series, present worth factor**, $\frac{P}{G}$ to determine the present worth of the series where $\frac{P}{G} = \left(\frac{A}{G}\right)\left(\frac{P}{A}\right)$, and $\left(\frac{P}{A}\right)$ is given by equation (14). The use of these factors will be illustrated in a subsequent example.

MINIMUM ATTRACTIVE RATE OF RETURN (MARR)

Private corporations require a **minimum attractive rate of return** (*MARR*) before considering investing in a project. This is an interest rate that may be a project-specific number, but it usually reflects the average return on investment for a particular corporation. Determining the appropriate *MARR* is a corporate policy matter. However, in an economist's point of view, an investment is attractive as long as the marginal rate of return is equal to or greater than the marginal cost of borrowed capital. However, a corporation and their investors usually require a substantially higher return than that which could be obtained by simply investing in a bank account.

The source of funds is a consideration when choosing a value for *MARR*. A private corporation can use funds from the owners, usually through the sale of stock, or from profits which are fed back into the corporation, or from capital recovery. The **cost of equity capital** is related to a company's policy on debt financing. High leverage makes equity in a business more risky, since equity capital must sustain any losses first before debt capital. Hence, investors require a higher return to compensate for the risk. Also, the minimum attractive rate of return generally represents the opportunity cost of money, since funds expended in one project are unavailable for others.

METHODS FOR PROJECT ANALYSIS

The present worth and annual worth methods of project analysis all assume estimations have been made regarding initial investment, cash flows, acceptable *MARR*, and the possibility of salvage value on equipment after the project has been completed. The equivalent uniform annual series equation used with these methods is described also.

Present-Worth Method

A method of economic decision analysis that converts cash flows into an equivalent present value for a certain minimum attractive rate of return as shown by equation (16) to evaluate the present worth (*PW*).

$$PW = \sum_{j=1}^{n} CF_j \left(\frac{P}{F} \right) \tag{16}$$

where $\left(\dfrac{P}{F} \right)$ is evaluated by equation (10) for the interest rate i and year j for the cash flow CF_j.

When a necessary expense is being planned (the installation of a new cooling system), then the present worth of costs is minimized. For maximizing profit, *MARR* is used, and the present worth has to be greater than zero. The following example illustrates the evaluation of present worth.

EXAMPLE : Consider the following series of payments and profits from ABC Pipes Inc.'s plan to institute a new product line of pipe fittings.

$150,000   initial cost for new equipment

$22,000   yearly after tax cash flow

$40,000   maintenance of equipment in year 10

$30,000   salvage value of equipment in 20 years

SOLUTION: To find the present worth of this project at a 10% *MARR*, equation (16) gives:

$$PW = \$150,000 + \$22,000 \left(\frac{P}{A}, 10\%, 20\right) - \$40,000 \left(\frac{P}{F}, 10\%, 10\right) + \$30,000$$

$$\left(\frac{P}{F}, 10\%, 20\right) = \$9,238$$

The present worth is greater than zero, and this project would be acceptable to a corporation requiring a 10% rate of return. A modification of this example shows another option that can be considered.

EXAMPLE: Instead of manufacturing the pipe fittings, ABC Corp. is also considering expanding its existing product line of pipes with a new state-of-the art plastic. The cash flows for this project are

$72,000	initial investment
6,000	profit in year 1
6,500	profit in year 2, and the profit continues to increase $500 every year for 20 years.

SOLUTION: Assuming zero maintenance and no salvage value, the present worth for this project is:

$$PW = -72,000 + 6,000 \left(\frac{P}{A}, 10\%, 20\right) + 500 \left(\frac{P}{G}, 10\%, 20\right)$$

$$= -72,000 + 6,000(8.514) + 500(55.407)$$

$$= -72,000 + 51,084 + 27,704$$

$$= \$6,787$$

Both courses of action would be acceptable to ABC, but if funds were not available for both, other considerations need to be addressed. The pipe fittings are more profitable; and, all things being equal, would be the clear choice. Yet all things are not equal. If the lower initial investment for the plastics would leave funds for another, even more profitable project, then that too must be evaluated. Another concern is competition. And what if the project lives of the two alternatives were unequal? The two would not be comparable if the equipment for plastics would have to be retired in 10 years. In instances when the analyst needs to compare projects with unequal lives, engineering judgment is required. In this instance, both could be considered over 10 years, or over 20 years, if additional new equipment would be purchased in year 10. The new plastics alternative would then look like this:

$$PW = 72,000 + 6,000 \left(\frac{P}{A}, 10\%, 20\right) + 500 \left(\frac{P}{G}, 10\%, 20\right) - 72,000$$

$$\left(\frac{P}{F}, 10\%, 10\right) = -\$20,969$$

An accurate approximation of project life is essential in economic analysis like these.

Annual-Worth Method

Also called equivalent uniform annual disbursements, annual worth method converts uneven cash flows into their equivalent uniform annual values using the equivalent uniform annual series as described below. The results are the same as in the present worth method; that is, the indicated best choice between alternatives will always be the same, but annual worth may be easier to grasp.

Equivalent Uniform Annual Series is an equation that converts the present value (present worth, *PW*) from a series of cash flows to a series of uniform annual payments. The present worth for the series of cash flows is computed using equation (16).

$$PW = \sum_{j=1}^{n} CF_j\left(\frac{P}{F}\right) \tag{16}$$

where $\left(\dfrac{P}{F}\right)$ is evaluated by equation (10) for the interest rate i and year j for the cash flow CF_j.

Then the present worth is converted to a uniform series (*EUAS*) using equation (13) for $\left(\dfrac{A}{P}\right)$, the uniform-series, capital-recovery factor, i.e.:

$$EUAS = PW\left(\frac{A}{P}\right) \tag{17}$$

The equivalent uniform annual series (*EUAS*) is also called the **annual worth** (*AW*). When the cash flows are negative, representing costs, then the above equation is called the **equivalent uniform annual cost** (*EUAC*) or **annual cost** (*AC*). The following example illustrates this procedure.

EXAMPLE: A company needs additional warehouse space for product as a result of plant expansion. The options include constructing a prefabricated steel building, a tilt-up concrete building, or renting space. The steel building has a cost of $150,000 and a service life of 25 years with annual maintenance and property taxes of $6,000 per year. The concrete building has a cost of $200,000 and a service life of 50 years with annual maintenance and property taxes of $4,000. Both buildings have no realizable salvage value, and the company uses a 15% minimum attractive rate of return. The company can rent suitable space for $32,000 per year. Basing the decision for additional warehouse space on the equivalent uniform annual cost, should the company construct the steel building, construct the concrete building or rent warehouse space?

SOLUTION: The equivalent uniform annual cost (*EUAC*) for the building is given by:

$$EUAC = -P(\frac{A}{P}, i\%, n) - [\text{maintenance and taxes}]$$

where $(\frac{A}{P}, i\%, n)$ is the uniform-series, capital-recovery factor and P is the cost of the building.

Steel Building: $EUAC = -150,000(\frac{A}{P}, 15\%, 25) - 6,000$

   $EUAC = -150,000(0.15470) - 6,000 = -\$29,205$

Concrete Building: $EUAC = -200,000(\frac{A}{P}, 15\%, 50) - 4,000$

   $EUAC = -200,000(0.15014) - 4,000 = -\$34,028$

Comparing the above annual rates with renting at $32,000 per year, the best decision is to build the steel building for $29,205 equivalent uniform annual cost.

The following example illustrates another way to use the equivalent uniform annual cost to determine the best alternative for an investment decision.

EXAMPLE: Refractory bricks lining a furnace have an installed cost of $35,000 and last six years. The furnace must be partially relined at the end of three years at a cost of $12,000. A new high temperature refractory material has been developed, and test results show that a lining with this material will last 15 years with no intermediate repair costs. What is the largest annual cost which can be justified economically for using this new material? The company uses a 25% minimum attractive rate of return.

SOLUTION: The equivalent uniform annual cost ($EUAC$) for the current and new refractory materials will be equated to determine the maximum initial cost of the new material using the company's minimum attractive rate of return of 25%.

current material: initial cost = $35,000 lasting 6 years
   repair cost = $12,000 after 3 years

$$EUAC = 35,000(\frac{A}{P}, 25\%, 6) + 12,000(\frac{P}{F}, 25\%, 3)(\frac{A}{P}, 25\%, 6)$$

   $= 35,000(0.33882) + 12,000(0.51001)(0.33882) = \$13,941$ per year

new material: The maximum equivalent uniform annual cost for new material lasting 15 years = $13,941

Let P = initial cost of new material and

$$13,941 = P(\frac{A}{P}, 25\%, 15)$$

$13,941 = P(0.25912)$

$P = \$53,801$

Thus, $53,801 is the largest initial cost that can be justified for this material.

PROFITABILITY ANALYSIS

The two standard methods used for profitability analysis by private corporations are net present value (*NPV*) and rate of return (*ROR*). Projects are ranked by these measures of return on investment to compete for the limited capital for plant improvements and new processes. The **net present value** is the sum of all of the cash flows for the project discounted to the present value, usually using the company's minimum attractive rate of return, *MARR*, and the capital investment required. The **rate of return** is the interest rate in the net present value calculation that give a zero net present value.

The net present value, is an estimate of profitability of a project. It has the advantage that the net present value for several projects can be added to obtain the net present value for all the projects.

The rate of return is sometimes called the discounted cash flow rate of return (*DCFRR*) and the internal rate of return (*IRR*). The rate of return has the advantage of being used to compare directly with alternate uses of money that have rates of return, such as bonds and certificates of deposit.

There are numerous similar measures of profitability, but all of these are variations of net present value and rate of return, except the payback period or payout time which has the flaw of neglecting the time value of money. One example is the net rate of return, except the payback period or payout time which has the flaw of neglecting the time value of money. One example is the net rate of return (*NRR*) which is the net present value divided by the product of the capital investment at year zero and the product life expressed as a percentage.

Net Present Value

To evaluate the net present value (*NPV*), the net annual cash flows, CF_j, are needed.

These cash flows are used in the following equation to compute the net present value (*NPV*) where CF_O is the initial capital investment for the project.

$$NPV = -CF_O + \sum_{j=1}^{n} CF_j (1+i)^{-j}$$

To determine the net present value, the interest rate, i, usually the minimum attractive rate of return, and the number of years, n, for the project are specified. There is no assumption about the signs of the cash flows, CF_j, but the equation has the initial cash flow, CF_O, being negative to represent the initial capital investment.

Referring to Figure 2 and equation (18), the individual annual cash flows would be discounted to the present and combined with the capital investment to estimate the net present value for the proposed new plant. The following simple example illustrates the calculation of the net present value.

EXAMPLE: A straight-run fuel oil stream in a refinery can be converted to a high octane fuel for blending into premium gasoline using hydrocracking. A proposal has been made to add a $15,000 bbl/day unit at a capital cost of $71.0 million. The annual net profit in million dollars is given below for the estimated life of the hydrocracking unit. The net present value is to be evaluated for interest rates of 15% and 25%, and the profitability compared. These results are shown in the following table.

TABLE 1
ANNUAL NET PROFIT FOR ESTIMATED
LIFE OF HYDROCRACKING UNIT

End of Year n	Annual Net Profit, F	$\frac{P}{F}$(15%) $(1.15)^{-n}$	Present Value	$\frac{P}{F}$(25%) $(1.25)^{-n}$	Present Value
1	32.0	0.8695	27.83	0.8000	25.60
2	28.0	0.7561	21.17	0.6400	17.92
3	22.0	0.6575	14.47	0.5120	11.26
4	17.0	0.5718	9.72	0.4091	6.96
5	15.0	0.4972	7.46	0.3277	4.92
Total	114.0		80.64		66.66

SOLUTION: Computing the net present value gives:

$NPV(15\%) = -71.0 + 80.84 = 9.64 \quad NPV(25\%) = -71.0 + 66.66 = -4.34$

The investment is marginally attractive with a positive net present value if funds are available at 15%, but the project is not considered with a negative net present value for funds available at 25%.

A convenient form of equation (18) is obtained if all of the cash flows, CF_j, are equal by using equation (14).

$$NPV = -CF_O + A\left[\frac{1-(1+i)^{-n}}{i}\right] \qquad (19)$$

where A is the uniform cash flow in the equation.

The use of this form of the equation for net present value is illustrated by the following example which also shows the use of the gradient series factor, equation (15).

EXAMPLE: Two projects are competing for an oil company's capital improvement funds. One is an additional distillation column for improved product quality, and the other is a new lubricant packaging system to reduce product packaging costs. The capital investment is $110,000 for each one. The cash flow for the distillation column is $38,000 for the first year, and then it declines by $4,000 for each subsequent year for the 10 year life of the project. The cash flow for the packaging system is $5,000 for the first year, and then it increases by $4,000 for each subsequent year for the 10 year life of the project. For a 15% minimum attractive rate of return compute the net present value for each proposed project.

SOLUTION: For both of the oil company's projects:

Capital investment, $CF_O = \$110,000$; $i = 15\%$ and $n = 10$ years.

Distillation column cash flows: Year 1 $CF_1 = \$38,000$

Year 2 – 10 – declines by \$4,000

Packaging system cash flows: Year 1 $CF_1 = \$5,000$

Year 2 – 10 – increases by \$4,000

The following equation gives the net present value (*NPV*) for these projects.

$$NPV = CF_O + CF_1(\frac{P}{A}, i\%, 10) + G(\frac{A}{G}, i\%, 10)(\frac{P}{A}, i\%, 10)$$

where: G = Gradient of the cash flow

$(\frac{A}{G}, i\%, n)$ = Gradient series factor

$(\frac{P}{A}, i\%, n)$ = Uniform series, present-worth factor

For the distillation column:

$$NPV = -110,000 + 38,000(\frac{P}{A}, 15\%, 10) - 4,000(\frac{A}{G}, 15\%, 10)(\frac{P}{A}, 15\%, 10)$$

$$= -110,000 + 38,000(5.0188) - 4,000(3.3832)(5.0188)$$

$$= -\$12,800$$

For the packaging system:

$$NPV = -110,000 + 5,000(5.0188) + 4,000(3.3832)(5.0188)$$

$$= -\$17,000$$

where compound interest factors were obtained from standard tables. The distillation column is a potentially attractive investment with a positive net present value, but the lubrication packaging system is not with a negative net present value.

Rate of Return

The rate of return (*ROR*) is the interest rate where the net present value is zero, i.e., from equation (18).

$$0 = -CF_O + \sum_{j=1}^{n} CF_j(1+i)^{-j} \tag{20}$$

To determine this interest rate, it is usually necessary to interpolate between two known values of the net present value. In the example on page 541 the net present value was 9.64 at an interest rate of 15% and –4.34 at 25%. Interpolating gives the rate of return for this case to be 21.9%. The following example gives an additional illustration of the evaluation of the rate of return.

EXAMPLE: A division of a company has been allocated $100,000 to invest at the start of the next fiscal year in cost-reduction projects. Three projects are under consideration and are summarized below.

TABLE 2
SAMPLE COST = REDUCTION PROJECTS

Project	Investment Required	Estimated Economic Life (years)	Net Annual Cash Flow
A	$50,000	9	$16,600
B	$50,000	8	$15,000
C	$100,000	6	$30,000

The minimum attractive rate of return for the company is 20% for projects with this economic life. Would recommendation based on the rate of return for these projects be (A) invest in A only, (B) invest in B only, (C) invest in A and B, (D) invest in C only, or (E) seek other alternatives?

SOLUTION: Alternatives are evaluated for investing $100,000 by comparing the rate of return for the projects with the minimum attractive rate of return of 20%. The rate of return (i) is the interest rate where the net present value is zero. For a uniform net annual cash flow (A), the equation for the net present value (NPV) is:

$$NPV = CF_O = A(\frac{P}{A}, i\%, n)$$

where CF_O is the capital investment and $(\frac{P}{A}, i\%, n) = \frac{\left[1-(1+i)^{-n}\right]}{i}$ is the uniform

series capital recovery factor.

For Project A:

$$0 = -50,000 + 16,600(\frac{P}{A}, i\%, 9) \text{ or } (\frac{P}{A}, i\%, 9) = 3.012$$

In addition to using tabulations of $(\frac{P}{A}, i\%, 9)$, from tables of compound interest

factors in standard texts, gives $i = 30.0\%$.

For Project B:

$$0 = -50,000 + 15,000 (\frac{P}{A}, i\%, 8) \text{ or } (\frac{P}{A}, i\%, 8) = 3.3333$$

Using standard tables of compound interest factors, gives $i = 25.0\%$.

For Project C:

$$0 = -100,000 + 30,000 (\frac{P}{A}, i\%, 6) \text{ or } (\frac{P}{A}, i\%, 6) = 3.33$$

Using standard table of compound interest factors, gives $i = 20.0\%$.

Summary:

Project	Rate of Return
A	30.0%
B	25.0%
C	20.0%

The investment decision is to select Projects A and B because their rate of return is greater than the minimum attractive rate of return; and all of the available capital is used.

It should be noted that the rate of return method is best when comparing independent alternatives. It is probably one of the most popular tools used in capital budgeting. Net present value is widely used to choose among dependent alternatives. A thorough economic analysis will use more than one method, and it will try to include as much information as can be made available.

The **Payback Period** is the time required to recover the capital investment from the net profit but neglecting the time value of money. The equation for the payback period is :

$$CF_O = \sum_{j=1}^{PBP} CF_j \tag{21}$$

and if the yearly net profits A are uniform, then the payback period (*PBP*) is given by the following equation.

$$PBP = \frac{CF_O}{A} \tag{22}$$

This is a simple and popular calculation, but it ignores the time value of money and should not be used for making economic decisions for that reason. It is sometimes called the **payout time**. However, there is a modification to the payout time called the **discounted payout time** that computes the number of years to have the cumulative discounted cash flows sum to zero, and this does include the time value of money.

To illustrate the calculation of the payback period using the example on page 541, the capital investment was to be $71.0 million, and the sum of the cash flows was $80.64 million for the five years. The payback period is determined by:

$$71.0 = 32.0 + 28.0 + 22.0(0.5) = 71$$

which gives a payback period of about 2.5 years. Although this time to recover the capital investment may sound attractive, it is not a good investment if the cost of money is at an interest rate of 25% as shown in the example.

Benefit-Cost Analysis

The benefit cost ratio (*BCR*) is used in municipal and government projects, and it is defined by the following equation.

$$BCR = \frac{B-D}{C} \qquad (23)$$

The benefit-cost ratio is the difference between the benefits, B, and the disbenefits, D, divided by the costs, C. For example, in a project to build a hydroelectric dam, the benefits would be electric power generation and possibly flood protection; and the disbenefits would be loss of productive farmland. The costs would include the construction and maintenance of the dam.

Although benefits and costs can be estimated without too much difficulty, it is usually difficult to measure the cost of the loss of wildlife habitat, scenic rivers, and land loss downstream in coastal marshes from lack of replenishing sediments. For projects to be considered by the U. S. Army Corp of Engineers, BCR should be about 2.0 or larger.

EQUIVALENCE

In comparing investment opportunities, there are times when two or more plans give the same result, the same net present value, for example, even though plans involve different interest rates and times. A plan that has a uniform annual series value of $8,000 for 25 years at an interest rate of 6% has a present value of $102,264. Another plan that has equal quarterly deposits of $156.07 for 40 years at a nominal interest rate of 6.0% has the same net present value of $102,264. This illustrates the concept of **equivalence**, i.e., being of equal value. The interest factors presented previously are called equivalence factors, also. Other simple examples of equivalence include the following:

A uniform annual payment of $655.56 for 20 years at an interest rate of 8% is equivalent to a single payment of $30,000 in year 20.

A single payment today of $30,000 would be equivalent to an annual payment of $6,436.45 over 20 years at an 8% interest rate.

Investing $6,436.45 for 20 years at a 15% rate, is equivalent to $105,342 today. A uniform series of payments of $1,028 for 20 years at a 15% rate is also equivalent to $105,342.

The first two examples show that, at 8%, $30,000 in 20 years is equal to $655.56 a year for 20 years, which is also equivalent to $6,436.45 in the present. The third example shows investing two different amounts of money are equivalent to $105,342.40 today.

TAXES AND DEPRECIATION

In any project analysis, it is important to distinguish between before tax cash flows and after tax cash flows (*BTCF* and *ATCF*). In general, project decisions are best made based on after tax cash flows, but before tax figures can often be used in the preliminary stages of analysis as an approximation for the desirability of a project. Typically, a corporation's first $75,000 to $100,000 in profits is tax exempt; but, in economic studies, it is usually assumed that this profit margin has been met, and that the corporation is operating and making decisions where marginal profits are taxed at the highest rate (currently about 38%). Although tax rates are constant for marginal profits, what makes after tax analysis more difficult than before tax analysis is the issue of depreciation.

To understand depreciation, you need to have an understanding of **capital**. Capital goods are those accumulated in order to produce other goods. Two kinds of capital can be distinguished, fixed and working. **Fixed capital** is that which cannot be readily converted into a different sort of asset. **Working capital** is the investment that puts the plant into production. It can be estimated to be a tenth to a fifth of fixed capital, or as a value of one month's worth of raw materials.

Total capital (fixed plus working) does not include the expenses of operation. These are other indirect expenses such as overhead, taxes, insurance, and **deprecia-tion**. The amount of depreciation in any given year depends on the amount of **fixed** capital only. Furthermore, income taxes are a function of depreciation.

Depreciation exists because anything valuable will lose some or all of that value with time. Depreciation is a tax allowance and is considered a cost of operation. It is a measure of declining value, and a means of building a fund to finance plant replacement.

The accounting of depreciation is also a way to plan for replacing equipment, except actual **replacement** of obsolete equipment is rare. New technology and new methods of production will usually require totally new equipment.

Depreciation appears on a financial statement as a cost of operation, but this is equally misleading. Capital costs are depreciated, and in that way are translated into yearly operating costs, but the actual disbursement was made at the time of equipment purchase. The reason for depreciating the amount over time is this: capital equipment is an asset to the plant; each year, the value of that asset declines and even though there is no out of pocket cost, the declining value is a real cost to the plant. However, since capital costs are incurred at the inception, depreciation costs are calculated yearly only for the calculation of taxes.

Depreciation as a tax allowance is the most important consideration when planning a project. The depreciation entry in the cost column serves to lessen taxable income, and can make the crucial difference in profitability. This entry may have little to do with the actual physical depreciation of the asset, for tax laws have increasingly standardized the ways in which depreciation is calculated, thus making the depreciation entry less a measure of an individual piece of equipment's useful life. Still, the expense should be viewed as a measure of useful life.

To determine the amount of depreciation for tax purposes, a few values need to be known. The **depreciable base** consists of fixed capital only. Land, however, is excepted from the depreciable base, because it is always thought to retain all of its original value, and so cannot be considered a depreciable item. The **write-off life** is the hypothetical life of the asset. The IRS has guidelines for write-off life of equipment, and plant equipment is depreciable over about ten years. Computers have a life of about five years. Next, various methods of calculating depreciation will be explained.

There are three methods of computing depreciation: straight line, accelerated, and decelerated. Straight line assumes a steady loss of value over time, while accelerated depreciates more in the early years of the asset, while decelerated depreciates more in later years.

Straight-Line Method

With the Straight-Line Method, the annual depreciation is constant. To evaluate the depreciation charge for the rth year, it is computed as the difference between the original cost P and the estimated salvage value S divided by the estimated service life n in years, i.e.,

$$D_r = \frac{(P-S)}{n} \tag{24}$$

Then the book value at the end of the rth year, BV_r is obtained by subtracting r times the depreciation D_r obtained from equation (24) as shown below.

$$BV_r = P - rD_r \tag{25}$$

For example, the original cost is \$20,000; and the service or write-off life is 12 years for a piece of equipment. With no salvage value, the depreciation for year 1 $(r = 1)$ is: $D_1 = \frac{(20,000-0)}{12} = \$1,667$; and the book value is: $BV_1 = 20,000 - (1)(1,667) = \$18,333$.

Sum-of-Years-Digits Method

The Sum-of-Years-Digits Method is an accelerated method using the sum-of-year-digits, SY, that depreciates about 75% of the cost in the first half of the service life. SY is given by the following equation.

$$SY = \sum j = 1 + 2 + ... + n = n\frac{(n+1)}{2} \tag{26}$$

where n is the service life. The amount of depreciation D_r is computed as follows:

$$C = \frac{(P-S)}{SY} \tag{27}$$

and
$$D_r = (n + 1 - r)C \tag{28}$$

Then the book value is obtained from the following equation.

$$BV_r = P - C\left[SY - (n-r)\frac{(n-r+1)}{2} \right] \tag{29}$$

Using the information from the straight line method, the sum-of-year-digits is:

$$SY = \frac{12(12+1)}{2} = 78,$$

and the value of

$$C = \frac{(20,000-0)}{78} = \$256.$$

Evaluating equation (28): $D_1 = (12+1-1)256 = \$3,072$, and the book value from equation (29) is:

$$BV_1 = 20,000 - 256\left[\frac{78-(12-1)(12-1+1)}{2}\right] = \$16,928.$$

Double-Rate Declining Balance Method

The Double-Rate Declining Balance Method computes the depreciation at double the straight line depreciation rate, and it is one way to do the declining balance method. If f is the double-declining balance rate given by:

$$f = \frac{2.0}{n} \tag{30}$$

then the book value is computed by the following equation.

$$BV_r = P(1-f)^r \tag{31}$$

The depreciation at year r is given by the following equation.

$$D_r = (BV_{r-1} - BV_r) = fBV_{r-1} \tag{32}$$

Using information from the straight line method, the double-declining balance rate is $f = \frac{2.0}{12} = 0.167$ by equation (30). Then the book value for year 1 ($r = 1$) is:

$BV_1 = (20,000)(1-0.167) = \$16,000$, and the depreciation is: $D_1 = (20,000-16,660) = \$3,340$.

In this method the net salvage value is not evaluated as part of the procedure. The method can have the book value become smaller than the net salvage value. However, IRS regulations require that the book value not be less than the salvage value, and the depreciation must stop at this point.

Sinking Fund Method

The Sinking Fund Method depreciates equipment with an imaginary sinking fund that is equivalent to the company making a series of equal annual deposits to have an amount equal to the cost of replacing the equipment at the end of its service life. The amount of depreciation in any given year is equal to the amount of the annual deposit into the sinking fund plus interest. This method is used when the replacement of the equipment is assumed to cost the same as the original. It is the only method in which depreciation increases in time. The accumulated depreciation in year r, AD_r is given by the following equation

$$AD_r = (P-S)\left(\frac{A}{F}, i\%, n\right)\left(\frac{F}{A}, i\%, r\right) \tag{33}$$

The depreciation for year r is computed by difference, i.e.:

$$D_r = AD_r - AD_{r-1} \qquad (34)$$

Then the equation to evaluate the book value is:

$$BV_r = P - (P - S)\left(\frac{A}{F}, i\%, n\right)\left(\frac{F}{A}, i\%, n\right)\left(\frac{F}{A}, i\%, r\right) \qquad (35)$$

Using the information from the straight line method and an interest rate of 8%, the accumulated depreciation for the first year is: $AD_1 = (20,000 - 0)(0.0527)(1) = \$1,054$ using equation (33), and the depreciation is: $D_1 = 1,054 - 0 = \$1,054$ using equation (34). Then the book value is $BV_r = 20,000 - 20,000 (0.0527)(1) = \$18,946$ from equation (35).

In this method the annual depreciation increases geometrically with time. Tax laws require that this method be used only with equipment that has to be replaced with equipment which cost at least as much as the original. This does not allow a company to take a total depreciation allowance that is greater than the equipment's current net adjusted cost.

Group Accounts

Often it will be more convenient to group assets, which are bought in the same year, for depreciation purposes. Items which have the same useful life can be pooled into a **group account**, regardless of whether they were bought in the beginning or the end of the year. A **classified account** groups items according to use, regardless of useful life, and a **composite account** includes items which have diverse lives and uses. With classified or composite accounts, the depreciation rate for any year is found by determining depreciation for each item or each group of similar items, and that total is divided by total cost. Any of the above depreciation methods will work.

Comparing the depreciation methods, the accelerated methods are those for which the accumulated depreciation exceeds the straight-line method. However, the sinking fund curve is below the straight-line, meaning it is decelerated. The amount of depreciation is ultimately the same for all, except in double declining balance, in which the depreciation total approaches, but never reaches, the full amount. In instances where the salvage value is significant, the double-declining rate's acceleration effect is enhanced, since salvage value is not considered in the calculation.

It might seem that, since the depreciation amount is ultimately the same, acceleration is simply a short-sighted way of grabbing tax credits in the shortest time possible. Not at all. The advantage of accelerated depreciation is very real. Because of money's value in time, the declining balance and sum of year's digits methods, because they are accelerated, yield a higher present worth for net income after taxes; the discount for the tax savings in earlier years is not as great as the discount in later years.

Sometimes, especially with an accelerated depreciation method, the depreciation charge exceeds the actual before tax cash flow in a given year. When this occurs, realize that, although each investment decision is treated separately in the economic study, the income generated and taxes paid are part of the cash flow for the entire firm. Hence, if the depreciation charge is $1,000 and income is $600 for that year, then the other $400 of depreciation is not lost, but is subtracted from the general before tax cash flow of the firm. For purposes of analysis, this extra benefit to the firm can be included in the economic study of the specific investment decision in the following manner, which shows an after tax profit greater than the actual before tax receipts.

Before Tax Cash Flow	$ 600
depreciation	1,000
taxable income	− 400
tax at 33%	− 132
After Tax Cash F low	$ 732

Every year, regulations on depreciation are being revised, and less choice is given to the corporation (or individual) in how they may claim depreciation deductions. The methods outlined above are important to know, since next year or thereafter they may be the basis of the newest law. Since 1981, however, formulas for depreciation have been set by the Accelerated Cost Recovery System (*ACRS*). With the *ACRS*, all assets can be grouped into one of four categories, each with a corresponding recovery period, and depreciation percentages are set for each year. This greatly simplifies matters, being very amenable to **group asset** depreciation.

Note that equipment is not subject to depreciation prior to the year it has been placed in service, that is, when it is not yet ready for use. A plant that takes 3 years to build is not depreciable until it is completed and placed in service. However, it makes no difference to the yearly deduction if the equipment was placed in service during the beginning of the year or in late December.

A company can reduce operating costs by $16,300 per year for 12 years by automating a process with new digital control computer. The cost of the computer and control system is $78,000, and straight line depreciation over 12 years is used with zero salvage value. If the applicable income tax rate is 38%, determine the rate of return after taxes.

For the new control system:

operating cost is reduced by $16,300 per year for 12 years

equipment cost is $78,000

straight line depreciation for 12 years with no salvage value income tax rate is 38%

The equation for the net present value (*NPV*) for uniform annual cash flow (*A*) is:

$$NPV = CF_O + A(\frac{P}{A}, i\%, n)$$

where CF_O = $78,000, the capital investment for the control system.

Change in the net profit = $16,300 − \dfrac{(\$78,000 - 0)}{12} = \$9,800$ per year before taxes

minus depreciation

Taxes on the change in net profit = $9,800 \times 0.38 = \$3,724$

A = Change in net annual income after taxes = $16,300 - \$3,724 = \$12,576$

Computing the rate of return (interest rate at zero net present value):

$$0 = -78,000 + 12,576(\frac{P}{A}, i\%, 12)$$

or

$$(\frac{P}{A}, i\%, 12) = 6.202$$

Using tables of compound interest factors at 12 years gives:

$$i = 12.0\%$$

which is the rate of return after taxes.

SENSITIVITY ANALYSIS

Sensitivity analysis is an important method of ensuring the reliability of the forecast. It will tell how "sensitive" a project is to some foreseeable change to any certain element that has been estimated, such as selling price, cost of materials, or the tax rate. A sensitive project is one whose desirability is highly affected by a small change in any certain variable. A project is insensitive to a variable if wide ranges of values do not alter the conclusions of the study. Typically, a base case is used, and the present value or rate of return is evaluated for variations in parameters such as product price, sales volume, plant cost and size, working capital, etc. For example, a plant design is based on a capacity that will produce 4,000 items per unit time with a net present value of $3.0 million based on a projected selling price of $0.50 per unit. The sensitivity of the net present value to selling price of $0.25 and $0.75 per unit is –$0.5 million and $4.2 million.

The sources of probable risk are any uncertainties, i.e., everything that is estimated: disbursements, receipts, the length of time the project will be in service, salvage value, and the tax rate. Again, a general rise in prices due to inflation is not a consideration in economic studies, since receipts should change along with disbursements. Differential price changes, however, when the price of a certain item varies in contrast to the general price level, should be included in analysis, if they can be foreseen. Differential prices may be foreseen if revenues rely on contract pricing which is not indexed for inflation, or if governmental price controls apply. Fluctuation in oil prices are another example.

Another source of differential price changes occurs regularly, because income tax deductions for depreciation are not indexed for current dollars. If a piece of equipment costing $100,000 is depreciated over a period when inflation is 10%, the costs and receipts will go up as well, but the deduction for tax purposes will not. Thus the corporation will be taxed on a higher percentage of income than would occur if there were no inflation, or if depreciation was indeed indexed.

RISK ANALYSIS

Sensitivity analysis is linked to risk analysis, which seeks to give the probability of the occurrence of certain scenarios. If the probability was high that the plant would not be used at capacity, then the desirability of the project is severely lessened. Sensitivity analysis and probability analysis together try to quantify elements of uncertainty.

Risk analysis attempts to quantitatively evaluate risks associated with research, development, economics, politics, natural disasters, and other possible uncertainties. Probability estimates for these risks can be obtained from past experience, simulations, expert estimation and experimental data, among others.

The objective is to use risk-weighted expected values of the profit to maximize the income in selecting among projects accounting for risk associated with them. Risk-weighted expected values are the sum of the product of the probabilities and the associated profit for that outcome. The profit can be measured by the net present value, annual worth, or other appropriate economic values.

A simple risk assessment for economic decision analysis involves the folowing steps. For each project under consideration, determine the range of possible outcomes that would effect the profit, e.g., the range in product selling price. Evaluate the profit over this range of possible outcomes. Separately, estimate the probability of occurrences of the possible outcomes, e.g., the probability the selling price will be at the low value when the plant is constructed (a very unfavorable situation). Then evaluate the weighted average of the profit by computing the sum of the profit and associated probabilities. This weighted average profit is an estimate of the expected value of the profit. These expected values for the projects are used to rank them in order of economic potential with risk incorporated in the comparison.

A simple illustration of the separate determination of profit and probability is given in this example. Consider having the opportunity of paying for a chance to bet on the flip of a coin. Each toss cost a quarter, but the payoff is $1.00. The expected value of the game is the sum of the probabilities of the possible outcomes, both 0.5, times the payoff, either $1.00 or 0. Thus the expected value is $(0.5 \times \$1.00 + 0.5 \times 0) = \$0.50$. Even though, for any one toss, you may not receive $0.50; this is a bargain at a quarter a game. However, if the price of playing the game was exactly fifty cents, then the odds would not be for you or against you.

Determining the expected value of any given project consists of summing the product of the probabilities of all possible events times their corresponding payoffs. For example, if it was determined that the supply of seafood is directly related to weather conditions, the probability that a seafood plant will be used to capacity in any one year could be predicted, using data from past years about what likely weather conditions will be and what the consequent seafood harvests will be. With this information, it is predicted that the probability of the plant being used at 75% capacity in any one year was 30%, the probability for a full capacity year was 45%, and the probability of usage of 125% of capacity was 25%. Using an estimate of the annual worth, calculations of expected worth are then simple to do.

TABLE 3
CALCULATION OF EXPECTED WORTH

% capacity	Annual Worth	Probability	Expected Value
75%	$ 22,314	0.30	$ 6,694.20
100%	124,045	0.45	55,820.25
125%	212,151	0.25	53,037.75
		Expected Value (AW)	$115,552.75

Note that the sum of the probabilities is equal to one. This will always be the case. This example gives the expected value of the annual worth, but any of the methods of project evaluation could be used.

Sometimes, if the decision maker does not know the probability of the success or failure of a specific scheme, risk analysis can still be used. For example, if an oil company is considering whether to invest in drilling a certain well, they might have these four alternatives

1. Drill the well with a 100% working interest (*WI*).

2. Drill with a 50% partner.

3. Farm out, but back in for a 50% working interest.

4. Do not drill the well and use the funds elsewhere.

The following table gives the payoffs for each alternative in the event of success or failure.

TABLE 4
PAYOFFS FOR EACH ALTERNATIVE

Possible Outcomes	Drill with 100% WI	Drill with 50% partner	Farm Out	Don't Drill
Dry hole (failure)	−500	−250	0	0
Producer (success)	3,000	1,500	1,500	0

If there are only two possible outcomes (success or failure), and if the chance for success is simply not known, a value P_S can be assigned to show a linear relationship between the chance of success and the expected value. If NPV_S is the net present value in case of success, then $(1-P_S)$ is the chance of failure with a net present value of NPV_S. The expected value becomes a function of the probability of success.

$$EV = P_S NPV_S + (1-P_S) NPV_f$$
$$= P_S (NPV_S - NPV_f) + NPV_f$$

Substituting values for the net present values for the four cases gives us the following equation:

$$EV_{100\%WI} = \$3,500\,P_S - \$350$$
$$EV_{50\%partner} = \$1,750\,P_S - \$250$$
$$EV_{farm\ out} = \$1,500\,P_S$$
$$EV_{do\ not\ drill} = 0$$

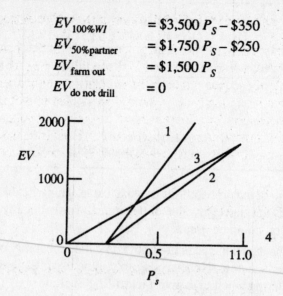

Figure 4. Expected value

These are linear equations, and the $EV_{100\%WI}$ is greater than the other two for values of P_S of more than 25%. If the analyst would hesitate to put a specific value to the probability of success, these relationships can at least illustrate how much latitude his estimates can contain. For any probability of success less than 25%, farming out would be the best alternative.

There are times when expected value is an inappropriate tool for decision making. Expected values are **long-run** values. If immediate outcome is crucial, such that any risk is unpalatable, then expected value is not a true measure of project worth. One common example is health insurance. People do not buy health insurance with the hope of making money. Rather, they are paying for the prevention of a financial disaster in case they are faced with a debilitating or long term illness. The value of being sure about preventing financial disaster is more important to them than the risk of losing money in the transaction. That the expected value of health insurance is lower than the price paid for it is evidenced by the fact that insurance companies are profit making businesses. To them, expected values and probability of illness are pertinent information; to the insurance policy holder, the knowledge that he is at risk is more important than knowing he's probably losing money.

What does this say to the engineering economist? When all risk is unacceptable, expected values should not be used. If an investment of $5,000 would yield an 8% return, or $400 a year, and a competing investment of $5,000 would yield a 75% return, but with only a 15% likelihood of occurrence, the investors would have to choose between a sure $400, or an expected value of $562.50. The smart choice that is, the economically sound choice, is the second investment, unless the chance of failure would mean disaster for the investor.

One Final Note about *MARR* (Minimum Attractive Rate of Return)

The concept of risk can add to a full comprehension of what the choice of an *MARR* entails. Three major elements can be seen to theoretically sum up the designated *MARR*. The first is pure interest, which is the amount an investor could make if he were to place his money in the bank to earn interest. Add to that an amount that represents the compensation for management, that is, whatever value the investor or company places on its involvement in the economy. A risk factor can then be added, which compensates implicitly for the level of risk the company or investor is willing to take.

REVIEW PROBLEMS

PROBLEM 1

What amount of money would have to be invested to have $4,000 at the end of 3 years at a 10% compound interest rate?

SOLUTION:

$$P = \frac{F}{(1+i)^n}$$

$$= \frac{4,000}{(1+0.10)^3}$$

$$= \$3,005$$

PROBLEM 2

If the inflation rate is 8%, and you invest $20,000 at an 11% simple interest rate, will you have retained your buying power at the end of a) 5 years? b) 10 years?

SOLUTION: a) The principal will grow to:

$$F = \$20,000[1 + 5(0.11)] = \$31,000$$

To see if the buying power is the same, the future amount must be discounted for inflation, so that $1.00 in the future can be converted to an equivalent amount today.

$$F_I = \frac{P_I}{(1+r)^n}$$

where F_I is the future worth, measured in today's dollars P_I, and where r is equal to the rate of inflation. Notice that discounting for inflation is the reverse of compound interest problems.

$$F_I = \frac{31,000}{(1+0.08)^5}$$

$$F_I = \$21,098$$

Buying power is retained, plus a little extra.

b)
$$F = \$20,000[1+10.11]$$
$$= \$42,000$$
$$F_I = \frac{42,000}{(1+0.08)^{10}}$$
$$F_I = \$19,454 < \$20,000$$

With an increase in time, buying power diminishes because of the compounding nature of inflation.

PROBLEM 3

$500,000 is borrowed at a nominal rate of 8%, compounded quarterly. If no payments are made in the first 3 years, how much will be owed? How much would the amount be if interest was compounded annually? Daily?

SOLUTION: For quarterly compounding, first find the effective interest rate from equation (3).

$$\left(1+\frac{0.08}{4}\right)^4 - 1 = 8.24\%$$

$$F = \$500,000 \ (\frac{F}{P}, 8.24\%, 12)$$
$$= \$636,064$$

Compounded annually

$$F = \$500,000 \ (\frac{F}{P}, 8\%, 3)$$
$$= \$629,856$$

Compounded daily

$$F = \$500,000(\frac{F}{P}, \frac{8}{365\%}, 1,095)$$
$$= \$635,608$$

PROBLEM 4

If you are 25 years old now, and want to retire at age 50 with an annuity account that would give you $5,000 a year for 30 years, how much do you need to deposit each month at 6% to be able to have enough in the account?

SOLUTION: First, find the present value of the annuity:

$$P = \$5,000(\frac{P}{A}, 6\%, 30)$$

$$= \$68,824$$

This is the future target amount. To get the annual deposit required:

$$A = \$68,824(\frac{A}{P}, 6\%, 25)$$

$$= \$1,254$$

PROBLEM 5

A company needs to decide whether to use some of their undistributed profits to automate certain of their procedures. The initial investment would cost $60,000 and the system would have a seven year lifetime. Savings from the automation are estimated at $18,000. Calculate the *AW*, the *PW*, and the *ROR* at 10%.

SOLUTION: $PW = \$27,631$

$AW = \$5,676$

$ROR = 22.92\%$

PROBLEM 6

Manufacturers of motors for small household equipment wonder if they should close one of their two plants (Plant A) and expand operations at Plant B. It seems there is not enough business to keep both plants running at capacity, yet expansion of Plant B would have to occur for it to handle double the usual load. Net income has been a steady $800,000 at Plant A, and $840,000 at Plant B. Due to savings in overhead costs, it is thought that, with the same level of business, net income from an upgraded Plant B alone would equal $1,780,000. Salvage value of Plant A and sale of the land is estimated at $120,000. The investment to expand Plant B would cost $1.1 million, and it would take three years before the level of production could be increased.

a) If the company has a required *MARR* of 11%, and operations are thought to be steady for the next 15 years, how desirable would this course of action be, if disassembly of Plant A were to begin immediately and the total cost of upgrading was incurred at time zero?

b) If disassembling Plant A was to occur at the end of year three, when the expansion is completed?

c) If the disbursement for upgrading was spread evenly over the first three years of construction?

SOLUTION: a) The first task is to separate all factors unique to the proposed course of action. If disassembling the plant were to occur immediately, that is, at time zero, a cost of $980,000 would be incurred, which is the expenditure for upgrading minus the income from sale and salvage of Plant A. The first three years of the project would incur a loss of $800,000 from Plant A, with no changes in income from Plant B. The next 15 years would see a positive cash flow of $140,000, which is the change in income caused by the project.

The internal rate of return is a negative 4.65%, and the return on investment is −2,198,866, clearly unacceptable.

b) If disassembling is delayed until the third year, the cost in year zero is $1.1 million, no income change occurs in years one and two, with a positive cash flow in year three from sale of land and Plant A. Income of $140,000 is then constant for 15 years. Delaying the plant disassembly is clearly the more practical choice, yet return on investment at 7.4%, still does not meet the requirement of the company. Net present value is −$276,150.

c) If the disbursements for upgrading were spread evenly over the three years, instead of a lump sum at time zero, then at the close of years one and two, −$366,667 is incurred, −246,667 at year three (366,667 − 120,000), and $140,000 for the 15 years thereafter. This makes the investment more attractive at a return on investment of 9.6%, but this is still not sufficient to meet an *MARR* of 11%. The net present value is −$72,180.

PROBLEM 7

A corporation has $225,000 in its capital budget to invest this year. There are several proposals for projects under consideration to either reduce costs or raise profits.

TABLE 5 PROPOSAL COMPARISON				
Project	Investment Required	Estimated Economic Life (Years)	Estimated Salvage Value	Annual After Tax Cash Flow
1	$18,000	5	$ 2,000	$ 4,000
2	70,000	8	10,000	12,000
3	60,000	12	0	14,000
4	45,000	10	2,000	4,800
5	52,000	8	5,000	18,000
6	10,000	5	1,000	3,400
7	85,000	15	5,000	1,800

a) Calculate the prospective return on investment for each project, and rank the projects accordingly. Which projects should be undertaken, if you were not to consider any other matter? What should be the company's *MARR* this year?

b) Suppose management is averse to the risk inherent in a long term and wants to consider profitability over only an 8 year period. How would the choice of projects, and the *MARR*, change?

SOLUTION: a) The choice of projects would occur in this order: 5, 6, 3, 7; and the *MARR* would correspond to the *ROR* of the last project chosen, i.e., 19.9%.

b) The *ROR* for projects 3, 4, and 7 change, and the choice of projects becomes 5, 6, 3, and 4. The *MARR* is lowered to 14.3%, and the $58,000 left of funds is either saved until another project can be found that meets that *MARR*, or some is invested in project 1, if the rate that can be earned in by investing in a highly liquid bank account is less than 6.6%.

PROBLEM 8

A paper company is building a paper mill, but is uncertain as yet how large to make it. The greater the capacity of the plant, the greater their income, but whether the extra investment will meet the company's *MARR* of 17% is unsure. The seven alternatives are outlined below.

TABLE 6
PAPER MILL ALTERNATIVES

	1	2	3	4	5	6	7
Investment	$50,000	55,000	60,000	65,000	70,000	75,000	88,000
After tax cash flow	10,000	15,000	18,000	20,000	21,000	22,000	25,000

a) Compute the return on investment for each alternative.

b) Calculate the marginal return on each increment of investment.

c) Which alternative is the best?

SOLUTION:

	1	2	3	4	5	6	7
a)*ROR*	5.47	16.19	19.91	20.93	19.91	19.01	17.75
b)Marginal *ROR*		98.36	55.81	32.66	5.47	5.47	8.16

c) Alternative four is best, because the marginal return on investment for alternative five is less than 17% and none of the successive alternatives can produce a marginal *ROR* of 17% or greater, when compared with alternative four.

PROBLEM 9

There are several models of a piece of equipment which would cut costs for a company. If their *MARR* is 20% before taxes, which of the five models should they choose if there is no salvage value and the equipment will last 8 years?

TABLE 7
COST-CUTTING MODELS

Model	Investment	Cost Reduction(annual)
1	$ 82,500	$18,010
2	93,200	22,020
3	98,066	22,252
4	104,400	30,003
5	110,110	32,580
6	112,000	33,008

SOLUTION:

TABLE 8
COMPARISON OF MODELS

	ROR	Increment of Investment	Increment of Cost Reduction	Before Tax ROR on Increment of Investment
1	14.38%			
2	16.81%			
3	19.6%			
4	23.39%			
5	24.44%	$5,710	$2,577	42.47%
6	24.3%	$1,890	$428	15.49%

Models 1, 2, and 3 do meet the *MARR*, so there is no need to calculate the marginal return on investment. Model four becomes the base case, by which model five is compared, and comes out ahead, becoming the new base case. Model six does not have a high enough marginal return on investment, so model five is best.

PROBLEM 10

If the tax rate is 33% and the equipment is depreciated as five year property under *ACRS*, and if the company's after-tax *MARR* is 15%,

a) Find the after-tax cash flows for each model and calculate the *ROI*.

b) Do an incremental analysis of cash flows and *ROI*, and decide which model is the best.

SOLUTION:

a)

	TABLE 9 AFTER TAX-FLOWS		

Year	Model 1 ATCF	Model 2 ATCF	Model 3 ATCF
1	$17,512	$20,905	$23,39
2	20,779	24,595	27,275
3	18,601	22,368	24,686
4	17,579	19,674	22,097
5	14,245	17,214	19,508
6	12,067	14,753	16,919
7	12,067	14,753	16,919
8	12,067	14,753	16,919
	ROR = 11.48%	ROR = 13.13%	ROR = 15.23%

	TABLE 10 AFTER TAX-FLOWS		

Year	Model 4 ATCF	Model 5 ATCF	Model 6 ATCF
1	$26,992	$29,096	$29,507
2	31,127	33,456	33,943
3	25,056	30,549	30,986
4	25,614	27,643	28,029
5	22,858	24,736	25,072
6	20,102	21,829	22,115
7	20,102	21,829	22,115
8	20,102	21,829	22,115
	ROR = 17.47%	ROR = 18.94%	ROR = 18.85%

b)

TABLE 11
INCREMENTAL ANALYSIS

Year	Comparison of Models 3 and 4	Comparison of Models 4 and 5	Comparison of Models 5 and 6
1	$3,601	$2,104	$411
2	3,852	2,329	487
3	370	5,493	437
4	3,517	2,029	386
5	3,350	1,878	336
6	3,183	1,727	286
7	3,183	1,727	286
8	3,183	1,727	286
	Increment of Investment = $6,334	Increment of investment = $5,710	Increment of investment = $1,890
	After-tax incremental ROR = 46.34%	After-tax incremental ROR = 43.18%	After-tax incremental ROR = 12.03%

* Model 5 is the choice in after tax analysis.

PROBLEM 11

A city is building an administrative building according to their increased needs for space. However, city planners see the need for even more space in a few years. Thus, the decision must be made about whether to include in the architectural design certain structural provisions for the addition of another story at a later date, which would add $90,000 on to the construction of the building which would cost $720,000 without these provisions. The second story would cost $400,000 if the provisions were built in, $540,000 if they were not. The building is thought to have a 40 year life, and maintenance costs are not thought to affect the decision. How soon must expansion occur in order to justify the more costly initial design? Assume a *MARR* of 8%. Taxes are not a consideration in this case, since it is a government building.

SOLUTION: 5.75 years.

CHAPTER THIRTEEN

Electronics

Chapter 13

ELECTRONICS

This chapter provides a review of electronics which is focused upon diodes, operational amplifiers, bipolar junction transistors, and the transistor hybrid model. As this chapter is intended to be a preparatory review for the FE Exam and not an introduction to electronics, emphasis is placed on application rather than theory.

DIODES

Ideal Diode Equation

The current-voltage relationship and the diode circuit symbol are shown in Figure 1. The ideal diode equation relates diode current and voltage as follows:

$$i_D = I_0 \left[e^{\frac{v_D}{\eta T}} - 1 \right]$$

where I_0 is the **saturation current** as shown in Figure 1 and η is an empirical constant. For the common semiconductor materials germanium and silicon, ε is equal to 1 and 2, respectively. The **volt-equivalent of the temperature**, V_T, is also required in the ideal diode equation above and is equal to

$$V_T = \frac{T}{11,600},$$

where T represents temperature in Kelvin.

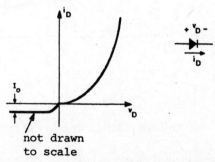

not drawn
to scale

Figure 1. Diode voltage-current relationship and circuit symbol

The inverse of the slope in the forward bias region of Figure 1 is the **forward resistance**, r_f. The forward resistance at room temperature is

$$r_f = \frac{0.026\text{mV}}{i_D}.$$

The **reverse resistance**, r_r, is the inverse of the slope in the reverse bias region and is usually assumed to be infinite in value.

The diode characteristic curve in Figure 1 provides a means of graphically analyzing diode circuits using the diode load line, as reviewed in the following example.

EXAMPLE: Each of the five load-lines in Figures 3 and 4 corresponds to the circuit in Figure 2. Find E_S (potential), R (resistance), I_R (current in the reverse direction), V_R (voltage in the reverse direction), and the mode in which the diode is operating.

SOLUTION: Start with the following method.

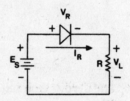

Figure 2. Circuit diagram

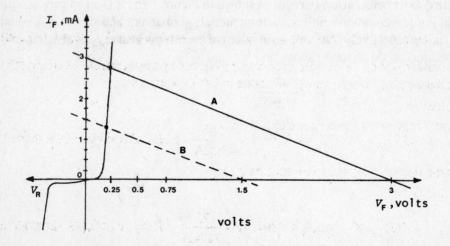

Figure 3. Current-voltage relationship in the forward direction

The endpoints of a load-line are given by

$$V = 0, \ I = \frac{E_S}{R}$$

and

$$I = 0, \ V = E_S.$$

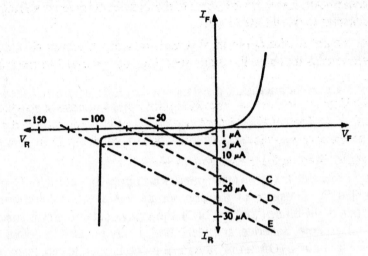

Figure 4. Current-voltage relationship in the reverse direction

For line *A*:

$$V_R = 0.25\text{V}, \ I_R = 2.75\text{mA}, \ E_S = 3\text{V}, \ R = \frac{3V}{3\text{mA}} = 1\text{k}\Omega \text{ and since } V_R > 0, \text{ the diode}$$

is operating in the forward mode.

Likewise for line *B*:

$$V_R + 0.25\text{V}, \ I_R = 1.25\text{mA}, \ E_S = 1.5\text{V}, \ R = \frac{1.4V}{1.5\text{mA}} = 1\text{k}\Omega \text{ and } V_R > 0, \text{ so the diode}$$

is in forward mode. Notice that reducing the current by more than 50% had virtually no effect on the voltage across the diode in the forward mode.

Line *C*:

$$V_R = -45\text{V}, \ I_R = -1 \text{ A}, \ E_S = -50\text{V}, \ R = \frac{-50V}{-10\mu\text{A}} = 5\text{M}\Omega. \text{ Since } V_R < 0, \text{ the diode is}$$

reverse-biased. Very little current flows.

Line *D*:

$$V_R = -70\text{V}, \ I_R = -1 \text{ A}, \ E_S = -75\text{V}, \ R = \frac{-75V}{-15\mu\text{A}} = 5\text{M}\Omega, \ V_R < 0, \text{ so the diode is}$$

reverse-biased. Note that increasing the voltage across the diode on reverse mode does not change the current.

Line E:

$$V_R = -100\text{V}, \ I_R = -5 \text{ A}, \ E_S = -125\text{V}, \ R = \frac{-125\text{V}}{25\mu\text{A}} = 5\text{M}\Omega. \text{ Now the diode has}$$

broken down. At this point, V_R will remain at -100V as long as $E_S < -100$V. The diode will burn out if the current is not limited by a large resistance, since power $= I_R V_R$ and V_R is so large.

Diode Equivalent Circuits

The **DC** or **static resistance** is the resistance of the diode at a particular operating point.

$$R_{DC} = \frac{V_D}{I_D}$$

If the static resistance is known at the **particular operating point,** or **Q-point,** it can be modeled by a resistor of the appropriate value. After determining the conducting state of the diode ("on" or "off"), the equivalent circuitry is substituted for the diode. Resistors, batteries, and ideal diodes may be used to obtain simple diode equivalent circuits. (Often only the ideal diode is used to represent the real diode.) The value of the battery reflects the conduction physics of a silicon diode—a silicon diode does not reach conduction state until 0.7 volts.

The open-circuit equivalent, or "off" state, of a diode results when a voltage with the polarity shown in Figure 5(a) is applied. The "off" state also results when a voltage less than the value required to reach conduction state (the back-bias voltage) is applied, as in Figure 5(b). This value is 0.3 volts for germanium diodes and, as mentioned previously, 0.7 volts for silicon diodes. The "on" state is shown in Figure 5(c) for an ideal diode and in Figure 5(d) for a silicon diode.

a) "Off" state of any diode

b) "Off" state of a silicon diode

$$V_D > 0$$

$$V_D \quad 0.7V$$

$$S_i$$

$$0.7V$$

c) "On" state of an ideal diode   d) "On" state of a silicon diode

Figure 5. Conduction states and equivalent circuits

EXAMPLE: A common application of silicon diodes is shown in Figure 6. Predict the current I_2 as a function of V_1.

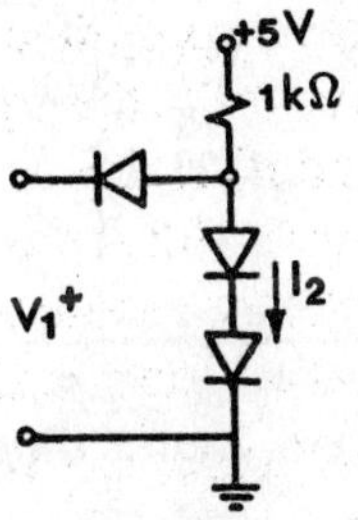

$$+5V$$
$$1k\Omega$$

$$V_1{}^+ \qquad \downarrow I_2$$

Figure 6. Silicon diodes

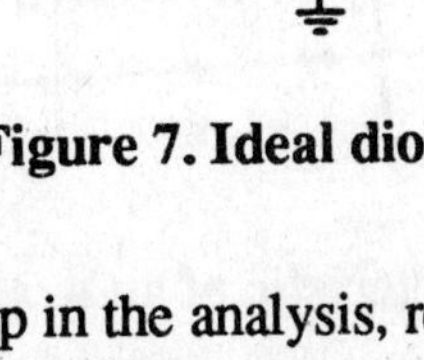

$$+5V$$

$$0.7 \quad D_1 \qquad P$$

$$V_1{}^+ \qquad I_1 \quad D_2 \quad \downarrow I_2$$

$$1.4V^+$$

Figure 7. Ideal diode

SOLUTION: As a first step in the analysis, replace the diodes by appropriate circuit models. The two diodes in series are represented by an ideal diode and a source of $2 \times 0.7 = 1.4V$. (See Figure 7.)

V_P cannot exceed 1.4V because at that voltage D_2 is forward biased and current I_2 flows readily. Also, V_P cannot exceed $V_1 + 0.7$ because at that voltage D_1 is forward-biased and current I_1 flows readily.

The critical value is $V_P = 1.4 = V_1 + 0.7$ or $V_1 = 1.4 - 0.7 = 0.7V$. For $0 < V_1 < 0.7V$, D_1 is forward biased and $I_2 = 0$. For $V_1 > 0.7V$, D_2 is forward-biased and $I_2 = \dfrac{V}{R} = \dfrac{(5-1.4)}{1,000} = 3.6\text{mA}$. The circuit thus provides a switchable constant current series.

Zener diodes operate differently in the reverse-bias region than the diodes discussed previously. While silicon diodes are modeled as open-circuits in the reverse-bias region, Zener diodes may be modeled by a short-circuit once an offset voltage is reached. The following example reviews Zener diode fundamentals.

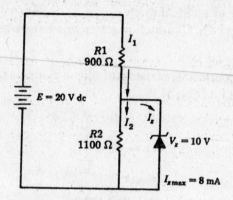

Figure 8. Zener diode

To determine if the Zener diode in Figure 8 is properly biased, the voltage-current relationship is examined. A Zener diode is basically a semiconductor diode which is designed to operate in the breakdown region (see Figure 9).

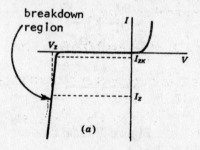

a) **The volt-ampere characteristic of an avalanche, or Zener, diode.**

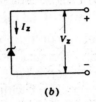

(b)

b) The symbol used for a breakdown diode
Figure 9. Breakdown region of Zener diode

The anode is connected to the negative terminal of the source; hence, its cathode is more positive than its anode. The Zener, therefore, has proper bias voltage polarity.

Assume that the Zener diode is open. Then the voltage across it is in the same proportion to the total as R_2 is to the total resistance $(R_1 + R_2)$. Writing this statement in equation form and solving it for V_{Z0} (voltage across the open Zener) yields the familiar voltage diode equation.

$$V_{Z0} = E\left(\frac{R_2}{(R_1 + R_2)}\right)$$

$$V_{Z0} = 20\left(\frac{1,100}{2,000}\right) = 11\text{V dc}$$

Note that $V_{Z0} > V_Z$ by 10 volts, and the Zener is properly voltage biased.

Because the Zener has proper voltage bias, the voltage across it is its V_Z of 10 volts. The same voltage is across R_2, making

$$I_2 = \frac{10\text{V}}{1,100} = 0.009\text{A or 9mA.}$$

The voltage across R_1 is the difference between the voltage across R_2 and the Zener, and the source voltage (E), or

$$V_{R_1} = E - V_2 = 20 - 10 = 10\text{V.}$$

Then, the current through R_1 is

$$I_1 = \frac{10\text{V}}{900} = 0.011\text{A or 11mA}$$

$$I_Z = I_1 - I_2 = 11\text{mA} - 9\text{mA} = 2\text{mA}$$

Therefore, the current through the Zener (I_Z) is less than I_{Zmax} of 8mA. The Zener has a negative open voltage greater than its V_Z and a current (I_Z) less than I_{Zmax}. Hence, it is properly biased.

Diode Applications

Diodes are commonly used as **full-wave** and **half-wave rectifiers**. A half-wave rectifier is shown in Figure 10 with its input and resulting output signals. The output voltage is explained for each half-cycle of the input voltage as follows.

1. **Positive half-cycle:** The polarity of the input voltage defines the "on" state for the ideal diode in the figure. The ideal diode may be represented by a short circuit. The output voltage is therefore equivalent to the input voltage for the first half of the period. (For a circuit with a silicon diode, the equivalent circuitry would be a 0.7 volts battery, and the output voltage would be equal to the input voltage—0.7 volts.)

2. **Negative half-cycle:** The polarity of the input voltage defines the "off" state of the ideal diode. This results in an open circuit equivalent for the negative half-cycle, and therefore zero voltage for the second half of the period.

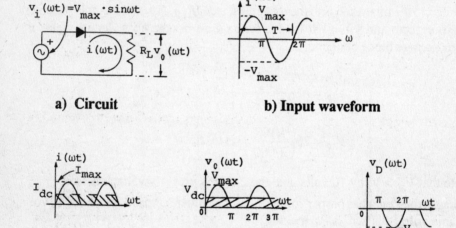

a) **Circuit** b) **Input waveform**

c) **Output Current** d) **Output voltage** e) **Voltage across the diode**

Figure 10. Half-wave rectifier

The load for the rectifier circuit in Figure 11, for the case of approximate U.S. line voltage ($C = 50 \times 10^{-6}$F, $R = 1,000\Omega$ and $V_1 = 165$ (sin $377t$)), can easily be drawn. For simplicity, assume that the diode is a perfect rectangle.

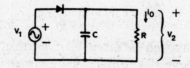

Figure 11. Rectifier circuit

First sketch the input voltage (Figure 12a). Initially, the capacitor is fully charged to 165V at time t_1. In the time interval from t_1 to $5t_1$, the input and the diode are effectively removed from the circuit. During this time the capacitor discharges into the load, supplying the load current, and the circuit reduces to a resistance in parallel with a capacitance. (See Figure 12b.) In RC circuits, the voltage V_C decays exponentially with time constant RC. The voltage is thus of the form

$V_C = 165 \exp\left(-\left(\dfrac{t-t_1}{RC}\right)\right)$. This equation is plotted in Figure 12(c).

The equation only describes the output up to the time t_C shown in the figure because the charging of the capacitor starts again at this instant and the cycle is repeated. Thus, the actual waveform is as given in Figure 12(d). The load current has the same form as the output voltage because $i_0 = \dfrac{V_C}{R}$.

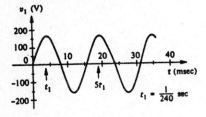

a) Input voltage

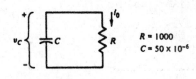

b) Resistance in parallel with a capacitance

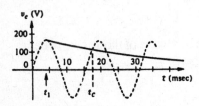

c) Voltage

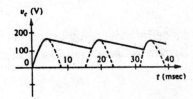

d) Actual waveform

Figure 12. Addition of a capacitor to the basic half-wave rectifier circuit

The average dc values for current and voltage shown in Figure 10 are defined as follows:

$$I_{dc} = \frac{I_{max}}{\pi} = 0.318\, I_{max}$$

$$V_{dc} = 0.318 I_{dc} \times R_L = 0.318 V_{max}$$

Other factors of interest concerning half-wave rectifier circuits are the ripple factor, efficiency, and the peak inverse voltage (PIV). The following example provides a review of these terms, and explains the procedure and equations needed to solve problems of this type.

EXAMPLE: For the half-wave rectifier shown in Figure 13, find (a) the I_{dc}, (b) the I_{rms}, (c) the ripple factor, r, (d) the rectifier efficiency η_r, and (e) the peak inverse.

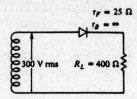

Figure 13. Half-wave rectifier

SOLUTION: (a) Since the input voltage is an rms (root mean square) value, V_m and I_m must be converted by

$$V_m = \frac{300}{0.707} = 425\text{V}$$

$$I_m = \frac{V_m}{(r_F + R_L)} = \frac{425}{425} = 1\text{A}$$

Therefore, for a half-wave rectifier,

$$I_{dc} = \frac{1}{2}\pi\int_0^\pi i\,d\theta = \frac{I_m}{\pi} = \frac{1}{\pi} = 0.32\text{A}.$$

(b) For a half-wave rectifier

$$I_{rms} = \frac{I_m}{2} = \frac{1}{2} = 0.5\text{A}.$$

(c) By definition, the ripple factor is the ratio of the rms value of the ac component of the current to the dc component of the current, where

$$i(t)_{ac} = i(t) - I_{dc}$$

or

$$I_{ac,rms}^2 = I_{rms}^2 - I_{dc}^2$$

so

$$r = \frac{(I_{ac,rms})}{(I_{dc})} = \frac{\sqrt{(I_{rms}^2 - I_{dc}^2)}}{I_{dc}} = \frac{\sqrt{(.5)^2 - (.32)^2}}{.32} = 1.21$$

(d) The rectifier efficiency is the ratio of the dc power absorbed by R_L to the average power supplied

or

$$\eta_r = \frac{P_{dc}}{P_i}$$

Since

$$P_i = I_{rms}^2(r_F + R_L)$$

and

$$P_{dc} = I_{dc}^2 R_L$$

it follows that

$$\eta_r = \frac{\left(I_{dc}^2 R_L\right)}{\left(I_{rms}^2 (r_F + R_L)\right)} = \frac{\left(0.32^2 (400)\right)}{\left(.5^2 (25 + 400)\right)} = 38\%.$$

(e) The peak reverse voltage is the maximum voltage the diode sees when reverse-biased. In this circuit, no current flows when the diode is reverse biased, so

$$PIV = V_m = 425V.$$

Full-wave rectification is accomplished using a circuit such as in Figure 14. The average dc values are shown in the figure and defined as follows:

$$I_{dc} = \frac{2 \times I_{max}}{\pi} = 0.636 I_{max}$$

$$V_{dc} = 0.636 V_{max}$$

These dc values are twice that of the half-wave circuit, as expected. The output signal is found using the same method as itemized for the half-wave rectifier circuit.

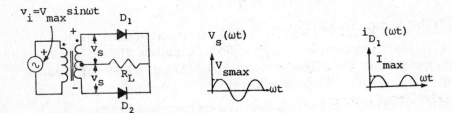

a) **Circuit** b) **Waveform across secondary winding** c) **Current in diode D_1**

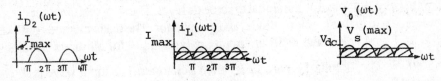

d) **Current in D_2**   e) **Load current waveform**   f) **Output voltage**

Figure 14. Full-wave rectifier

Other common diode applications are **clipper** and **clamper networks**. Clipper circuits "clip" off part of the input signal without affecting the remaining part of the input signal. For example, the half-wave rectifier circuit discussed previously was a simple clipper circuit because it clipped off the negative half of the voltage without affecting the positive cycle.

An example of a symmetrical clipper is shown in Figure 15. It clips the positive and negative portions of the input sine wave, V_i, so the output, V_o, approximates a trapezoidal waveform. (See Figure 16.)

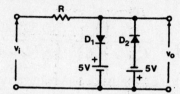

Figure 15. Symmetrical clipper

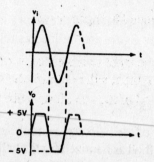

Figure 16. Reverse-biased diodes

In Figure 16, when V_i is less than 5V, both diodes are reverse-biased, and $V_o = V_i$. As soon as V_i is slightly more positive than 5V, diode D_1 is forward biased (diode D_2 remains reverse biased). The output is clamped (maintained) at +5V and remains clamped at that value until V_i becomes less than 5V.

When V_i is between +5V and –5V, both diodes are reverse-biased again, and $V_o = -V_i$. As soon as V_i becomes more negative than –5V, D_2 conducts (D_1 is reverse-biased), and the output is now clamped to –5V. It remains clamped at this value until V_i is less negative than –5V.

Clamper circuits "clamp" the signal to some dc level. The simplest clamper circuit must include a resistive element, diode, and a capacitor. The basic methods outlined with the full- and half-wave rectifier circuit analysis apply for all diode circuits.

EXAMPLE: The circuit of Figure 17 is called a clamper. The input wave is shown on Figure 18.

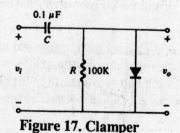

Figure 17. Clamper

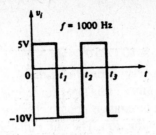

Figure 18. Input wave

SOLUTION: At the instant the input switches to the +5V state, the circuit will appear as shown in Figure 19. The input will remain in the +5V state for an interval of time equal to one-half the period of the waveform since the time interval $0 \rightarrow t_1$ is equal to the interval $t_1 \rightarrow t_2$.

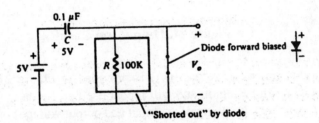

Figure 19. Input in +5V state

The period at V_1 is $T = \dfrac{1}{f} = \dfrac{1}{1,000} = 1\text{ms}$ and the time interval is the +5V state is

$\dfrac{T}{2} = 0.5\text{ms}$.

Since the output is taken from directly across the diode, it is 0V for this interval of time. The capacitor, however, will rapidly charge to 5V, since the time constant of the network is not $\tau = RC \sim OC = 0$.

When the input switches to –10V, the circuit of Figure 20 will result.

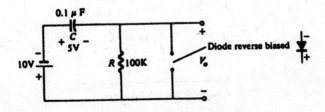

Figure 20. Input in –10V state

The time constant for the circuit of Figure 20 is

$$RC = 100 \times 10^8 \times 0.1 \times 10^{-6} = 10\text{ms}$$

Since it takes approximately 5 time constants, or 50ms, for a capacitor to discharge, and the input is only in this state for 0.5ms, to assume that the voltage across the capacitor does not change appreciably during this interval of time is certainly a reasonable approximation. The output is therefore

$$V_0 = -10 - 5 = -15\text{V}.$$

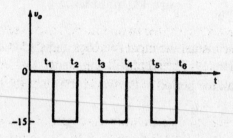

Figure 21. Output waveform

The resulting output waveform (V_0) is provided in Figure 21. The output is clamped to the negative region and will repeat itself at the same frequency as the input signal. Note that the swing of the input and output voltages is the same: 15V. For all clamper circuits the voltage swing of the input and output waveforms will be the same. This is in contrast to clipping circuits.

Diodes are also used in **logic gates**, such as the **AND** and **OR** gate. The basic diode AND gate is shown in Figure 22 with the corresponding truth table. Although Boolean algebra is not reviewed in this chapter, recall that an AND gate produces an output of 1 only if both inputs are 1. The inputs A and B are varied in Figure 22(d) for the four combinations shown in the truth table. For example, notice that the second configuration ($A = 0$, $B = 1$, $F = 0$) has a voltage source at input B, but does not show a voltage source for the A input. As with the previous diode applications, analysis is based upon the "off" or "on" state of the diodes.

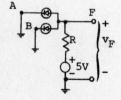

a) Absence of voltage

b) Terminal *B* is high, *A* is low

c) **Terminal A is high, B is low**   d) **Terminals A and B are high**

Figure 22. The diode "AND" gate

An absence of voltage sources tied to terminals A and B results in a low output (first entry in truth table). When terminal B is high, terminal A is low, a low output results. When terminal A is high and terminal B is low, low output results. When terminals A and B are high, high output results, as shown in the last entry of the truth table.

An OR gate, such as in Figure 23, produces an output of 1 if any or all of the inputs are 1. Although the different input variations of the circuit are not shown in Figure 23, the addition of a voltage source at the appropriate input(s) would result in the truth table configurations shown. For example, a voltage source at the A input only ($A = 1, B = 0, C = 0$) would produce an output $F = 1$, as shown in the fifth entry of the truth table. The following example illustrates an OR diode logic (DL) circuit.

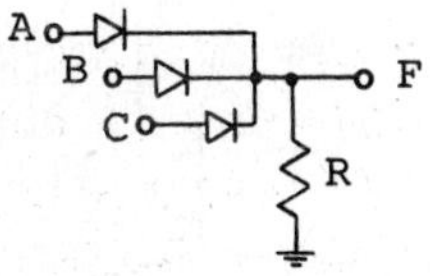

Inputs			Output
A	B	C	F
0	0	0	0
0	0	1	1
0	1	0	1
0	1	1	1
1	0	0	1
1	0	1	1
1	1	0	1
1	1	1	1

a) **The circuit**   b) **Associated truth table**

Figure 23. A diode "OR" gate

EXAMPLE: Verify that the **DL** gate shown can perform the exclusive OR logic.

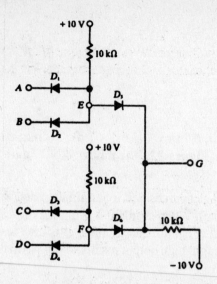

Figure 24. DL gate

SOLUTION: If either A or B is low, then D_1 or D_2 will conduct, and point E will be at V_D, low. If both are high, D_1 and D_2 are both off, and point E is high. Thus, D_1 and D_2 form the AND function $E = AB$ while similarly $F = CD$. The directions of D_5 and D_6 are reversed and they form an OR, $G = E + F$. (A high at E or F forward biases D_5 or D_6, the voltage at E or F is carried through to G, since the diode voltage drop is small.) Thus,

$$G = E + F = AB + CD.$$

With

$$B = D \text{ and } C = A$$
$$G = AD + AD = A \times D.$$

D_5 and D_6 form the final OR function besides providing cancellation of the 0.7V drop in the input diodes.

BIPOLAR JUNCTION TRANSISTORS

Common Base Configuration

The **bipolar junction transistor** will be reviewed in terms of the common base and common emitter configurations. As expected, the base of the common base transistor as shown in Figure 25(a) is common to both input (emitter) and output (collector). Figure 25(b) illustrates the active, saturation, and cutoff regions of the output characteristics. The collector junction is reverse-biased in the active region, and the emitter junction is forward-biased. The **collector** and **emitter currents** are approximately equivalent in the active region.

$$I_C \simeq I_E$$

Both collector and emitter junctions are reverse-biased in the cutoff region and forward-biased in the saturation region. The input characteristics are shown in Figure 25(c).

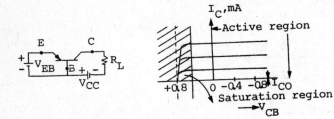

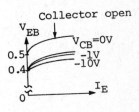

a) BJT common base configuration

b) output characteristics

c) input characteristics

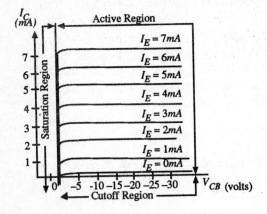

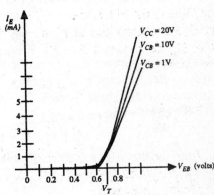

d) Output characteristics

e) Input characteristics

Figure 25. Transistor configuration

EXAMPLE: In Figure 26, I_E is chosen to be 2.0mA. If $V_{EE} = 12.0$V, $V_{CC} = 12.0$V, and $R_L = 5.0$ kΩ. Find (a) R_E, (b) I_C, (c) $I_C R_L$, and (d) V_{CB}.

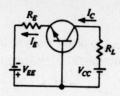

Figure 26. Transistor circuit

SOLUTION: (a) $R_E \dfrac{V_{EE}}{I_E} = \dfrac{12}{2(10^{-3})} = 6k\Omega.$

(b) $I_C \sim I_E \sim 2mA.$

(c) $I_C R_L = 2.0(10^{-3})(5.0)(10^3) = 10V.$

(d) $V_{CB} = V_{CC} - I_C R_L = 12 - 10 = 2.0V.$

Common Emitter Configuration

The common emitter configuration of Figure 27 produces a collector current of:

$$\beta = \frac{\alpha}{1-\alpha}, \; I_c = (1+\beta)I_{co} + \beta I_B$$

where α is the approximate ratio of collector current to emitter current, and β (the common-emitter forward-current amplification factor) is approximately the ratio of collector current to base current. The collector current equation above was derived from **Kirchoff's Current Law:**

$$I_E = I_C + I_B,$$

and from the majority and minority carrier relationship of the collector current:

$$I_C = I_{C(\text{majority})} + I_{CO(\text{minority})}.$$

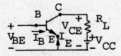

a) Emitter configuration

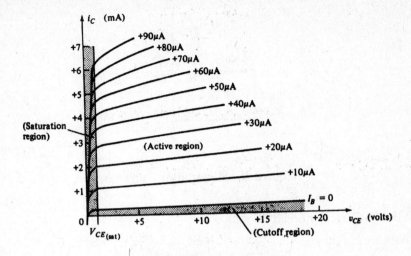

b) Output characteristics

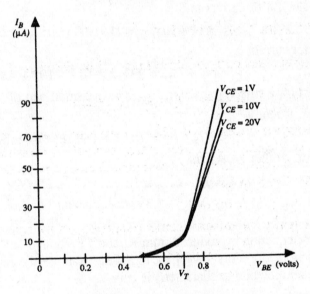

c) Output characteristics
Figure 27. Common emitter configuration

EXAMPLE: A silicon npn transistor with $\alpha = 0.99$ and $I_{CBO} = 10^{-11}$ A is shown in Figure 28. Predict the quantities i_C, i_E, and v_{CE}. (Note: It is convenient and customary in drawing electronic circuits to omit the battery which is assumed to be connected between the +10V terminal and ground.)

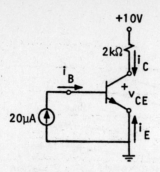

Figure 28. Silicon npn transistor

SOLUTION: For this transistor,

$$\beta = \frac{\alpha}{1-\alpha} = \frac{0.99}{1-0.99} = 99$$

and the collector cutoff current is

$$I_{CEO} = (1+\beta)\,I_{CBO} = (1+99)\,10^{-11} = 10^{-9}\text{A}.$$

The collector current is

$$i_C = \beta i_B + I_{CEO} = 99 \times 2 \times 10^{-5} + 10^{-9} \cong 1.98\text{mA}.$$

As expected for a silicon transistor, I_{CEO} is a very small part of i_C.
The emitter current is

$$i_E = -(i_B + i_C) = -(0.02 + 1.98)10^{-3} = -2\text{mA}.$$

The collector emitter voltage is

$$v_{CE} = 10 - i_C R_C \cong 10 - 2\text{mA} \times 2\text{k}\Omega = 6\text{V}.$$

Since $v_{CB} = v_{CE} - v_{BE} \cong 6 - 0.7 = +5.3\text{V}$, the collector-base junction is reverse biased.

The load line equation is produced from Kirchoff's Voltage or Current Law for the transistor output loop. As shown in the following example, the load line equation is then graphed on the output characteristic curves.

EXAMPLE: The transistor shown in Figure 29 has the *CE* output characteristics drawn in Figure 30. If $I_B = 50\mu A$ with $V_{cc} = 12$ and $R_E = 1k\Omega$, then calculate I_C and V_{CE}.

SOLUTION:

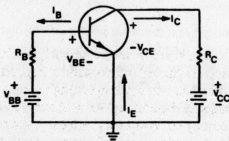

Figure 29. Transistor

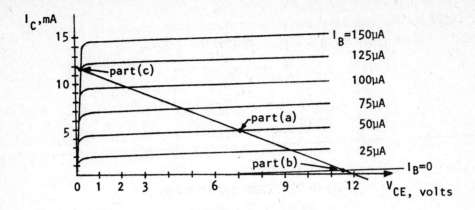

Figure 30. Load-line relationship

The load-line equation is

$$12V = I_C \times 1k\Omega + V_{CE}$$

and is plotted as shown in Figure 30. The intersection of this load line and the $I_B = 50\mu A$ curve is thus $I_C = 50mA$ and $V_{CE} = 7V$. The voltage across the 1k Ω resistor is $5mA \times 1k\Omega = 5V$. These values can be checked by substituting into the load-line equation. In this case, the operating point lies in the active region of the transistor characteristics.

Similarly, the intersection of the load line and the $I_B = 0$ curve gives the new operating point as $I_C \simeq 0$ and $V_{CE} = 12V$. Since the transistor is in cutoff, only a small leakage (I_{CEO}) flows and the voltage dropped across the series resistor is essentially zero. Thus, $V_{CE} = V_{CC}$.

The intersection of the load line and the $I_B = 125$ A curve gives the operating point as $I_C = 11.8mA$ and $V_{CE} = 0.2V$. this operating point lies in the saturation region of the characteristics.

The operating point using $I_V = 150\mu A$ is the same as for $I_B = 125\mu$ A, since the different curves merge in the saturation region. Thus, $I_C = 11.8mA$ and $V_{CE} = 0.2V$ is still the operating point. In fact, further increases in I_B will still result in the same I_C and V_{CE}. The extra base current does not produce any increase in collector once the transistor is saturated. Notice that V_{CE} is very small in the saturation region.

DC Biasing

The purpose of this section is to review basic **DC biasing circuits**. Transistors are biased in order to obtain a particular current and voltage relationship, or Q-point. The Q-point is a specific point within the operating region of the transistor, generally between the saturation and cutoff regions as shown in Figures 25 and 27. Biasing analysis techniques may be performed graphically, mathematically, or using computer programs. The following example illustrates graphical determination of the Q-point.

EXAMPLE: The common emitter (CE) characteristics for the transistor used in the circuit of Figure 31 are given in Figure 32. Find the Q-point.

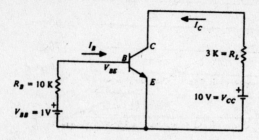

Figure 31. Transistor

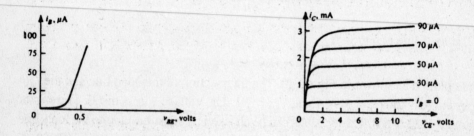

a) **Input characteristics**   b) **Output characteristics**

Figure 32. Common emitter (CE) characteristics

SOLUTION: For the input side from Figure 31, when $I_B = 0$, $V_{BE} = 1$V. With $V_{EE} = 0$, $I_B = \dfrac{(1V)}{(10k)} = 100\mu A$. The load line is then superimposed on the input (base) characteristic as shown in Figure 34. At their point of intersection, it is found that $I_B = 50\mu A$ and $V_{BE} = 0.5$V

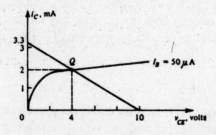

Figure 33. Collector characteristics

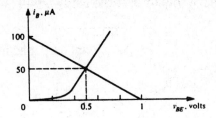

Figure 34. Load line superimposed on the input

On the output side, setting $I_C=0$ yields $V_{CE}=10V$. For $V_C=0$, $I_C=\dfrac{10V}{3k}=3.3mA$.

With the load line superimposed on the collector characteristic as shown in Figure 33, the line intersects the base curve $I_B=50\mu A$, and the Q-point is $I_C=2mA$ and $V_{CE}=4V$. The emitter current is given by

$$I_E=I_C+I_B$$

Therefore, $2+0.05=2.05mA$.

The design of a BJT circuit which may be used in an environment with fluctuating temperature employs the use of a factor known as the **stability factor**, S. The collector current will vary for fluctuations in temperature because of the temperature dependence of the **reverse saturation current** I_{CO}, the **base-emitter voltage** V_{BE}, and the **amplification factor (gain)** β. The stability factor is therefore defined in terms of these three parameters:

$$S(I_{CO})=\frac{\Delta I_C}{\Delta I_{CO}} \quad S(V_{BE})=\frac{\Delta I_C}{\Delta V_{BE}} \quad S(\beta)=\frac{\Delta I_C}{\Delta \beta}$$

TRANSISTOR MODELING

The small-signal AC response of BJTs will be analyzed using **hybrid equivalent modeling**. The hybrid model is founded upon two-port theory, in which a network is represented as shown in Figure 35(a). The hybrid equivalent circuit is shown in Figure 35(b). The voltage and current equations, or hybrid equations, are:

$$\left. \begin{array}{l} v_1=h_i\times i_1+h_r\times v_2 \\ i_2=h_f\bullet\times i_1+h_o\times v_2 \end{array} \right\} \quad \begin{array}{l} v_1=h_{11}\times i_1+h_{12}\times v_2 \\ i_2=h_{21}\times i_1+h_{22}\times v_2 \end{array}$$

and make use of the hybrid parameters, or h-parameters, defined as:

$$h_i=\left.\frac{v_1}{i_1}\right|_{v_2=0}=\text{short-circuit input impedance}$$

$$h_r=\left.\frac{v_1}{v_2}\right|_{i_1=0}=\text{open-circuit reverse voltage gain}$$

$$h_f = \frac{i_2}{i_1}\bigg|_{v_2=0} = \text{short-circuit forward current gain}$$

$$h_0 = \frac{i_2}{v_2}\bigg|_{i_1=0} = \text{open-circuit output admittance}$$

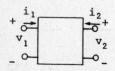

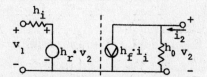

a) Two-part network   **b) Equivalent circuit of two-part network**
Figure 35. General two-port network

Problems involving the use of hybrid equivalent models for small-signal AC analysis generally involve the determination of input and output impedance and current and voltage gains. The output impedance is determined for zero input voltage. Recall that voltage gain, A_v, and current gain, A_i, are defined as

$$A_v = \frac{V_{out}}{V_{in}} \qquad A_i = \frac{I_{out}}{I_{in}}.$$

EXAMPLE: Examine the circuit in Figure 36.

FIG her

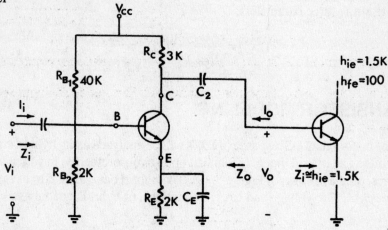

Figure 36. Hybrid circuit diagram

Find the following:

(a) $A_i = \dfrac{I_o}{I_i}$ (b) $A_v = \dfrac{V_o}{V_i}$ (c) Z_i (d) Z_o.

SOLUTION: Replacing the dc supplies and capacitors by short circuits and substituting the appropriate hybrid equivalent circuit will result in the configuration of Figure 37. Note that the 2k emitter resistor has been "shorted out" by the capacitor CE.

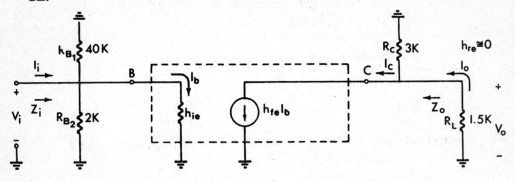

Figure 37. Replacing and substituting hybrids and short circuits

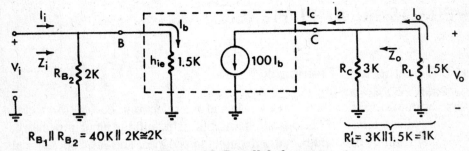

Figure 38. Parallel elements

Considering parallel elements will result in the configuration of Figure 38, $I_2 = 100$ I_B. That is $A_1 = \dfrac{I_C}{I_B} = h_{fe}$. Applying the current divider rule to the input and output circuits, we get

$$I_B = \frac{(2k)(I_i)}{(2k + 1.5k)} = 0.571 I_i,$$

and

$$I_0 = \frac{(3k)(I_2)}{(3k + 1.5k)} = 0.667 I_2.$$

Substituting,

$$A_i = \frac{I_o}{I_i} = \left(\frac{I_o}{I_2}\right)\left(\frac{I_2}{I_i}\right) = \left(\frac{I_o}{I_2}\right)\left(\frac{I_2}{I_b}\right)\left(\frac{I_b}{I_i}\right)$$
$$= (0.667)(100)(0.571)$$
$$= 38.1$$

and likewise

$$A_v = \frac{V_o}{V_i}$$

$$V_i = h_{ie} I_B$$
$$V_o = -h_{fe} I_B R_L'$$

so

$$A_v = \frac{V_o}{V_i} = h_{fe} \frac{R_L'}{h_{ie}}$$

$$= \frac{-(100)(1 \times 10^3)}{(1.5 \times 10^3)}$$

$$= 66.7$$

OPERATIONAL AMPLIFIERS

Basic Op-Amp with Feedback

Operational amplifiers make use of BJTs and other electronic devices to provide current or voltage gain. An ideal operational amplifier has infinite input impedance, infinite gain, infinite bandwidth, and zero output impedance. The basic operational amplifier with feedback is shown in Figure 39.

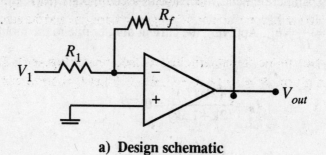

a) Design schematic

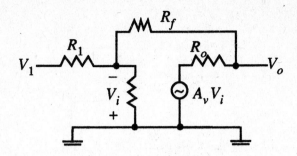

b) Circuit diagram
Figure 39. Basic operational amplifier with feedback.

Derived using the principle of superposition and the definition of voltage gain, the input voltage is

$$V_i = \frac{R_f}{R_f(1+A_v)R_1}V_1.$$

The input voltage is approximately equal to

$$V_i \approx \frac{R_f}{A_v R_1}V_1$$

when

$$A_v >> 1$$
$$R_1 >> R_f$$

The following example derives the output voltage equation for the basic operation amplifier with feedback. This circuit is also referred to as an inverting constant-gain circuit, as the sign of the input signal is changed at the output and the gain is a constant multiplier.

EXAMPLE: Determine the output voltage of the circuit in Figure 40. The following parameters are given: $R_i = 10^5\Omega$, $R_o = 0$, $A = 10^5$. Let $R_1 = 10^5\Omega$ and $R_2 = 10^7\Omega$.

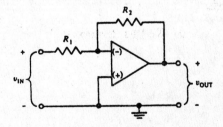

Figure 40. Circuit

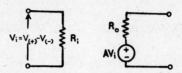

Figure 41. Model for op-amp

SOLUTION: By replacing the op-amp with the model of Figure 41, the result is the circuit in Figure 42. In order to keep track of the negative (–) and positive (+) input terminals, their locations in the new circuit have been shown.

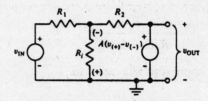

Figure 42. Resulting circuit

The voltage at the (–) input terminal $V_{(-)}$ is found by the node equation. Summing the currents leaving the (–) node:

$$\frac{V_{(-)} - V_{in}}{R_1} + \frac{V_{(-)}}{R_i} + \frac{V_{(-)} - A\left(V_{(+)} - V_{(-)}\right)}{R_2} = 0.$$

$V_{(+)} = 0$, so the equation is rewritten as

$$V_{(-)}\left[\frac{1}{R_1} + \frac{1}{R_i} + \frac{(1+A)}{R_2}\right] = \frac{V_{in}}{R_1}.$$

Referring to the values of R_1, R_i, R_2, and A, the last term in the square brackets is clearly the largest, so the first two terms in the brackets may be neglected. Moreover, $(1 + A)$ is nearly equal to A. Thus,

$$V_{(-)}\left[\frac{A}{R_2}\right] \approx \frac{V_{in}}{R_1}$$

$$V_{(-)} \approx \frac{R_2}{R_1}\frac{V_{in}}{A}.$$

The output voltage is equal to the voltage of the dependent voltage source, which is $-AV_{(-)}$. Thus,

$$V_{out} \simeq -\frac{R_L}{R_1}V_{in}.$$

Op-AMP Applications

Common operation amplifier applications include circuits which function as **adders, integrators, differentiators,** and **simple phase changers.** A noninverting constant-gain circuit is shown in Figure 43. The gain for this circuit is

$$A_v = \frac{R_f + R_1}{R_1}.$$

As expected, the noninverting amplifier does not change the sign of the input signal.

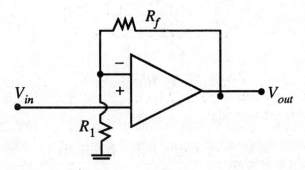

Figure 43. Noninverting constant-gain operation amplifier

A summing amplifier, or adder, is shown in Figure 44. This operational amplifier configuration provides a means to sum voltages, in this case, the three voltages V_1, V_2, and V_3. The output voltage is

$$V_o = -R'\left(\frac{V_1}{R_1} + \frac{V_2}{R_2} + \frac{V_3}{R_3}\right),$$

or

$$V_o = \frac{-R'}{R}(V_1 + V_2 + V_3), \text{ if } R_1 = R_2 = R_3 = R.$$

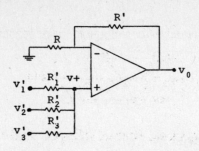

Figure 44. Summing amplifier

The output of the operational amplifier in Figure 45 is the integral of the input. This integrator circuit produces an output voltage as follows:

$$V_o = \frac{-1}{RC} \int V dt$$

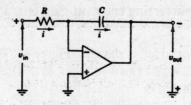

Figure 45. Integrator

The input to the integrator shown in Figure 46 is the +/− 10V, 250 Hz square waveform shown in Figure 47. Knowing that $C = 1\mu F$ and $R = 100k\Omega$, the output voltage waveform is determined.

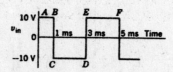

Figure 46. Integrator

Figure 47. Hz square waveform

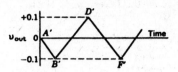

Figure 48. Output voltage waveform

The time constant of the circuit, RC, is

$$RC = 10^5 \Omega \times 1 \times 10^{-5} \text{F} = 0.1\text{s}.$$

Then

$$V_{out} = \frac{-1}{RC} \int V_{in} dt = -10 \int V_{in} dt.$$

The area under the curve is $\int V_{in} dt$. Starting from A toward B on the square wave, the area increases linearly. The final area at B is

$$\int V_{in} dt = (100) \times (0.001\text{s}) = 0.01\text{s}$$

then

$$V_{out} = -10 \int V_{in} dt = (-10) \times (0.01) = -0.1\text{V}.$$

Thus, the voltage changes linearly from zero at A' to -0.1V at B'.

The area from C to D is negative and the magnitude of the area increases linearly from C to D. the total area from C to D is

$$(-10V) \times (0.002s) = -0.02s$$

$$V_{out} = -10 \int V_{in} dt = (-10) \times (-0.02) = +0.2V$$

At B', V_{out} is -0.1V. The linear change of Vout from B' to D' is $+0.2V$. Thus the voltage at D' is $-0.1 + 0.2 = +0.1V$.

The area under the input voltage waveform increases linearly from E to F. This increasing area causes the voltage to fall linearly from D' to F'.

Thus, if a square wave is fed into the input of an integrator, the output voltage is the triangular wave form shown in Figure 48.

A differentiator circuit is shown in Figure 49, which as expected, is very similar to the integrator circuit of Figure 48. the output voltage is

$$V_o = -RC \frac{dV}{dt}.$$

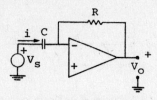

Figure 49. Differentiator

REVIEW PROBLEMS

PROBLEM 1

For the diode circuit of Figure 50, assume that $R_F = 50\Omega$, $R_R = \infty$, and $V_T = 0V$. Determine the current and dissipated power in the diode and load resistance, R_L, when

(a) Point A is positive with respect to B.

(b) Point A is negative with respect to B.

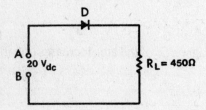

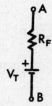

Figure 50. Forward biased diode   **Figure 51. Diode model**

SOLUTION: (a) When A is positive with respect to B, the diode is forward-biased. Representing the diode by the model of Figure 51 with $V_T = 0$, the circuit appears as in Figure 52. Current I is $\dfrac{20}{(50+450)} = \dfrac{20}{500} = 0.04A$. The power dissipated in the diode is $P_D = I^2 R_F = (0.04)^2 \times 50 = 0.08W = 80mW$. In R_L, the power dissipated is $(0.04)^2 \times 450 = 0.72W = 720mW$.

(b) When A is negative with respect to B, the diode is reverse-biased. Because $R_R = \infty$, it is represented by an open switch in Figure 53. Current $I = 0$, and the dissipated power in the diode and load resistance is also equal to zero.

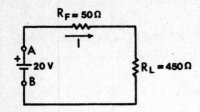

Figure 52. Circuit

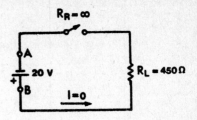

Figure 53. Open switch

PROBLEM 2

In Figure 54, $V_S = 5$V. a) Determine I_F for $R_L = 100\Omega$. b) Find an R_L that will give $I_F = 30$mA. The diode characteristics are given in Figure 55.

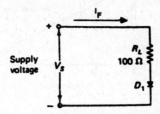

Figure 54. Diode circuit diagram

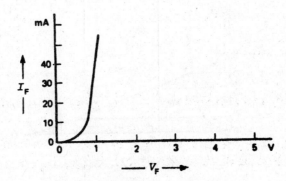

Figure 55. Diode characteristics

SOLUTION: (a) By *KVL*,

$$V_S = I_F R_L + V_F.$$

When $I_F = 0$,

$$V_S = 0 + V_F.$$

Therefore, the diode voltage is

$$V_F = V_S = 5\text{V}$$

We plot point A on the diode characteristics at $I_F = 0$ and $V_F = 5\text{V}$. When $V_F = 0$,

$$V_S = I_F R_L + 0$$

$$I_F = \frac{V_S}{R_L} = \frac{5\text{V}}{100\Omega} = 50\text{mA}$$

We plot point B on the diode characteristics at $I_F = 50\text{mA}$ and $V_F = 0$. Now we draw the dc load line through points A and B. The dc load line intersects the diode characteristics at point Q, where $I_F = 40\text{mA}$.

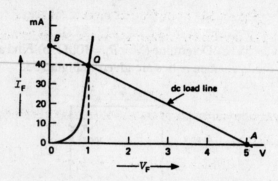

Figure 56. Load line when $R_L = 100\Omega$

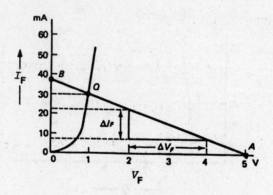

Figure 57. Load line when $I_F = 30\text{mA}$

(b) Our load line now passes through points A and Q in Figure 57. R_L is the reciprocal of the slope of the load line.

$$R_L = \frac{\Delta V_F}{\Delta I_F} = \frac{2\text{V}}{15\text{mA}} = 133\Omega$$

PROBLEM 3

For the circuit shown, determine the direct current in the diode and the dc output voltage.

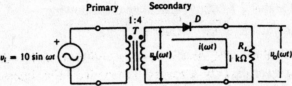

Figure 58. Transformer circuit diagram

SOLUTION: The maximum voltage across the secondary of the transformer, $V_{s(max)}$, is

$$V_{s(max)} = 4V_{p(max)} = 4 \times 10 = 40V$$

where $V_{p(max)}$ is the maximum voltage across the primary winding. While the current is positive, it is given by

$$i(\omega t) = I_{max} \sin \omega t \quad 0 < \omega t < \pi$$

It cannot go negative because of the diode, so

$$i(\omega t) = 0 \quad \pi \le \omega t \le 2\pi$$

Then the dc (average) value of $i(\omega t)$ is

$$I_{dc} = I_{max} \frac{1}{2\pi} \left[\int_0^\pi \sin(\omega t) d(\omega t) + \int_\pi^{2\pi} o \, d(\omega t) \right]$$

$$= \frac{I_{max}}{\pi} = 0.318 I_{max}$$

$$I_{dc} = \frac{0.318 V_{s(max)}}{R_L} = \frac{0.318 \times 40}{1} = 12.72 \text{mA}$$

Similarly,

$$V_{dc} = \frac{V_{max}}{\pi}$$

$$V_{dc} = 0.318 \times 40 = 12.72 \text{V}$$

The dc output also can be determined from the following relationship:

$$V_{dc} = I_{dc(HW)} R_L$$

$$= 12.72 \times 1 = 12.72 \text{V}$$

PROBLEM 4

For the circuit shown, the following data applies: ideal diodes, input voltage to primary coil of the transformer, E_i, is 115V RMS; turns ratio of the transformer is 2.3:1; $R_L = 500\Omega$ and $C = 1,000\mu F$. Calculate the following:

(a) RMS ripple voltage

(b) DC component of load voltage

(c) Total load power, P_L

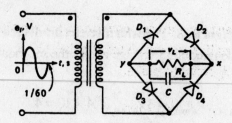

Figure 59. Circuit diagram

SOLUTION: The ratio of the voltage at the secondary coil transformer to the voltage at the primary coil of the transformer equals the turns ratio; therefore, the secondary voltage is $\dfrac{115}{2.3} = 50\text{V RMS}$. The peak load voltage, V_{LM}, is then $50\sqrt{2} = 70.7\text{V}$.

(a) The peak-to-peak ripple voltage for the full-wave 60 Hz signal input is

$$V_{R(PP)} = \frac{V_{LM}}{120R_L C}$$

$$V_{R(PP)} = \frac{70.7}{120(500 \times 10^3)} = 1.18\text{V}$$

The RMS ripple voltage is

$$V_{L(AC)} = \frac{V_{R(PP)}}{2\sqrt{3}}$$

$$V_{L(AC)} = \frac{1.18}{2\sqrt{3}} = 0.341\text{V}$$

(b) The DC component is

$$V_{L(DC)} = V_{LM} - \frac{1}{2}V_{R(PP)}$$

$$V_{L(DC)} = 70.7 - 0.59 \quad 70.1\text{V}$$

(c) To calculate the total load power, we can assume that the ripple voltage is negligible since it is small compared to $V_{L(DC)}$.

$$P_L = V_L I_L$$

$$P_L = 70.1\left(\frac{70.1}{500}\right) = 9.85\text{W}$$

PROBLEM 5

Given the circuit and the transistor collector characteristics of Figure 60 and 61,

(a) Plot the dc load line and obtain the Q-point.

(b) Find V_{CE} and I_C from the graph.

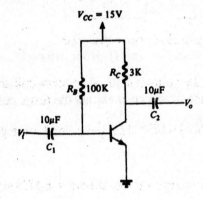

Figure 60. Circuit characteristics

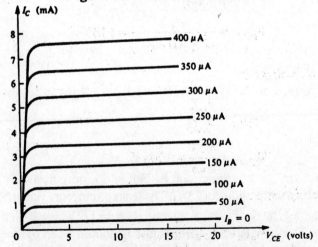

Figure 61. Transistor collector characteristics

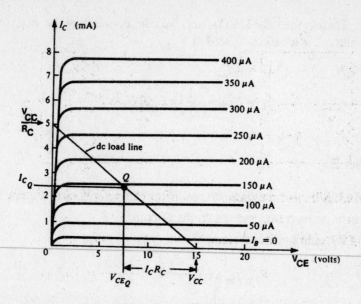

Figure 62. DC load line and Q-point intersection

SOLUTION: (a) Draw the dc load line as shown in Figure 62. The two points for the load line are

a. at $I_C = 0$, $V_{CE} = V_{CC} = 15$V.

b. at $V_{CE} = 0$, $I_C = \dfrac{V_{CC}}{R_C} = \dfrac{15\text{V}}{3\text{K}} = 5\text{mA}$.

Calculate the base current using

$$I_B = \frac{V_{CC} - V_{BE}}{R_B} = \frac{15 - 0.7}{100\text{K}} = 143\mu\text{A}$$

The intersection of the load line and the transistor curve for $I_B = 143\mu$A defines the quiescent operating point.

(b) At the Q-point, we find $V_{CE_Q} = 7.2$V and $I_{C_Q} = 2.4$mA.

PROBLEM 6

A silicon transistor, used in the circuit of Figure 63, may have any value of β between 36 and 90 at a temperature of 25°C. The leakage current I_{CO} can be neglected at room temperature. Find R_e, R_1 and R_2 given: $R_C = 4$ kΩ, $V_{CC} = 20$V; the nominal bias point is to be at $V_{CE} = 10$V, $I_C = 2$mA; and I_C should be in the range 1.75 to 2.24mA as β varies from 36 to 90.

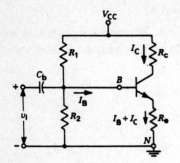

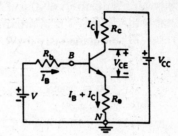

Figure 63. Silicon transistor   **Figure 64. Collector circuit**

SOLUTION: From the collector circuit (with $I_C \gg I_B$)

$$R_c + R_e = \frac{V_{CC} - V_{CE}}{I_C} = \frac{20 - 10}{2} = 5k\Omega$$

Hence, $R_e = 5 - 4 = 1k\Omega$.

Knowing that for two different β and the same V_{CE}

$$I_{C_2}\left[\frac{R_b + R_e(1+\beta_2)}{\beta_2}\right] = I_{C_1}\left[\frac{R_b + R_e(1+\beta_1)}{\beta_1}\right]$$

We have

$$2.25\left[\frac{R_b + 1(1+90)}{90}\right] = 1.75\left[\frac{R_b + 1(1+36)}{36}\right]$$

$$0.025R_b + 2.275 = 0.049R_b + 1.7986$$

$$R_b = 19.9k\Omega$$

To find R_1 and R_2, we use the Thevenin resistor formula

$$R_b = \frac{R_1 R_2}{R_1 + R_2}$$

and the voltage division formula

$$V_i = \frac{R_1}{R_1 + R_2} V_{CC}$$

From Figure 64, we can use *KVL*

$$V = I_B R_b + V_{BE} + (I_B + I_C)R_e$$

Where $I_B = \dfrac{I_C}{\beta}$

$$V = \frac{(I_C R_b)}{\beta} + V_{BE} \frac{(\beta + 1)I_C R_e}{\beta}$$

For $\beta = 36$ and $I_C = 1.75\text{mA}$, and $V_{BE} = 0.7\text{V}$, $V = 3.47$.

The same answer results for $\beta = 90$ and $I_C = 2.25\text{mA}$. Because $I_B R_b$ is small, we can assume that V in Figure 64 is approximately the dc bias voltage V_i and solve equations (2) and (3) simultaneously for R_1 and R_2.

$$R_1 = 115\text{k}\Omega$$

$$R_2 = 24\text{k}\Omega$$

PROBLEM 7

The silicon transistor in the amplifier circuit of Figure 65 has $h_{fe} = h_{FE} = 50$. Assume all bypass and coupling capacitors have zero reactance at the signal frequency. Determine

(a) Quiescent conditions.

(b) The small-signal equivalent circuit, neglecting h_{oe} and h_{re}.

(c) The current gain, $A_i = \frac{i_L}{i_i}$.

(d) The input impedance seen by the signal current source, i_i.

(e) The output impedance seen by the $1 - \text{k}\Omega$ load.

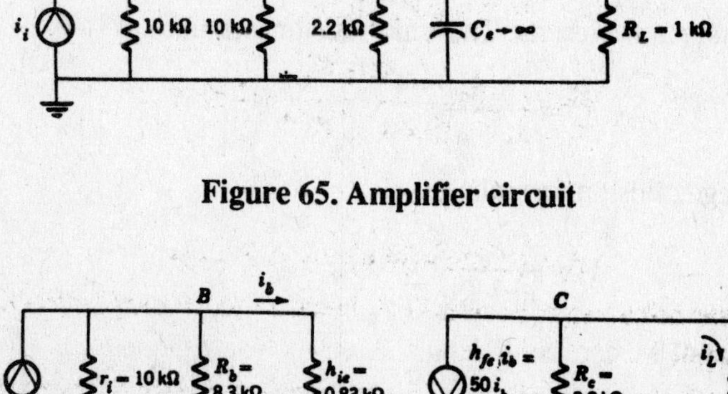

Figure 65. Amplifier circuit

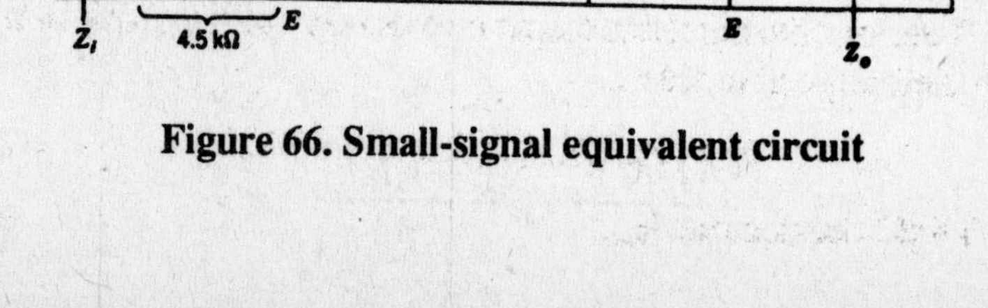

Figure 66. Small-signal equivalent circuit

SOLUTION: (a) Using voltage division across the 10kΩ and 50kΩ resistors,

$$V_{BB} = \left(\frac{10}{10+50}\right)(24) = 4\text{V}$$

R_b is the parallel combination of the 10kΩ and 50kΩ resistors.

$$R_b = \frac{(10)(50)}{10+50}\text{k}\Omega = 8.3\text{k}\Omega$$

Note that $R_b \ll h_{FE}R_e$, so that the resulting bias is insensitive to Q-point variations. Neglecting I_{BQ}, the emitter voltage V_{EQ} is

$$V_{EQ} \approx V_{BB} - V_{BEQ} = 4 - 0.7 = 3.3\text{V}$$

and

$$I_{EQ} = \frac{3.3\text{V}}{2.2\text{k}\Omega} \approx 1.5\text{mA}$$

and

$$V_{CEQ} \approx 24 - I_{EQ}(R_c + R_e) = 15\text{V}$$

(b) $h_{ie} \approx \dfrac{25h_{fe} \times 10^{-3}}{I_{EQ}} = \dfrac{(25)(50)(10^{-3})}{(1.5)(10^{-3})} = 833\Omega$

Thus, $h_{ie} \ll R_b$. The resulting small-signal equivalent circuit is shown in Figure 66.

(c) $A_i = \dfrac{i_L}{i_i} = \left(\dfrac{i_b}{i_i}\right)\left(\dfrac{i_L}{i_b}\right)$

We find, using current division (see Figure 66), that

$$\frac{i_b}{i_i} = \frac{(4.5)(10^3)}{(4.5+0.83)(10^3)} = 0.85$$

$$\frac{i_L}{i_b} = (-50)\frac{(3.8)(10^3)}{(3.8+1)(10^3)} = -39.6$$

Thus,

$$A_i = (0.85)(-39.6) = -33.7 \approx -34$$

Note that the minus sign occurs in A_i because the positive direction for i_L is opposite to that of i_c.

(d) Note that the left part of Figure 66 is not connected to the right in a circuit sense. The only connection is that the current source on the right is controlled by a current on the left. Therefore, the input and output impedances are calculated using only the resistors on their respective sides.

$$z_i = 10\text{k}\Omega\|8.3\text{k}\Omega\|0.83\text{k}\Omega \approx 700\Omega$$

(e) $z_o = 3.8\text{k}\Omega$ neglecting h_{oe}.

PROBLEM 8

The following characteristics apply to the transistor of the circuit shown in Figure 67: $h_{fe} = 100, h_{ie} = 2,000\Omega$, h_{re} is negligible, $h_{oe} = 10^{-5}$ mho. Estimate to a reasonable degree the values of A_v, A_i, R_i, and R_o.

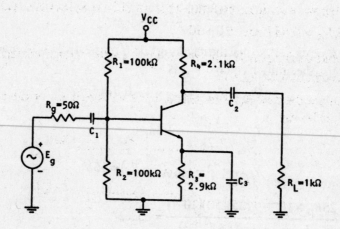

Figure 67. Circuit

SOLUTION: Figure 68 and Figure 69 show the AC equivalent and the *h*-parameter models of the circuit, respectively.

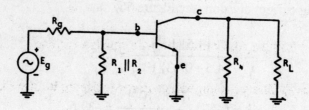

Figure 68. AC equivalent model

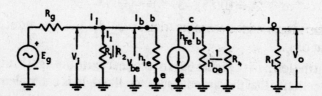

Figure 69. *H*-Parameter model

The resistance looking into the base will just be h_{ie} since h_{re} is disregarded. The resistance looking into the input of the amplifier is $R_1 \| R_2 \| h_{ie}$.

PROBLEM 9

Figure 70 shows a common base amplifier circuit, whose transistor parameters are $h_{ib} = 27.6\Omega, h_{fb} = 0.987$, and $h_{ob} = 10^{-6}\,S$.

(a) Calculate the input and output impedance, and the voltage, current, and power gains for the circuit. Use $h_{rb} \approx 0$.

(b) Calculate the new input impedance and voltage gain for the circuit when capacitor C_B is removed.

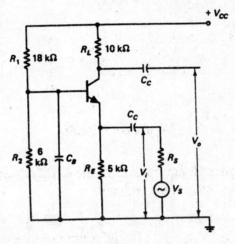

Figure 70. Common base amplifier circuit

SOLUTION: (a) The input impedance of the transistor is (see Figure 71)

$$Z_i = \frac{V_i}{I_i}. \tag{1}$$

Since $h_{rb} = 0$, this is just

$$Z_i = h_{ib} \tag{2}$$

The input impedance of the circuit is then

$$Z_i' = Z_i \| R_E \tag{3}$$

Thus,

$$Z_i \approx 27.6\Omega$$

and the circuit input impedance is

$$Z_i' \approx 27.6\Omega \| 5k\Omega \approx 27.4\Omega$$

Because h_{ob} is so small, the output impedance of the transistor is

$$Z_o = \frac{1}{h_{ob}}.$$ (4)

The circuit output impedance is then

$$Z_o' = Z_o \| R_L$$ (5)

Thus,

$$Z_o \approx \frac{1}{10^{-6}} \approx 1M\Omega$$

and the circuit output impedance is

$$Z_o' = R_L \| Z_o \approx 10k\Omega.$$

The voltage gain is

$$A_v = \frac{V_o}{V_i}.$$ (6)

We have, from Figure 71,

$$V_o = I_o R_L.$$ (7)

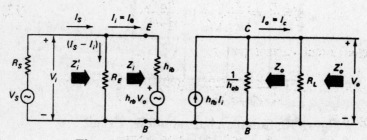

Figure 71. Input impedance of the transistor

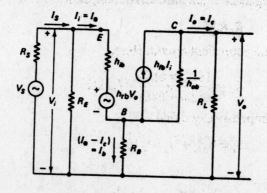

Figure 72. Equivalent circuit

and

$$V_i = h_{ib}I_i + h_{rb}V_o = h_{ib}I_i. \tag{8}$$

Since the current gain of the transistor is

$$A_i = \frac{I_o}{I_i} = h_{fb} \tag{9}$$

we have

$$A_v = \frac{V_o}{V_i} = \frac{I_o R_L}{I_i h_{ib}} = \frac{h_{fb} R_L}{h_{ib}}. \tag{10}$$

Thus, $A_i = 0.987$ and

$$A_v = \frac{0.987 \times 10k\Omega}{27.6\Omega} = 358$$

Since current divides between R_E and the transistor, we have

$$I_i = \frac{R_E}{z_i + R_E} I_s. \tag{11}$$

The actual circuit current gain is

$$A_i' = \frac{I_o}{I_s} = \left(\frac{I_o}{I_i}\right)\left(\frac{I_i}{I_s}\right) = \frac{h_{fb} R_E}{z_i + R_E} \tag{12}$$

and here the actual circuit current gain is

$$A_i' = \frac{0.987 \times 5k\Omega}{27.6\Omega + 5k\Omega} = 0.98.$$

The power gain is

$$A_p = A_v \times A_i' = 358 \times 0.98 = 351.$$

(b) Without the capacitor C_B, a base resistor appears,

$$R_B R_1 \| R_2$$

Figure 72 shows the equivalent circuit. Here,

$$R_B = \frac{18k\Omega \times 6k\Omega}{18k\Omega + 6k\Omega} = 4.5k\Omega$$

To find the input impedance, we find

$$V_i = h_{ib}I_i + (I_i - I_c)R_B \tag{13}$$

or

$$V_i = \left[h_{ib} + (1 - h_{fb})R_B\right]I_i \tag{14}$$

so

$$z_i = h_{ib} + (1 - h_{fb})R_B \tag{15}$$

Equation 3 still gives z_i', so we have

$$z_i = 27.6\Omega + 4.5k\Omega(1 - 0.987) = 86.1\Omega$$

and the circuit input impedance is

$$z_i' = 86.1\Omega \| 5k\Omega = 84.6\Omega.$$

Using Equations 7, 9, and 15, the new voltage gain is

$$A_v = \frac{V_o}{V_i} = \frac{I_o R_L}{I_i\left[h_{ib} + \left(1 - h_{fb}\right)R_B\right]}$$

$$= \frac{h_{fb} R_L}{h_{ib} + \left(1 - h_{fb}\right)R_B} \tag{16}$$

Here,

$$A_v = \frac{0.987 \times 10k\Omega}{27.6\Omega + 4.5k\Omega(1 - 0.987)} = 115$$

PROBLEM 10

For the differentiator of Figure 73, $R = 10k\Omega$, $C = 0.1\mu F$, R_c and C_c are of the appropriate size. The input is a triangular waveform shown in Figure 74. What is the output?

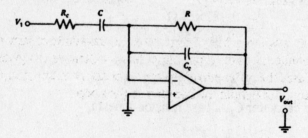

Figure 73. Differentiator

SOLUTION: First express the waveform as a function of time over the period of interest. Since this is a repetitive waveform that is symmetric about t_1, we need only solve one half period. The output for the next half period will look the same but the polarity will be reversed. We see the voltage rises linearly to 2V during 0.5ms, so we can write

$$V_{out} = \frac{2V}{0.5ms}t = \left(\frac{4 \times 10^{3V}}{s}\right)t$$

where t is time in seconds. Since the differentiator reacts only to changes in voltage, we can neglect the dc component of the input signal. We can now solve for the output:

$$V_{out} = -RC\frac{dV_{in}}{dt}$$

$$V_{out} = -RC\frac{d(4\times10^3)t}{dt}$$

$$V_{out} = -RC\left(\frac{4\times10^3\,\text{V}}{\text{s}}\right)$$

$$V_{out} = -(10\text{k}\Omega)(0.1\mu\text{F})\left(\frac{4\times10^3\,\text{V}}{\text{s}}\right)$$

$$V_{out} = -(0.001\text{s})\left(\frac{4\times10^3\,\text{V}}{\text{s}}\right) = -4\text{V}$$

FI

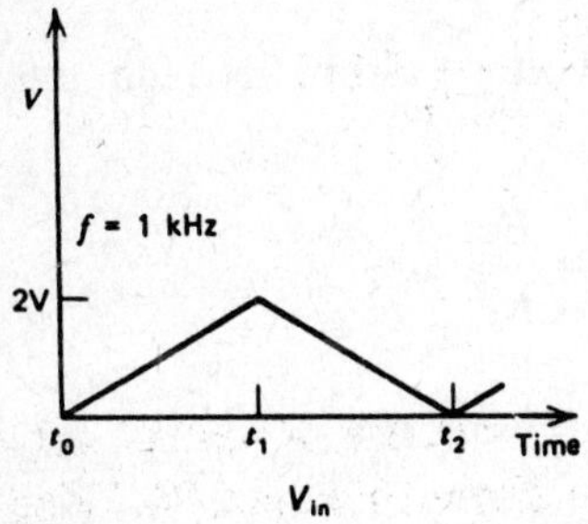

Figure 74. Triangular waveform

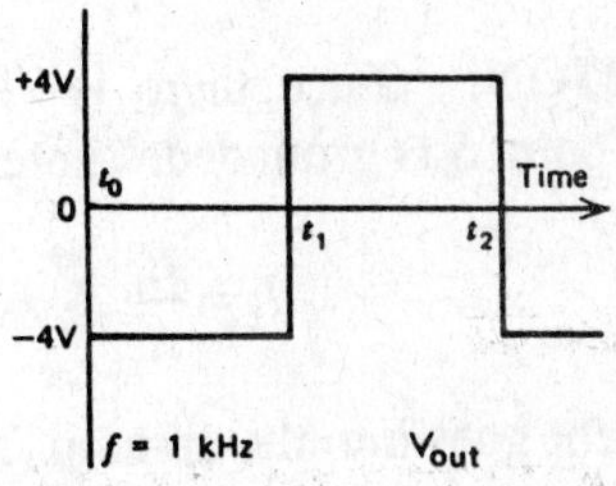

Figure 75. Square waveform

The output is then a square wave of 4Vp, 8Vp–p, amplitude as shown in Figure 75 with the same frequency as the input. From this problem we can generalize that any linear ramp causes the differentiator to have a constant output, proportional to the slope of the ramp, throughout the duration of the ramp.

PROBLEM 11

Describe the action of the op-amp circuit shown. Then, if $R_1 = R_2 = R_3 = R = 100\text{k}\Omega$, $R_f = \dfrac{100\text{k}\Omega}{3} = 33\text{k}\Omega$, $E_1 = +5\text{V}$, $E_2 = +5\text{V}$, and $E_3 = -1\text{V}$, find V_0.

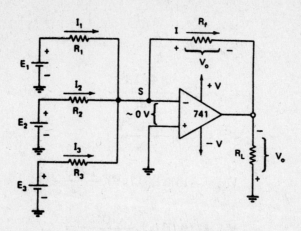

Figure 76. Op-amp circuit

SOLUTION: Since there is virtually no voltage drop between the input terminals, point S is grounded. Therefore,

$$I_1 = \frac{E_1}{R_1}, I_2 = \frac{E_2}{R_2}, \text{ and } I_3 = \frac{E_3}{R_3}.$$

No current goes into the op-amp. Therefore,

$$I = I_1 + I_2 + I_3.$$

The output voltage is then

$$V_0 = -R_f I = -R_f \left[\frac{E_1}{R_1} + \frac{E_2}{R_2} + \frac{E_3}{R_3} \right].$$

If, as in this case, $R_1 = R_2 = R_3$ and $R_f = \frac{R}{3}$, then

$$V_0 = \frac{-\left(E_1 + E_2 + E_3 \right)}{3}$$

and the circuit is an averager. For the values given,

$$V_0 = -\left[\frac{5V + 5V + (-1V)}{3} \right] = -\left(\frac{9V}{3} \right) = -3V.$$

APPENDIX

VARIABLES

a	=	acceleration
a_t	=	tangential acceleration
a_r	=	radial acceleration
d	=	distance
e	=	coefficient of restitution
f	=	frequency
F	=	force
g	=	gravity = 32.2 ft/sec^2 or 9.81 m/sec^2
h	=	height
I	=	mass inertia
k	=	spring constant, radius of gyration
KE	=	kinetic energy
m	=	mass
M	=	moment
PE	=	potential energy
r	=	radius
s	=	position
t	=	time
T	=	tension, torsion, period
v	=	velocity
w	=	weight
x	=	horizontal position
y	=	vertical position
α	=	angular acceleration
ω	=	angular velocity
θ	=	angle
μ	=	coefficient of friction

EQUATIONS

Kinematics

Linear Particle Motion

Constant velocity

$$s = s_o + vt$$

Constant acceleration

$$v = v_o + at$$

$$s = s_o + v_o t + \left(\frac{1}{2}\right) a t^2$$

$$v^2 = v_o^2 + 2a(s - s_o)$$

Projectile Motion

$$x = x_o + v_x t$$

$$v_y = v_{yo} - gt$$

$$y = y_o + v_{yo}t - \left(\frac{1}{2}\right)gt^2$$

$$v_y^2 = v_{yo}^2 - 2g\,(y - y_o)$$

Rotational Motion

Constant rotational velocity

$$\theta = \theta_o + \omega t$$

Constant angular acceleration

$$\omega = \omega_o + \alpha t$$

$$\theta = \theta_o + \omega_o t + \left(\frac{1}{2}\right)\alpha t^2$$

$$\omega^2 = \omega_o^2 + 2\alpha\,(\theta - \theta_o)$$

Tangential velocity

$$v_t = r\omega$$

Tangential acceleration

$$a_t = r\alpha$$

Radial acceleration

$$a_r = r\omega^2 = \frac{v_t^2}{r}$$

Polar coordinates

$$a_r = \frac{d^2 r}{dt^2} - r\left(\frac{d\theta}{dt}\right)^2 = \frac{d^2 r}{dt^2} - r\omega^2$$

$$a_\theta = r\left(\frac{d^2\theta}{dt^2}\right) + 2\left(\frac{dr}{dt}\right)\left(\frac{d\theta}{dt}\right) = r\alpha + 2\left(\frac{dr}{dt}\right)\omega$$

$$v_r = \frac{dr}{dt}$$

$$v_\theta = r\left(\frac{d\theta}{dt}\right) = r\omega$$

Relative and Related Motion

Acceleration

$$a_A = a_B + a_{A/B}$$

Velocity

$$v_A = v_B + v_{A/B}$$

Position

$$x_A = x_B + x_{A/B}$$

Kinetics

$$w = mg$$

$$F = ma$$

$$F_c = ma_n = \frac{mv_t^2}{r}$$

$$F_f = \mu N$$

Kinetic Energy

$$KE = \left(\frac{1}{2}\right)mv^2$$

Work of a force $= \int F ds$

$$KE_1 + \text{Work}_{1-2} = KE_2$$

Potential Energy

Spring $PE = \left(\frac{1}{2}\right)kx^2$

Weight $PE = wy$

$$KE_1 + PE_1 = KE_2 + PE_2$$

Power

Linear power $P = Fv$

Torsional or rotational power $P = T\omega$

Impulse-Momentum

$$mv_1 + \int F dt = mv_2$$

Impact

$$m_A v_{A1} + m_B v_{B1} = m_A v_{A2} + m_B v_{B2}$$

$$e = \frac{v_{B2} - v_{A2}}{v_{A1} - v_{B1}}$$

Perfectly plastic impact ($e = 0$)

$$m_A v_{A1} + m_B v_{B1} = (m_A + m_B)v'$$

One mass is infinite

$$v_2 = e v_1$$

Inertia

Beam $\quad I_A = \left(\dfrac{1}{12}\right)ml^2 + m\left(\dfrac{1}{2}\right)^2 = \left(\dfrac{1}{3}\right)ml^2$

Plate $\quad I_A = \left(\dfrac{1}{12}\right)m(a^2 + b^2) + m\left[\left(\dfrac{a}{b}\right)^2 + \left(\dfrac{b}{2}\right)^2\right] = \left(\dfrac{1}{3}\right)m(a^2 + b^2)$

Wheel $\quad I_A = mk^2 + mr^2$

Two-Dimensional Rigid Body Motion

$$F_x = ma_x$$
$$F_y = ma_y$$
$$M_A = I_A\alpha = I_{cg}\alpha + m(a)d$$

Rolling Resistance

$$F_r = \frac{mga}{r}$$

Energy Methods for Rigid Body Motion

$$KE_1 + \text{Work}_{1-2} = KE_2$$

$$\text{Work} = \int F ds + \int M d\theta$$

Mechanical Vibration

Differential equation

$$\frac{md^2x}{dt^2} + kx = 0$$

Position

$$x = x_m \sin\left[\sqrt{\frac{k}{m}}\,t + \theta\right]$$

Velocity

$$v = \frac{dx}{dt} = x_m \sqrt{\frac{k}{m}} \cos\left[\sqrt{\frac{k}{m}}\,t + \theta\right]$$

Acceleration

$$a = \frac{d^2x}{dt^2} = -x_m \left(\frac{k}{m}\right) \sin\left[\sqrt{\frac{k}{m}}\, t + \theta\right]$$

Maximum values

$$x = x_m, v = x_m \sqrt{\frac{k}{m}},\ a = -x_m\left(\frac{k}{m}\right)$$

Period

$$T = \frac{2\pi}{\left(\sqrt{\frac{k}{m}}\right)}$$

Frequency

$$f = \frac{1}{T} = \frac{\sqrt{\frac{k}{m}}}{2\pi}$$

Springs in parallel

$$k = k_1 + k_2$$

Springs in series

$$\frac{1}{k} = \frac{1}{k_1} + \frac{1}{k_2}$$

AREA UNDER NORMAL CURVE

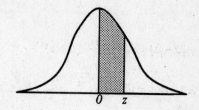

Z	0	1	2	3	4	5	6	7	8	9
0.0	.0000	.0040	.0080	.0120	.0160	.0199	0239	.0279	.0319	.0359
0.1	.0398	.0438	.0478	.0517	.0557	.0596	.0636	.0675	.0714	.0754
0.2	.0793	.0832	.0871	.0910	.0948	.0987	.1026	.1064	.1103	.1141
0.3	.1179	.1217	.1255	.1293	.1331	.1368	.1406	.1443	.1480	.1517
0.4	.1554	.1591	.1628	.1664	.1700	.1736	.1772	.1808	.1844	.1879
0.5	.1915	.1950	.1985	.2019	.2054	.2088	.2123	.2157	.2190	.2224
0.6	.2258	.2291	.2324	.2357	.2389	.2422	.2454	.2486	.2518	.2549
0.7	.2580	.2612	.2642	.2673	.2704	.2734	.2764	.2794	.2823	.2852
0.8	.2881	.2910	.2939	.2967	.2996	.3023	.3051	.3078	.3106	.3133
0.9	.3159	.3186	.3212	.3238	.3264	.3289	.3315	.3340	.3365	.3389
1.0	.3413	.3438	.3461	.3485	.3508	.3531	.3554	.3577	.3599	.3621
1.1	.3643	.3665	.3686	.3708	.3729	.3749	.3770	.3790	.3810	.3830
1.2	.3849	.3869	.3888	.3907	.3925	.3944	.3962	.3980	.3997	.4015
1.3	.4032	.4049	.4066	.4082	.4099	.4115	.4131	.4147	.4162	.4177
1.4	.4192	.4207	.4222	.4236	.4251	.4265	.4279	.4292	.4306	.4319
1.5	.4332	.4345	.4357	.4370	.4382	.4394	.4406	.4418	.4429	.4441
1.6	.4452	.4463	.4474	.4484	.4495	.4505	.4515	.4525	.4535	.4545
1.7	.4554	.4564	.4573	.4582	.4591	.4599	.4608	.4616	.4625	.4633
1.8	.4641	.4649	.4656	.4664	.4671	.4678	.4686	.4693	.4699	.4706
1.9	.4713	.4719	.4726	.4732	.4738	.4744	.4750	.4756	.4761	.4767
2.0	.4772	.4778	.4783	.4788	.4793	.4798	.4803	.4808	.4812	.4817
2.1	.4821	.4826	.4830	.4834	.4838	.4842	.4846	.4850	.4854	.4857
2.2	.4861	.4864	.4868	.4871	.4875	.4878	.4881	.4884	.4887	.4890
2.3	.4893	.4896	.4898	.4901	.4904	.4906	.4909	.4911	.4913	.4916
2.4	.4918	.4920	.4922	.4925	.4927	.4929	.4931	.4932	.4934	.4936
2.5	.4938	.4940	.4941	.4943	.4945	.4946	.4948	.4949	.4951	.4952
2.6	.4953	.4955	.4956	.4957	.4959	.4960	.4961	.4962	.4963	.4964
2.7	.4965	.4966	.4967	.4968	.4969	.4970	.4971	.4972	.4973	.4974
2.8	.4974	.4975	.4976	.4977	.4977	.4978	.4979	.4979	.4980	.4981
2.9	.4981	.4982	.4982	.4983	.4984	.4984	.4985	.4985	.4986	.4986
3.0	.4987	.4987	.4987	.4988	.4988	.4989	.4989	.4989	.4990	.4990
3.1	.4990	.4991	.4991	.4991	.4992	.4992	.4992	.4992	.4993	.4993
3.2	.4993	.4993	.4994	.4994	.4994	.4994	.4994	.4995	.4995	.4995
3.3	.4995	.4995	.4995	.4996	.4996	.4996	.4996	.4996	.4996	.4997
3.4	.4997	.4997	.4997	.4997	.4997	.4997	.4997	.4997	.4997	.4998
3.5	.4998	.4998	.4998	.4998	.4998	.4998	.4998	.4998	.4998	.4998
3.6	.4998	.4998	.4999	.4999	.4999	.4999	.4999	.4999	.4999	.4999
3.7	.4999	.4999	.4999	.4999	.4999	.4999	.4999	.4999	.4999	.4999
3.8	.4999	.4999	.4999	.4999	.4999	.4999	.4999	.4999	.4999	.4999
3.9	.5000	.5000	.5000	.5000	.5000	.5000	.5000	.5000	.5000	.5000

POWER SERIES FOR ELEMENTARY FUNCTIONS

$$\frac{1}{x}=1-(x-1)+(x-1)^2-(x-1)^3+(x-1)^4-...+(-1)^n(x-1)^n+..., \quad 0<x<2$$

$$\frac{1}{1+x}=1-x+x^2-x^3+x^4-x^5+...+(-1)^n x^n+..., \qquad -1<x<1$$

$$\ln x=(x-1)-\frac{(x-1)^2}{2}+\frac{(x-1)^3}{3}-\frac{(x-1)^4}{4}+...+\frac{(-1)^{n-1}(x-1)^n}{n}+..., \quad 0<x\le2$$

$$e^x=1+x+\frac{x^2}{2!}+\frac{x^3}{3!}+\frac{x^4}{4!}+\frac{x^5}{5!}+...+\frac{x^n}{n!}+..., \qquad -\infty<x<\infty$$

$$\sin x=x-\frac{x^3}{3!}+\frac{x^5}{5!}-\frac{x^7}{7!}+\frac{x^9}{9!}-...+\frac{(-1)^n x^{2n+1}}{(2n+1)!}+..., \qquad -\infty<x<\infty$$

$$\cos x=x-\frac{x^2}{2!}+\frac{x^4}{4!}-\frac{x^6}{6!}+\frac{x^8}{8!}-...+\frac{(-1)^n x^{2n}}{(2n)!}+..., \qquad -\infty<x<\infty$$

$$\arctan x=x-\frac{x^3}{3}+\frac{x^5}{5}-\frac{x^7}{7}+\frac{x^9}{9}-...+\frac{(-1)^n x^{2n+1}}{2n+1}+..., \qquad -1\le x\le1$$

$$\arcsin x=x+\frac{x^3}{2\cdot3}+\frac{1\cdot3x^5}{2\cdot4\cdot5}+\frac{1\cdot3\cdot5x^7}{2\cdot4\cdot6\cdot7}+...+\frac{(2n)!\,x^{2n+1}}{\left(2^n n!\right)^2(2n+1)}+..., \quad -1\le x\le1$$

$$(1+x)^k=1+kx+\frac{k(k-1)x^2}{2!}+\frac{k(k-1)(k-2)x^3}{3!}+\frac{k(k-1)(k-2)(k-3)x^4}{4!}+...,$$
$$-1<x<1$$

$$(1+x)^{-k}=1-kx+\frac{k(k+1)x^2}{2!}-\frac{k(k+1)(k+2)x^3}{3!}+\frac{k(k+1)(k+2)(k+3)x^4}{4!}-...,$$
$$-1<x<1$$

TABLE OF MORE COMMON LAPLACE TRANSFORMS

$f(t) = \mathcal{L}^{-1}\{F(s)\}$	$F(s) = \mathcal{L}\{f(t)\}$
1	$\dfrac{1}{s}$
t	$\dfrac{1}{s^2}$
$\dfrac{t^{n-1}}{(n-1)!}; n=1,2,\ldots$	$\dfrac{1}{s^n}$
e^{at}	$\dfrac{1}{s-a}$
$t\,e^{at}$	$\dfrac{1}{(s-a)^2}$
$\dfrac{t^{n-1}e^{-at}}{(n-1)!}$	$\dfrac{1}{(s+a)^n}; n=1,2,\ldots$
$\dfrac{e^{-at}-e^{-bt}}{b-a}; a\neq b$	$\dfrac{1}{(s+a)(s+b)}$
$\dfrac{a\,e^{-at}-b\,e^{-bt}}{a-b}; a\neq b$	$\dfrac{s}{(s+a)(s+b)}$
$\sin st$	$\dfrac{a}{s^2+a^2}$
$\cos at$	$\dfrac{s}{s^2+a^2}$
$\sinh at$	$\dfrac{a}{s^2-a^2}$

$f(t) = \mathcal{L}^{-1}\{F(s)\}$	$F(s) = \mathcal{L}\{f(t)\}$
$\cosh at$	$\dfrac{s}{s^2 - a^2}$
$\dfrac{1}{a^2}(1 - \cos at)$	$\dfrac{1}{s(s^2 + a^2)}$
$\dfrac{1}{a^3}(at - \sin at)$	$\dfrac{1}{s(s^2 + a^2)}$
$\dfrac{t}{2a}\sin at$	$\dfrac{s}{(s^2 + a^2)^2}$
$\dfrac{1}{b}e^{-at}\sin bt$	$\dfrac{1}{(s + a)^2 + b^2}$
$e^{-at}\cos bt$	$\dfrac{s + a}{(s + a)^2 + b^2}$
$h_1(t - a)$	$\dfrac{1}{s}e^{-as}$
$h_1(t) - h_1(t - a)$	$\dfrac{1 - e^{-as}}{s}$
$\dfrac{1}{t}\sin kt$	$\arctan\dfrac{k}{s}$

CONVERSION FACTORS - GENERAL

To convert from	To	Multiply by
Acres	Square feet	43,560
Acres	Square meters	4074
Acres	Square miles	0.001563
Acre-feet	Cubic meters	1233
Ampere-hours (absolute)	Coulombs (absolute)	3600
Angstrom units	Inches	3.937×10^{-9}
Angstrom units	Meters	1×10^{-10}
Angstrom units	Microns	1×10^{-4}
Atmospheres	Millimeters of mercury at 32°F.	760
Atmospheres	Dynes per square centimeter	1.0133×10^{6}
Atmospheres	Newtons per square meter	101,325
Atmospheres	Feet of water at 39.1°F.	33.90
Atmospheres	Grams per square centimeter	1033.3
Atmospheres	Inches of mercury at 32°F.	29.921
Atmospheres	Pounds per square foot	2116.3
Atmospheres	Pounds per square inch	14.696
Bags (cement)	Pounds (cement)	94
Barrels (cement)	Pounds (cement)	376
Barrels (oil)	Cubic meters	0.15899
Barrels (oil)	Gallons	42
Barrels (U.S. liquid)	Cubic meters	0.11924
Barrels (U.S. liquid)	Gallons	31.5
Barrels per day	Gallons per minute	0.02917
Bars	Atmospheres	0.9869
Bars	Newtons per square meter	1×10^{5}
Bars	Pounds per square inch	14.504
Board feet	Cubic feet	$1/12$
Boiler horsepower	B.t.u. per hour	33,480
Boiler horsepower	Kilowatts	9.803
B.t.u.	Calories (gram)	252
B.t.u.	Centigrade heat units (c.h.u. or p.c.u.)	0.55556
B.t.u.	Foot-pounds	777.9
B.t.u.	Horsepower-hours	3.929×10^{-4}
B.t.u.	Joules	1055.1
B.t.u.	Liter-atmospheres	10.41
B.t.u.	Pounds carbon to CO_2	6.88×10^{-5}
B.t.u.	Pounds water evaporated from and at 212°F.	0.001036
B.t.u.	Cubic foot-atmospheres	0.3676
B.t.u.	Kilowatt-hours	2.930×10^{-4}
B.t.u. per cubic foot	Joules per cubic meter	37,260
B.t.u. per hour	Watts	0.29307
B.t.u. per minute	Horsepower	0.02357
B.t.u. per pound	Joules per kilogram	2326
B.t.u. per pound per degree Fahrenheit	Calories per gram per degree centigrade	1
B.t.u. per pound per degree Fahrenheit	Joules per kilogram per degree Kelvin	4186.8
B.t.u. per second	Watts	1054.4
B.t.u. per square foot per hour	Joules per square meter per second	3.1546
B.t.u. per square foot per minute	Kilowatts per square foot	0.1758
B.t.u. per square foot per second for a temperature gradient of 1°F. per inch	Calories, gram (15°C.), per square centimeter per second for a temperature gradient of 1°C. per centimeter	1.2405

(continued)

To convert from	To	Multiply by
B.t.u. (60°F.) per degree Fahrenheit	Calories per degree centigrade	453.6
Bushels (U.S. dry)	Cubic feet	1.2444
Bushels (U.S. dry)	Cubic meters	0.03524
Calories, gram	B.t.u.	3.968×10^{-3}
Calories, gram	Foot-pounds	3.087
Calories, gram	Joules	4.1868
Calories, gram	Liter-atmospheres	4.130×10^{-2}
Calories, gram	Horsepower-hours	1.5591×10^{-6}
Calories. gram, per gram per degree C.	Joules per kilogram per degree Kelvin	4186.8
Calories, kilogram	Kilowatt-hours	0.0011626
Calories, kilogram per second	Kilowatts	4.185
Candle power (spherical)	Lumens	12.556
Carats (metric)	Grams	0.2
Centigrade heat units	B.t.u.	1.8
Centimeters	Angstrom units	1×10^{8}
Centimeters	Feet	0.03281
Centimeters	Inches	0.3937
Centimeters	Meters	0.01
Centimeters	Microns	10,000
Centimeters of mercury at 0°C.	Atmospheres	0.013158
Centimeters of mercury at 0°C.	Feet of water at 39.1°F.	0.4460
Centimeters of mercury at 0°C.	Newtons per square meter	1333.2
Centimeters of mercury at 0°C.	Pounds per square foot	27.845
Centimeters of mercury at 0°C.	Pounds per square inch	0.19337
Centimeters per second	Feet per minute	1.9685
Centimeters of water at 4°C.	Newtons per square meter	98.064
Centistokes	Square meters per second	1×10^{-6}
Circular mils	Square centimeters	5.067×10^{-6}
Circular mils	Square inches	7.854×10^{-7}
Circular mils	Square mils	0.7854
Cords	Cubic feet	128
Cubic centimeters	Cubic feet	3.532×10^{-5}
Cubic centimeters	Gallons	2.6417×10^{-4}
Cubic centimeters	Ounces (U.S. fluid)	0.03381
Cubic centimeters	Quarts (U.S. fluid)	0.0010567
Cubic feet	Bushels (U.S.)	0.8036
Cubic feet	Cubic centimeters	28,317
Cubic feet	Cubic meters	0.028317
Cubic feet	Cubic yards	0.03704
Cubic feet	Gallons	7.481
Cubic feet	Liters	28.316
Cubic foot-atmospheres	Foot-pounds	2116.3
Cubic foot-atmospheres	Liter-atmospheres	28.316
Cubic feet of water (60°F.)	Pounds	62.37
Cubic feet per minute	Cubic centimeters per second	472.0
Cubic feet per minute	Gallons per second	0.1247
Cubic feet per second	Gallons per minute	448.8
Cubic feet per second	Million gallons per day	0.64632
Cubic inches	Cubic meters	1.6387×10^{-5}
Cubic yards	Cubic meters	0.76456
Curies	Disintegrations per minute	2.2×10^{12}
Curies	Coulombs per minute	1.1×10^{12}
Degrees	Radians	0.017453
Drams (apothecaries' or troy)	Grams	3.888

(continued)

To convert from	To	Multiply by
Drams (avoirdupois)	Grams	1.7719
Dynes	Newtons	1×10^{-5}
Ergs	Joules	1×10^{-7}
Faradays	Coulombs (abs.)	96,500
Fathoms	Feet	6
Feet	Meters	0.3048
Feet per minute	Centimeters per second	0.5080
Feet per minute	Miles per hour	0.011364
Feet per (second)^2	Meters per (second)^2	0.3048
Feet of water at 39.2°F.	Newtons per square meter	2989
Foot-poundals	B.t.u.	3.995×10^{-5}
Foot-poundals	Joules	0.04214
Foot-poundals	Liter-atmospheres	4.159×10^{-4}
Foot-pounds	B.t.u.	0.0012856
Foot-pounds	Calories, gram	0.3239
Foot-pounds	Foot-poundals	32.174
Foot-pounds	Horsepower-hours	5.051×10^{-7}
Foot-pounds	Kilowatt-hours	3.766×10^{-7}
Foot-pounds	Liter-atmospheres	0.013381
Foot-pounds force	Joules	1.3558
Foot-pounds per second	Horsepower	0.0018182
Foot-pounds per second	Kilowatts	0.0013558
Furlongs	Miles	0.125
Gallons (U.S. liquid)	Barrels (U.S. liquid)	0.03175
Gallons	Cubic meters	0.003785
Gallons	Cubic feet	0.13368
Gallons	Gallons (Imperial)	0.8327
Gallons	Liters	3.785
Gallons	Ounces (U.S. fluid)	128
Gallons per minute	Cubic feet per hour	8.021
Gallons per minute	Cubic feet per second	0.002228
Grains	Grams	0.06480
Grains	Pounds	$\frac{1}{7000}$
Grains per cubic foot	Grams per cubic meter	2.2884
Grains per gallon	Parts per million	17.118
Grams	Drams (avoirdupois)	0.5644
Grams	Drams (troy)	0.2572
Grams	Grains	15.432
Grams	Kilograms	0.001
Grams	Pounds (avoirdupois)	0.0022046
Grams	Pounds (troy)	0.002679
Grams per cubic centimeter	Pounds per cubic foot	62.43
Grams per cubic centimeter	Pounds per gallon	8.345
Grams per liter	Grains per gallon	58.42
Grams per liter	Pounds per cubic foot	0.0624
Grams per square centimeter	Pounds per square foot	2.0482
Grams per square centimeter	Pounds per square inch	0.014223
Hectares	Acres	2.471
Hectares	Square meters	10,000
Horsepower (British)	B.t.u. per minute	42.42
Horsepower (British)	B.t.u. per hour	2545
Horsepower (British)	Foot-pounds per minute	33,000
Horsepower (British)	Foot-pounds per second	550
Horsepower (British)	Watts	745.7
Horsepower (British)	Horsepower (metric)	1.0139
Horsepower (British)	Pounds carbon to CO_2 per hour	0.175

(continued)

To convert from	To	Multiply by
Horsepower (British)	Pounds water evaporated per hour at 212°F	2.64
Horsepower (metric)	Foot-pounds per second	542.47
Horsepower (metric)	Kilogram-meters per second	7.5
Hours (mean solar)	Seconds	3600
Inches	Meters	0.0254
Inches of mercury at 60°F	Newtons per square meter	3376.9
Inches of water at 60°F	Newtons per square meter	248.84
Joules (absolute)	B.t.u. (mean)	9.480×10^{-4}
Joules (absolute)	Calories, gram (mean)	0.2389
Joules (absolute)	Cubic foot-atmospheres	0.3485
Joules (absolute)	Foot-pounds	0.7376
Joules (absolute)	Kilowatt-hours	2.7778×10^{-7}
Joules (absolute)	Liter-atmospheres	0.009869
Kilocalories	Joules	4186.8
Kilograms	Pounds (avoirdupois)	2.2046
Kilograms force	Newtons	9.807
Kilograms per square centimeter	Pounds per square inch	14.223
Kilometers	Miles	0.6214
Kilowatt-hours	B.t.u.	3414
Kilowatt-hours	Foot-pounds	2.6552×10^{6}
Kilowatts	Horsepower	1.3410
Knots (international)	Meters per second	0.5144
Knots (nautical miles per hour)	Miles per hour	1.1516
Lamberts	Candles per square inch	2.054
Liter-atmospheres	Cubic foot-atmospheres	0.03532
Liter-atmospheres	Foot-pounds	74.74
Liters	Cubic feet	0.03532
Liters	Cubic meters	0.001
Liters	Gallons	0.26418
Lumens	Watts	0.001496
Micromicrons	Microns	1×10^{-6}
Microns	Angstrom units	1×10^{4}
Microns	Meters	1×10^{-6}
Miles (nautical)	Feet	6080
Miles (nautical)	Miles (U.S. statute)	1.1516
Miles	Feet	5280
Miles	Meters	1609.3
Miles per hour	Feet per second	1.4667
Miles per hour	Meters per second	0.4470
Milliliters	Cubic centimeters	1
Millimeters	Meters	0.001
Millimeters of mercury at 0°C.	Newtons per square meter	133.32
Millimicrons	Microns	0.001
Mils	Inches	0.001
Mils	Meters	2.54×10^{-5}
Minims (U.S.)	Cubic centimeters	0.06161
Minutes (angle)	Radians	2.909×10^{-4}
Minutes (mean solar)	Seconds	60
Newtons	Kilograms	0.10197
Ounces (avoirdupois)	Kilograms	0.02835
Ounces (avoirdupois)	Ounces (troy)	0.9115
Ounces (U.S. fluid)	Cubic meters	2.957×10^{-5}
Ounces (troy)	Ounces (apothecaries')	1.000
Pints (U.S. liquid)	Cubic meters	4.732×10^{-4}
Poundals	Newtons	0.13826

(continued)

To convert from	To	Multiply by
Pounds (avoirdupois)	Grains	7000
Pounds (avoirdupois)	Kilograms	0.45359
Pounds (avoirdupois)	Pounds (troy)	1.2153
Pounds per cubic foot	Grams per cubic centimeter	0.016018
Pounds per cubic foot	Kilograms per cubic meter	16.018
Pounds per square foot	Atmospheres	4.725×10^{-4}
Pounds per square foot	Kilograms per square meter	4.882
Pounds per square inch	Atmospheres	0.06805
Pounds per square inch	Kilograms per square centimeter	0.07031
Pounds per square inch	Newtons per square meter	6894.8
Pounds force	Newtons	4.4482
Pounds force per square foot	Newtons per square meter	47.88
Pounds water evaporated from and at 212°F.	Horsepower-hours	0.379
Pound-centigrade units (p.c.u.)	B.t.u.	1.8
Quarts (U.S. liquid)	Cubic meters	9.464×10^{-4}
Radians	Degrees	57.30
Revolutions per minute	Radians per second	0.10472
Seconds (angle)	Radians	4.848×10^{-6}
Slugs	Gee pounds	1
Slugs	Kilograms	14.594
Slugs	Pounds	32.17
Square centimeters	Square feet	0.0010764
Square feet	Square meters	0.0929
Square feet per hour	Square meters per second	2.581×10^{-5}
Square inches	Square centimeters	6.452
Square inches	Square meters	6.452×10^{-4}
Square yards	Square meters	0.8361
Stokes	Square meters per second	1×10^{-4}
Tons (long)	Kilograms	1016
Tons (long)	Pounds	2240
Tons (metric)	Kilograms	1000
Tons (metric)	Pounds	2204.6
Tons (metric)	Tons (short)	1.1023
Tons (short)	Kilograms	907.18
Tons (short)	Pounds	2000
Tons (refrigeration)	B.t.u. per hour	12,000
Tons (British shipping)	Cubic feet	42.00
Tons (U.S. shipping)	Cubic feet	40.00
Torr (mm. mercury, 0°C.)	Newtons per square meter	133.32
Watts	B.t.u. per hour	3.413
Watts	Joules per second	1
Watts	Kilogram-meters per second	0.10197
Watt-hours	Joules	3600
Yards	Meters	0.9144

TEMPERATURE FACTORS

$$°F = 9/5 \,(°C) + 32$$

Fahrenheit temperature = 1.8 (temperature in kelvins) −459.67

$$°C = 5/9 \,[(°F) - 32]$$

Celsius temperature = temperature in kelvins −273.15

Fahrenheit temperature = 1.8 (Celsius temperature) +32